Cultivating Food Justice

Food, Health, and the Environment
Series editor: Robert Gottlieb, Henry R. Luce Professor of Urban and
Environmental Policy, Occidental College

Keith Douglass Warner, *Agroecology in Action: Extending Alternative
Agriculture through Social Networks*

Christopher M. Bacon, V. Ernesto Méndez, Stephen R. Gliessman,
David Goodman, and Jonathan A. Fox, eds., *Confronting the Coffee
Crisis: Fair Trade, Sustainable Livelihoods and Ecosystems in Mexico
and Central America*

Thomas A. Lyson, G. W. Stevenson, and Rick Welsh, eds., *Food and
the Mid-Level Farm: Renewing an Agriculture of the Middle*

Jennifer Clapp and Doris Fuchs, eds., *Corporate Power in Global
Agrifood Governance*

Robert Gottlieb and Anupama Joshi, *Food Justice*

Jill Lindsey Harrison, *Pesticide Drift and the Pursuit of Environmental
Justice*

Alison Hope Alkon and Julian Agyeman, eds., *Cultivating Food
Justice: Race, Class, and Sustainability*

Cultivating Food Justice
Race, Class, and Sustainability

edited by Alison Hope Alkon and Julian Agyeman

The MIT Press
Cambridge, Massachusetts
London, England

For information about special quantity discounts, please email special_sales@ mitpress.mit.edu.

This book was set in Sabon by Toppan Best-set Premedia Limited. Printed and bound in the United States of America.

Library of Congress Cataloging-in-Publication Data

Cultivating food justice : race, class, and sustainability / edited by Alison Hope Alkon and Julian Agyeman.
    p.   cm. — (Food, health, and the environment)
Includes bibliographical references and index.
ISBN 978-0-262-01626-1 (hardcover : alk. paper)—ISBN 978-0-262-51632-7 (pbk. : alk. paper)
1. Food consumption—United States. 2. Minorities—Nutrition—United States. 3. Poor—Nutrition—United States. 4. African American—Nutrition. 5. Discrimination—United States. 6. Social justice—United States. I. Alkon, Alison Hope. II. Agyeman, Julian.
HD9005.C88 2011
363.80973—dc22
2011002082
10 9 8 7 6 5 4

This book is dedicated to those who work to create a more just and sustainable food system, and through it, a more just and sustainable world.

Contents

Series Foreword

I am pleased to present the seventh book in the Food, Health, and the Environment series. This series explores the global and local dimensions of food systems and examines issues of access, justice, and environmental and community well-being. It includes books that focus on the way food is grown, processed, manufactured, distributed, sold, and consumed. Among the matters addressed are what foods are available to communities and individuals, how those foods are obtained, and what health and environmental factors are embedded in food-system choices and outcomes. The series focuses not only on food security and well-being but also on regional, state, national, and international policy decisions and economic and cultural forces. Food, Health, and the Environment books provide a window into the public debates, theoretical considerations, and multidisciplinary perspectives that have made food systems and their connections to health and environment important subjects of study.

Robert Gottlieb, Occidental College
Series Editor

Preface

In this book, a key concept is positionality, the understanding that our life experiences and practices are deeply entangled with the ways we see the world. Given the importance of this idea to our work, it seems appropriate to share something about the journeys that have led to the production of this book while acknowledging those who have helped cultivate our thinking.

From Alison

My path began during my undergraduate education, where many of my courses probed the exploitations brought on by corporate globalization. I found opposition to that exploitation, as well as a deep sense of joy, in a place-based environmentalism emphasizing connection to the natural world. At the same time, my coursework highlighted resistance among poor communities of color while criticizing well-intentioned outsiders' problematic reform efforts. It seemed that lessening my own resource use might decrease the need for exploitation while respecting, but not interfering with, communities' struggles.

I learned to do this while living in an intentional community as a graduate student at UC Davis. I "put my hands in the soil" of our organic gardens and learned to cultivate food. Evenings were filled with shared meals, often followed by conversations lasting into the night. I'm grateful to have experienced firsthand the desire for ecological sustainability and deep social bonds that can arise from growing and sharing food. I also acknowledge that I, and those with whom I shared this experience, hailed predominantly from white and middle-class backgrounds. Even then, I knew that there was something about this way of life that rarely worked for people of color.

My early research arose from my personal and political interest in place, examining its relationship with meaning-making processes and power. This interest grew as I learned about the environmental justice movement. In contrast to movement leaders' claims that its emphasis on place was in conflict with the environmental movement, place was essential to the kind of environmentalism I had been trying to live. I wanted to understand how activists could combine place-based environmentalism with antiracism. I moved to Oakland, where I had heard such efforts were emerging.

There I met a group of activists willing to share their work with me who, while not the subjects of this book, have deeply shaped my thinking. I am grateful to Jason Harvey (who loaned me his phrase "will work for food justice"), David Roach, Dana Harvey, Leroy Musgrave, Will Scott and the Scott Family, Charlotte Coleman, Ted Dixon, Xan West, and Jada White.

Additionally, I have found supportive communities of scholars including my dissertation committee (Tom Beamish, Jim Cramer, Julie Sze, and Kimberly Nettles), the Environmental Justice Project (Jonathan London, Marisol Cortez, Raoul Lievanos, and Tracy Perkins), the UC Multicampus Research Group on Food and the Body (especially Julie Guthman, Melanie DuPuis, Carolyn de la Peña, Laura-Anne Minkoff-Zern, and Christie McCullen), my writing group (Joan S. M. Meyers, Dina Biscotti, Julie Collins-Dogrul, Jen Gregson, Macky Yamaguchi, Lori Freeman, Julie Setele, and Jaime Becker), and my colleagues at the University of the Pacific (Marcia Hernandez, Ethel Nicdao, George Lewis, and Ken Albala). Kari Norgaard has been an incredible mentor, colleague, and coauthor, helping me think through the framework for this book. Additionally, it is rare that a well-known scholar reaches out to a graduate student, but Julian Agyeman has been extremely supportive, eventually suggesting we collaborate on this project. I am deeply grateful for his guidance and positive attitude throughout. Thanks also to Natalia Skolnik and my partner Aaron Simon for their genuine interest in this work, as well as my family, Penny, Michael, and Matty Alkon, for their love, good humor, and encouragement.

From Julian

In the early 1980s, I was a geography teacher in Carlisle, England. Carlisle sits at the northern edge of one of Britain's most beautiful regions, the Lake District. This is Wordsworth country, the place of England's highest mountain, Scafell Pike, and its biggest lake, Lake Windermere. It is a place of amazing beauty in every season but it's a place that also perked my interest in race, ethnicity, and space. I always wondered why, when leading field trips to engage students in the very visible glacial history of the area, people would stop and stare. I also wondered why there were so few other visitors (and even fewer residents) that looked like me, a person of color in charge of a white group of students. What I didn't know was that at the same time, Guyanese-born British photographer Ingrid Pollard, who became a good friend, was having similar experiences in the Lake District, and other rural spaces.

I chose to set up an organization to investigate these exclusive English "white spaces" in 1988. The organization, the Black Environment Network, of which Ingrid was also a founder, is still active today, and is Britain's only minority-founded environmental organization. I also began to investigate this theme through my popular writings and TV programs during the 1980s and early 1990s. I took a more academic slant in my 1997 book chapter, written with then PhD student Rachel Spooner, called "Ethnicity and the Rural Environment" in *Contested Countryside Cultures*, edited by P. Cloke and J. Little. In 2006, Sarah Neal and I coedited a book-length investigation of what had become known as "rural racism" in *The New Countryside? Ethnicity, Nation and Exclusion in Contemporary Rural Britain.* Rural racism became the Black Environment Network's equivalent of "environmental racism" in the United States. It was the first real framing of environmental (in)justice in Britain.

Ingrid Pollard's way of expressing her feelings about her experiences was to produce a series of celebrated photographs entitled *Pastoral Interludes* of herself and others, including me, in rural English landscapes. In doing this, she helped to transgress the notion that we, as people of color, were "out of place" in the English countryside. Her work draws us in to view a moment of apparent disjuncture, of rupture between the black body and the (English) rural setting. With her inclusion of written text in her pictures, Pollard challenges the notion of a disjuncture *and* details the experience of living the disjuncture.

This, in a nutshell, was the beginning of my academic career. This was my experience with environmental (in)justice, which I've developed in different ways in all my writings since the 1980s, culminating in the concept of "just sustainability," which Bob Bullard, Bob Evans, and I described in our 2003 book *Just Sustainabilities: Development in an Unequal World* (Cambridge, MA: MIT Press) and which I describe more fully in my 2005 book *Sustainable Communities and the Challenge of*

Environmental Justice. It is also the concept that brought Alison and me together. One of the great joys of being an academic is having bright, keen students chew over and reshape things that you have shaped. I'd heard about Alison's PhD research at UC Davis and the fact that she was using the "just sustainability" framework to investigate two farmers markets in the Bay Area. We finally met at the 2007 American Association of Geographers conference in San Francisco and began to hatch this project. I thank Alison for opening my eyes to the world of food justice as that bridge between social justice and environmental protection that I call "just sustainability."

In closing, we'd both like to thank all of the chapter authors, series editor Robert Gottlieb, Clay Morgan at the MIT Press, our editorial assistant Tufts student Laura Tolkoff, and our anonymous reviewers, each of whose hard work is reflected in this book.

1

Introduction

The Food Movement as Polyculture

Alison Hope Alkon and Julian Agyeman

Picture a field of corn stretching out into the horizon. Each evenly spaced stalk is genetically identical. Each needs exactly the same amount of water, fertilizer, sunlight, and time as every other. And each is ready for harvest at exactly the same moment. For this reason, the cultivation and collection of this field can be entirely mechanized. Heavy machinery has crossed this field many times, laying seed, applying fertilizer, and eventually gathering many tons of corn. Mechanized production is necessary on a farm of this size, as hand cultivation would be prohibitively expensive. This approach toward farming, called a monoculture, ensures that the finished product will always meet the specifications required for processing, which will transform it into everything from animal feed to Coca-Cola.

In the past twenty-five years, a variety of food movements have crafted strong and coherent opposition to this type of industrial monoculture. Such monoculture, they claim, requires energy-intensive chemical fertilizers, pesticides and machinery, all of which necessitate up-front investments. These circumstances favor large farms with available capital, and have led to the increased consolidation and corporate ownership of agriculture (Bell 2004; Magdoff, Foster, and Buttel 2000; Buttel, Larson, and Gillespie 1990).[1] Ecologists have also found that chemical inputs deplete soils of nutrients and pollute nearby waterways (Altieri 2000), and food movement activists have drawn on such studies to argue that industrial agriculture is environmentally harmful. And increasingly, as public attention to diet-related health diseases has increased, food movement activists[2] have argued that responsibility lies with industrial monocultures, whose processed results are often used as inexpensive fillers in high-calorie foods (Nestle 2002).

The food movement encourages eaters to turn away from the industrially produced and processed goods created from monocultures—what

New York times bestselling author Michael Pollan (2008) refers to as "edible food-like substances"—and instead to choose fresh, local, and often organic offerings supplied by local small farms. Food movement activists routinely trumpet such foods as healthier and tastier than their industrial and processed counterparts. But for the movement, eating is not only personal but also political. The purchase of local and organic food is cast as a "vote with your fork," to quote a common movement refrain. It is a vote for environmental sustainability, as local, organic producers cultivate biologically diverse polycultures and avoid the use of synthetic pesticides and fertilizers. It is also a vote for small, family-owned farms, as opposed to their large, corporate counterparts, and for creating local communities filled with rich interpersonal interactions. The food movement narrative also argues that linking producers and consumers will endow both with a sense of connection to local place, which allows eaters to better understand the social and environmental processes through which their food is produced. By transforming our food practices, the movement tells us, we can live healthier, more authentic lives while supporting positive social and environmental change. In this way, the food movement is responding to popular anxieties that modern life is alienating and antisocial, and an American mythology that locates the good life in romanticized small towns.[3]

The food movement's vision has become widely popular. In addition to Pollan, authors such as Barbara Kingsolver (2007), Eric Schlosser (2001), and Marion Nestle (2006) have drawn upon, rehearsed, and expanded the movement narrative. Additionally, in 2007, the *Oxford English Dictionary* named *locavore*, meaning one who eats local food, as its "word of the year" (Prentice 2007). The food movement has also played an important role in driving the economic success of farmers markets and organic food, both of which have experienced enormous growth, and have even held their market value during the economic crisis of the late 2000s (Chu 2009). The food movement, and its narrative linking the production and consumption of local organic food to positive economic, environmental, and social changes, is one force behind such a lucrative market.

But such a consistent narrative, along with the movement's predominantly white and middle-class character, suggests that it may itself be something of a monoculture. It consists of a group of "like-minded" people, with similar backgrounds, values, and proclivities, who have come to similar conclusions about how our food system should change. Moreover, those active in the food movement tend to have the wealth necessary to participate in its dominant social change strategy—the purchase of local organic food—or at least the cultural cachet necessary to obtain such foods through barter, work trades, or other forms of exchange. Of course, those who buy local organic food are choosing to spend money in this way. But amid rising economic insecurity brought on by decades of neoliberal policies, including the privatization of many public goods and the weakening of entitlement programs, increasing numbers of people struggle to meet basic needs. It is worth recognizing that food movement adherents must be willing, and also able, to participate by purchasing. The similar lifestyles held by food movement adherents are reflected in its movement narrative.

Feminist social scientists use the term *positionality* to refer to the understanding that our lived experiences, particularly those of race, class, and gender, shape our worldviews. The food movement narrative is largely created by, and resonates most deeply with, white and middle-class individuals. For example, Michael Pollan's recently offered list of food rules (2007) is intended to guide consumers toward eating practices aligned with the food movement. However, when Pollan begins his first rule by telling us not to "eat anything your great-grandmother wouldn't recognize as food," he ignores the fact that "our" great-grandmothers come from a wide variety of social and economic contexts that may have informed their perceptions of food quite differently. Some were enslaved, transported across the ocean, and forced to subsist on the overflow from the master's table. Others were forcibly sent to state-mandated boarding schools, in which they were taught to despise, and even forget, any foods they would previously have recognized. And those who have emigrated from various parts of the global south in the past few generations may have great-grandmothers who saw the foods they recognized demeaned, or even forbidden, by those who claimed their lands. Of course it is not these histories that Pollan intends to invoke when he urges readers to choose fruits and vegetables over processed foods. But because of his privileged positionality, Pollan fails to consider the effects of race on food access and the alternative meanings his words may hold for people of color in the United States. In this same way, whites in the food movement often simply do not see the subtle exclusivities that are woven into its narrative.

This is not to say that some of the food movement's insights, particularly those regarding the destructive nature of industrial agriculture, are not important. Indeed, many of the authors included in this volume are supporters and consumers of local and organic food. Among us are individual and community gardeners, former farm apprentices, members of community-supported agriculture projects, and farmers market shoppers. Nevertheless, for many of us, our involvement with the food movement, along with our academic training, has contributed to the belief that the dominant narrative described earlier, compelling as it may be to some, might drown out other stories. In these additional stories, food is not only linked to ecological sustainability, community, and health but also to racial, economic, and environmental justice.

Our goal in highlighting these additional stories is not to chastise the food movement, but to work toward building a stronger and deeper critique of industrialized agriculture, which includes injustice along with environmental and social degradation. If activists in the food movement are to go beyond providing alternatives and truly challenge agribusiness's destructive power, they will need a broad coalition of supporters. We argue that such support can best be found in the low-income communities and communities of color that have been, and are currently, most deeply harmed by the food system. But this alliance will require that the food movement reach beyond its own dominant narrative to understand the experiences and perspectives of its potential allies.

Situating Food Justice

In this volume, we offer a variety of accounts that explore the ways that race and class are enmeshed in the food system. Such stories are rooted in the low-income communities and communities of color that are all too often absent from the dominant food movement narrative, and are disproportionately harmed by the current food system. From these communities' silenced histories, and from the framings deployed by activists seeking redress, we learn that race and class play a central role in organizing the production, distribution, and consumption of food. Time and again, communities of color have been subject to laws and policies that have taken away their ability to own and manage land for food production, though members of these communities continue to be exploited as farm laborers. Moreover, low-income communities and communities of color often lack access to locally available healthy food, and what food is available is often more expensive than similar purchases in wealthier areas (Winne 2008).

In these ways, the food system is implicated in many of what Omi and Winant (1994) call *racial projects*, political and economic undertakings through which racial hierarchies are established and racialized subjectivities are created. Federal immigration laws, for example, act as racial projects when they define who is a legitimate subject deserving of workplace protections, and who is regarded as an alien "other." Similarly, urban planning and mortgage lending policies become racial projects when they serve to shape built environments that lack access to basic amenities, and to restrict communities of color to those areas. In another example, the appropriation of Native American lands by white settlers and past and present-day forced assimilation are examples of racial projects that have deprived some Native American tribes of both material wealth and cultural sovereignty. While these communities' circumstances are widely divergent, various chapters in this volume demonstrate the ways that racial projects have led to widespread hunger, exploitation, and environmental degradation.

In response, communities of color are beginning to engage in *food justice* activism in order to provide food for themselves while imagining new ecological and social relationships. The divergent stories represented in this volume warrant a broad definition of food justice. According to veteran organization Just Food (2010), food justice is "communities exercising their right to grow, sell, and eat [food that is] fresh, nutritious, affordable, culturally appropriate, and grown locally with care for the well-being of the land, workers, and animals." Detroit's D-Town Farmers additionally emphasize that those communities that have been most marginalized by the agribusiness system need to "lead the movement to provide food for the members of their community" (White 2010, 204). Essential to the food justice movement is an analysis that recognizes the food system itself as a racial project and problematizes the influence of race and class on the production, distribution, and consumption of food. Communities of color and poor communities have time and again been denied access to the means of food production, and, due to both price and store location, often cannot access the diet advocated by the food movement. Through food justice activism, low-income communities and communities of color seek to create local food systems that meet their own food needs.

The movement for food justice has begun to take root between the cracks in busted sidewalks in some of the poorest neighborhoods in the United States. Areas such as West Oakland, California; the South Bronx, New York; Detroit, Michigan; and Milwaukee, Wisconsin, are hardly the places in which one would expect to hear the food movement narrative. And yet these neighborhoods are host to a growing array of projects encouraging low-income people and people of color to create local, sustainable food systems using many of the same tools—farmers markets, community gardens, and the like—as are found in wealthier locales. Because local and state governments have traditionally ignored food in their policy and planning goals (Pothukuchi and Kaufman 2000) and because food movements tend to view the federal government as a staunch supporter of industrial agriculture (Buttel, Larson, and Gillespie 1990), food justice projects tend to operate through grassroots, community-based organizations. According to food justice activists, local food systems can help to create social justice in the form of stable, meaningful jobs in communities that have been decimated by the decline of the manufacturing industry (Jones 2008). They can also contribute to sustainability by providing access to environmental benefits, such as healthy food, green space, and outdoor activities, that are often missing from the places inhabited by low-income people and people of color. Indeed, food justice activists often refer to these neighborhoods as *food deserts* because it is so difficult to find fresh food there.

The food justice movement combines an analysis of racial and economic injustice with practical support for environmentally sustainable alternatives that can provide economic empowerment and access to environmental benefits in marginalized communities. Its race- and class-conscious analysis expands that of the food movement to include not only ecological sustainability but also social justice. The commonly celebrated notion of *sustainability* indicates the mutual dependence and necessity of economic growth, environmental protection, and social equity (Campbell 1996). In practice, however, the equity component has been systematically marginalized, leading Agyeman, Bullard, and Evans to argue for a *just sustainability* that can "ensure a better quality of life for all, now, and into the future, in a just and equitable manner, while living within the limits of supporting ecosystems" (2003, 2). Common examples of just sustainability activism include campaigns for transit-oriented land-use planning, waste and toxic materials reduction, and retrofitting affordable housing units for energy efficiency (Agyeman 2005). Each of these projects holds benefits for low-income people and people of color while helping to reduce resource use. The vision espoused by many food justice activists goes beyond one in which wealthy consumers vote with their forks in favor of a more environmentally sustainable food system to imagine that all communities, regardless of race or income, can have both increased access to healthy food and the power to influence a food system that prioritizes environmental and human needs over agribusiness profits. This vision clearly weaves together justice and sustainability.

Cultivating Food Justice reflects the aims of the food justice movement. Our goal is to highlight the alternative imaginings eclipsed by the pervasiveness of the food movement narrative, to offer other voices and visions of how past and present human communities relate food, agriculture, and ecology. Not all of the community-based efforts described in this volume embrace the relatively new label of food justice. However, each of these efforts contains two important attributes: they are firmly rooted in the low-income communities and communities of color that most suffer from inequalities embedded in the food system; and they articulate their claims through a frame combining the environmental justice movement's emphasis on access to environmental benefits and decision-making power with the food movement's focus on creating environmentally sustainable alternatives.

Food Justice and Environmental Justice

The environmental justice movement cohered around the well-substantiated claim that low-income people and people of color bear a disproportionate share of the burden of environmental degradation. Low-income people and people of color are more likely to live in neighborhoods dominated by toxic industries and diesel emissions, or in rural areas burdened by the pesticides and dust that result from agribusiness (United Church of Christ 1987, 2007). Many environmental justice studies are largely quantitative and epidemiological in nature, aiming to prove disproportionate exposure (Pastor, Morello-Frosch, and Sadd 2006; Mohai and Saha 2003), and to link such exposure to public health outcomes (Israel et al. 2005; Petersen et al. 2006).

The academic literature has also followed communities as they have mobilized against incinerators (Cole and Foster 2000), petrochemical facilities (Allen 2003), and toxic waste (LaDuke 2004). The environmental justice movement's early tactics reveal its roots in the civil rights movement (Taylor 2000). For example, one of its first notable campaigns, which took place in predominantly African American Warren County, North Carolina, in 1982, occurred when the local government issued permits for the construction of a hazardous waste landfill despite the concerns of local residents already beleaguered by other toxic land uses. While the activists' campaign was ultimately unsuccessful, the sympathetic publicity garnered by more than five hundred protestor arrests spurred a growing social movement (Bullard 1990). In addition to opposing the permitting of locally unwanted land uses, communities have also organized for procedural justice, which encompasses the rights of all affected communities to be included in environmental decision making (Fletcher 2004; Shrader-Frechette 2002). Although definitions of environmental justice vary, the two most common components are equal protection from environmental pollution and procedural justice.

The food justice movement mirrors these two key concerns through the concepts of *food access* and *food sovereignty*. Food access is the ability to produce and consume healthy food. While the environmental justice movement is primarily concerned with preventing disproportionate exposure to toxic environmental burdens, the food justice movement works to ensure equal access to the environmental benefit of healthy food. Food sovereignty is a community's "right to define their own food and agriculture systems" (Via Campesina 2002). Like procedural justice, food sovereignty moves beyond the distribution of benefits and burdens to call for a greater distribution of power in the management of food and environmental systems.

In addition, the food justice and environmental justice movements both rely on an institutional concept of racism consistent with those of antiracist activists and the academic literature. Popular approaches to racism in the United States tend to assume that an individual must consciously make a biased decision based on race. In contrast, *institutional racism* occurs when institutions such as government agencies, the military, or the prison system adopt policies that exclude or target people of color either overtly or in their effects. From the perspective of environmental justice scholars and activists, the disproportionate burden of toxics borne by low-income people and people of color need not be linked to intentional discrimination (though the U.S. Environmental Protection Agency has tried to define it in this limited way) if the outcome is a disproportionate burden borne by communities of color. Similarly, food justice activists point to a variety of institutional policies, such as the U.S. Department of Agriculture's discrimination against African American farmers (Gilbert, Sharp, and Felin 2002) or the supermarket industry's practice of charging lower prices in suburban versus urban locations, through which communities of color have been systematically disadvantaged. An institutional approach claims that racial and economic inequalities are built into the zoning ordinances, mortgage requirements, and other policies that determine how industries, human communities, and goods and services come to exist in particular places (Massey and Denton 1998; Lipsitz 1998; Pellow 2002). Such institutional understandings of racism do not see it separable from class, but rather as producing and being produced by economic inequalities. Activists from both movements tend to highlight issues of race, although they do so with the knowledge that communities of color are disproportionately poor.

Issues of food, particularly urban agriculture, have been of concern to environmental justice activists since at least the 1980s (Pinderhughes 2003). The Second National People of Color Environmental Justice Summit, held in 2002, included three panels on sustainable agriculture and the publication of a preconference resource paper (Peña 2002). The paper identified a wide variety of concerns, many of which are addressed in this volume. Important issues include land loss among farmers of color, farmworker health and organizing rights, the enclosure of tribal lands for the public domain, lack of access by communities of color to healthy food, and the inability of the food movement to address social justice concerns. Attention to food, however, was nascent compared to the environmental justice movement's more established concerns such as air pollution and waste management.

Despite the previously described theoretical similarities and fledgling activism, academics studying environmental justice have paid little attention to food. Few scholars have responded to Gottlieb and Fisher's (1996) call, made more than fifteen years ago, to explore the nexus between these movements. When the environmental justice literature has addressed food, it has most often employed the traditional environmental justice approach focusing on disproportionate burden. For example, lands cultivated by people of color are often exposed to toxic chemicals and (often immigrant) farmworkers are disproportionately exposed to pesticides, fertilizers, and other noxious chemicals (Harrison 2006; Moses 1993). Other analyses have addressed political organizing among farmworkers (Pulido 1998), or have highlighted the importance of foods such as salmon for Native American cultural practices (Dupris, Hill, and Rodgers 2006; Wilkinson 2005). Such approaches, however, do not contextualize the disproportionate burden of toxics and access to resources experienced by low-income people of color within a broad critique of the environmental and social harms of industrial agriculture.

The body of work contained in this volume begins to fill an important omission located at the nexus of food, agriculture, and inequality. It locates instances of food injustice in the wider political, economic, and cultural systems that produce both environmental degradation and racial and economic inequality. It also highlights the ways that communities of color resist these conditions by creating local and organic food systems and practices. The volume's focus on sustainable alternatives ties the food justice movement not only to environmental justice but also to just sustainability.

Food and (Poly)culture

In addition to the pragmatic work of highlighting the development of alternatives, food justice scholarship can also help the environmental justice literature to incorporate more theoretically sophisticated understandings of race. Pulido (1996) convincingly argues that many debates within the environmental justice literature—most notably those weighing the various impacts of race and class (Anderton et al. 1994; Mohai and Bryant 1992) and those determining whether communities of color preceded local toxic land uses (Pastor, Sadd, and Hipp 2001)—are based on essentialist notions of race and racism that ignore the socially constructed and lived dynamics of racial identities and oppressions. While such studies importantly do battle with those forces seeking to undermine the claims of the environmental justice movement, they presume that race is a biological absolute, rather than examining the ways that racial identities are formed through a variety of social processes including exposure to environmental toxins or a dearth of environmental benefits (Sze 2006).

Studies of food justice offer an excellent opportunity to bring the environmental justice and just sustainability literatures together with contemporary social science approaches to race because food is deeply intertwined with both personal and cultural identities. Winson (1993) refers to food as an "intimate commodity" that is literally taken inside the body and imbued with heightened significance. Not only is it a physiological necessity, but food practices—what scholars often call *foodways*—are manifestations and symbols of cultural histories and proclivities. As individuals participate in culturally defined proper ways of eating, they perform their own identities and memberships in particular groups (Douglass 1996). Food informs individuals' identities, including their racial identities, in ways that other environmental justice and sustainability issues—energy, water, garbage and so on—do not. Many of the food justice projects described in part II of this book attempt to tap into this deep relationship between food and cultural identity in order to develop explicitly race-conscious approaches to food access and food sovereignty.

Processes of individual and collective identity formation, however, are hardly static, but develop fluidly in accordance with particular cultural moments and movements. Food is culturally important, yet the relationship between food and culture is not deterministic. For example, Avakian (1997) noted that American-born descendents of Jewish and Chinese immigrants continue to cherish stigmatized foods such as matzoh or boiled chicken feet, respectively. In contrast, however, others have found that second-generation Jewish Americans in New York developed a love for Chinese food, much as the Chinese in Hong Kong embrace McDonalds fast food, precisely because they see such foods as distinct from their cultural background (Tuchman and Levine 1993; Watson 1997). Food justice activists tend to highlight particular foods as both healthy and indicative of a racial or cultural group. In doing so, they both reflect and reproduce notions of cultural identity.

Furthermore, competing political projects may espouse varying food practices as constituting a particular, proper subject. Wit's *Black Hunger*, for example, demonstrates how Black Muslims condemned soul food as "the diet of a slave mentality, an unclean, unhealthy practice of racial genocide" while the Black Panther Party "valorized [it] as an expression of pride in the cultural forms created out of and articulated through a history of black oppression" (2004, 260). By embracing or disdaining soul food, both Black Muslims and Black Panthers performed and pushed for their competing, contested understandings of black identity.

These examples also demonstrate the lack of any essentialist link between racial identities and particular foods. Some contemporary food justice efforts in low-income African American communities attempt to reframe soul food once more, this time eliminating pork and chicken and instead emphasizing vegetarian staples such as collard greens and sweet potatoes. Activists cast such foodways as a form of racial empowerment, as well as personal health and environmental sustainability, and in doing so, project a new notion of black identity (Terry 2009; Harper 2010). Additionally, food scholars have demonstrated that the cultivation and distribution of culturally identified foods can provide opportunities for economic growth, as was the case for the women depicted in Williams-Forson's 2006 book *Building Houses out of Chicken Legs*. Many food justice advocates similarly argued that links between food and cultural identity can serve as sources of wealth for their communities, providing culturally relevant underpinnings for green economic growth.

Although food can be a source of material and cultural empowerment, it can also reflect, and even create, social and economic hierarchies. It is not surprising to observe, as Goody (1998) did, that ancient kings ate much more meat than their subjects did, or that women and children disproportionately bear the burden of contemporary food shortages (Poppendieck 1999). Moreover, when people use food to perform identities, they simultaneously create exclusions. Jews who observe traditional dietary laws called Kashrut, for example, often cast their less observant counterparts as somehow improperly or not as fully Jewish in the same way that the aforementioned Black Muslims labeled pork eaters as improperly or not fully black. As particular ways of eating come to be associated with cultural identities, those who eat differently become marked as less worthy others.

By combining a prescient critique of industrial agriculture with a consumption-driven response, the food movement marks a particular set of foodways (organic, local, and slow foods) as right and proper, and condemns what Michael Pollan (2006) calls "industrial eaters" as less worthy others. This is one reason that the food movement is sometimes interpreted as elitist. Pollan's analysis presumes that foodways are individual choices removed from their social and economic constraints. In a critique rooted in the environmental justice movement, food justice activists demonstrate that institutional racism, in its intersections with economic inequality, has stripped communities of color of their local food sovereignty, preventing many of them from eating in the way the food movement describes as proper. By aiming their critique at the structural barriers communities of color face to accessing local and organic food, the food justice movement hopes to craft collective racial and cultural identities through the celebration of particular foods while reaching out to those whose foodways still reflect the dominance of American agribusiness.

Broadly, scholars have established a relationship among food, social status, and cultural identity that is extremely fluid and certainly not proscriptive or straightforward. This complexity, however, is precisely the reason that studies of food have so much to offer to the environmental justice and just sustainability literatures, which have thus far rarely engaged with social science understandings of race that conceptualize it not as biologically determined, but as learned, performed, and practiced through social interaction (cf. Pulido 1996, 2000; Park and Pellow 2004; Sze 2006). Moreover, the field of food studies has devoted significant attention to the relationship between food and cultural identity, but has rarely examined the structural context in which these relationships occur. In the contemporary United States, this context consists of an environmentally and socially destructive centralized agribusiness system in which race and class inform inequalities of material resources and decision-making power. Neither has the food studies literature addressed the power of deeply held and meaningful foodways to inspire culturally resonant social movements against this agribusiness system. It is only in this newly emergent body of work on food justice that the racialized political economy of food production and distribution meets the cultural politics of food consumption. In this context, some of the key possibilities of cultivating food justice arise.

Overview

This book is divided into four sections, each of which is unified by an overarching theme. In part I, The Production of Unequal Access, the authors analyze the social and economic factors that prevent low-income people and people of color from producing healthy, affordable, and culturally appropriate food. In the first of these chapters, Norgaard, Reed, and Van Horn explore the *racialized environmental history* through which white settlement and the appropriation of Native American lands forced tribes to abandon sustainable, traditional foodways. Using a case study of the Karuk Tribe of Northern California, the authors illuminate how institutional racism can create both ecological degradation and human hunger. Next, Green, Green, and Kleiner investigate the severe decline in African American farming during the twentieth century, highlighting the plantation political economy, sharecropping systems, limits on civil rights, and discrimination by the U.S. Department of Agriculture. Finally, Minkoff-Zern, Peluso, Sowerwine, and Getz examine how U.S. federal policies have affected both access to material resources and the construction of racial identities among Chinese, Japanese, and Hmong immigrant farmers and farmworkers. The authors find that diverse U.S. policies—ranging from immigration exclusion to labor laws—assume (and therefore create) a white farmer as their subject. These processes can criminalize the agricultural practices common among past and present Asian immigrant farmers. In each of these chapters, institutional racism prevents communities of color from participating in both agribusiness and sustainable agriculture.

Part II, Consumption Denied, contains two chapters that employ the lens of political ecology to explore the reasons many people of color cannot purchase what they and their ancestors once produced. McClintock traces how Oakland, California's predominantly black and Latino/a "food desert" landscapes were created through a combination of industrial location, urban planning, and racist mortgage lending practices.

Additionally, Brown and Getz explore the paradox of farmworker food insecurity in California's Central Valley, where those who do the bulk of cultivation in the nation's most bountiful agricultural region often go hungry. These chapters locate food desert residents' and farmworkers' hunger within the context of local and national policies that foster the ability of uneven capital development, rather than human needs, to shape urban and rural environments.

Part III, Will Work for Food Justice, offers a series of case studies of marginalized peoples working to create food systems that are both environmentally sustainable and socially just. The stories through which these communities give meaning to their work are often quite different from the narratives that inform the food movement. This section begins with a chapter by Morales exploring an initiative called Growing Food and Justice for All. Led largely by people of color, this initiative provides support to grassroots organizations using sustainable food systems as a tool to dismantle racism. Morales's chapter presents an overview of the initiative's approach and spotlights organizations working within it. The following two chapters present examples of communities of color engaged in sustainable agricultural production. McCutcheon demonstrates how Black Nationalist religions in Georgia and South Carolina use food and health to address hunger, as well as religious and race-based goals. She argues that in marginalized communities, food self-reliance can become a means toward just sustainability. Mares and Peña then draw on ethnographic data from two predominantly Latino/a urban community garden projects to analyze how food and farming can connect growers to local and extralocal landscapes. The authors conclude that these gardeners, many of whom are indigenous people, create an autotopography linking their life experiences to a deep sense of place. Next, Harper examines how color-blind and class-blind approaches assuming race and class are unrelated to food practices emerge in the vegan and animal rights movements. In contrast to previous scholarship aimed at understanding how such discourses affects whites, Harper examines the effect of whitened discourses on vegan activists of color. Part III concludes with a chapter by McEntee, who argues that by presenting alternative food practices as a way of life rather than a set of individual choices, the food movement misses an opportunity to appeal to working-class rural populations that embrace *traditional* rather than *contemporary* notions of localism. Movements for food justice, he argues, need a more compassionate approach that understands the constraints of poverty, and that take low-income rural peoples' lived experiences as a point of departure.

The first three parts of the book fit within and extend the analysis offered by the food justice movement. Part IV, Future Directions, contains three chapters critiquing omissions and slippages in current food justice scholarship and activism. First, Guthman draws upon her extensive work to describe how the white volunteers who support many food justice programs remain inspired by the food movement's narrative. Thus, even the explicitly antiracist ethic of food justice remains complicatedly entangled with white cultural discourses. DuPuis, Harrison, and Goodman offer a reflection on why food movements have largely failed to emphasize issues of justice. Any movement for food justice, the authors argue, must contend with the multiple and contradictory meanings that have historically been embedded in this ideal, and they offer several suggestions for how food justice activists can do so. Finally, Holt-Giménez highlights the potential overlap between U.S. communities' demands for food justice and the international peasants movement's calls for food sovereignty. He argues that a food justice framework can help the U.S. food movement shift to a more transformational approach to food politics, through which it might become a powerful ally in a worldwide movement to replace industrial agriculture with a more just and sustainable food system. Taken together, these chapters push the food justice movement to challenge the food movement, and in doing so, highlight important future directions for theory and practice.

The individual chapters in *Cultivating Food Justice* advance important arguments as to how racial and economic inequalities have shaped both the food system and the food movement. The chapter authors use a variety of theoretical lenses including environmental justice, political ecology, and critical race theory. Collectively, the contributions reveal three critical insights. First, the chapters highlight a variety of vehicles, including state policies and market-exchange relations, through which race and class affect the production, distribution, and consumption of food. Second, the book demonstrates how communities of color are weaving together new narratives and foodways to create just and sustainable local food systems. Last, the chapters suggest that issues of meaning and identity formation are paramount to communities' engagements with alternative food systems, and that those seeking to create interracial coalitions on these issues would do well to begin by listening to one another's stories. We hope that this volume can contribute to such a dialogue by making visible diverse accounts of the relationships between food, environmentalism, justice, race, and identity. We believe that this emergent body of literature can help to nurture fertile soil in which a polyculture of approaches to a just and sustainable agriculture can thrive.

Notes

1. In California, where many of the chapters in the book are set, mechanization proceeded white settlement, and the industrial farm has been dominant since that time.

2. We use the term *food movement* to indicate a broad range of proponents including those of organic, local, and slow food. Allen et al. (2003) use the term *alternative agrifood movement*, but we prefer the simpler, though admittedly less descriptive *food movement*.

3. Understandings of modernity as opposed to community and authenticity can be found in the works of early sociologists including Emile Durkheim's concept of *anomie* ([1893] 1997), Ferdinand Tönnies's *gemeinschaft* and *gesellschaft* ([1887] 2005), and Georg Simmel's "The Metropolis in Mental Life" (1950). More contemporary scholars such as Anselm Strauss (1961) and Raymond Williams (1973) have observed the continuation of this trend.

References

Agyeman, Julian. 2005. *Sustainable Communities and the Challenge of Environmental Justice*. New York: New York University Press.

Agyeman, Julian, Robert Bullard, and Bob Evans. 2003. *Just Sustainabilities: Development in an Unequal World*. Cambridge, MA: MIT Press.

Allen, Barbara L. 2003. *Uneasy Alchemy: Citizens and Experts in Louisiana's Chemical Corridor Disputes*. Cambridge, MA: MIT Press.

Allen, Patricia, Margaret FitzSimmons, Michael Goodman, and Keith Warner. 2003. Shifting Plates in the Agrifood Landscape: The Tectonics of Alternative Agrifood Initiatives in California. *Journal of Rural Studies* 19 (1): 61–75.

Altieri, Miguel. 2000. Ecological Impacts of Industrial Agriculture and the Possibilities for Truly Sustainable Farming. In *Hungry for Profit*, ed. F. Magdoff, J. B. Foster, and F. H. Buttel, 77–92. New York: Monthly Review Press.

Anderton, Douglas, Andy B. Anderson, Peter Rossi, John Oakes, Fraser Eleanor Weber Michael, and Edward Calabrese. 1994. Hazardous Waste Facilities: Environmental Equity Issues in Metropolitan Areas. *Evaluation Review* 18 (2): 123–140.

Avakian, Arlene Voski, ed. 1997. *Through the Kitchen Window: Women Explore the Intimate Meanings of Food and Cooking*. Boston: Beacon Press.

Bell, M. 2004. *Farming for Us All: Practical Agriculture and the Cultivation of Sustainability*. University Park: Pennsylvania State University Press.

Bullard, Robert. 1990. *Dumping in Dixie*. Boulder, CO: Westview Press.

Buttel, Fred, Olaf F. Larson, and Gilbert W. Gillespie Jr. 1990. *The Sociology of Agriculture*. New York: Greenwood Press.

Campbell, S. 1996. Green Cities, Growing Cities, Just Cities: Urban Planning and the Contradictions of Sustainable Development. *Journal of the American Planning Association* 62 (3): 296–312.

Chu, Jeff. 2009. Are Fair Trade Goods Recession Proof? Wal-mart, Starbucks and Cadbury Hope So. <http://www.fastcompany.com/blog/jeff-chu/inquisition/fair-trade-%20recession-proof> (accessed July 13, 2009).

Cole, Luke, and Sheila Foster. 2000. *From the Ground Up: Environmental Racism and the Rise of the Environmental Justice Movement*. New York: NYU Press.

Douglass, Mary. 1996. *Purity and Danger: An Analysis of the Concepts of Purity and Taboo*. New York: Taylor.

Dupris, Joseph, Kathleen S. Hill, and William H. Rodgers. 2006. *The Si'lailo Way: Indians, Salmon, and Law on the Columbia River*. Durham, NC: Carolina Academic Press.

Durkheim, Emile. [1893] 1997. *Suicide: A Study in Sociology*. Washington, DC: Free Press.

Fletcher, Thomas H. 2004. *From Love Canal to Environmental Justice: The Politics of Hazardous Waste on the Canada–U.S. Border*. Ontario, Canada: Broadview Press.

Gilbert, J., G. Sharp, and S. Felin. 2002. The Loss and Persistence of Black-owned Farms and Farmland: A Review of the Research Literature and Its Implications. *Southern Rural Sociology* 18: 1–30.

Goody, Jack. 1998. *Food and Love: A Cultural History of East and West*. London: Verso.

Gottlieb, R., and A. Fisher. 1996. "First Feed the Face": Environmental Justice and Community Food Security. *Antipode* 29 (2): 193–203.

Harper, A. Breeze. 2010. *Sistah Vegan: Food, Identity, Health and Society: Black Female Vegans Speak*. New York: Lantern.

Harrison, Jill. 2006. Accidents and Invisibilities: Scaled Discourse and the Naturalization of Regulatory Neglect in California's Pesticide Drift Conflict. *Political Geography* 25: 506–529.

Israel, B. A., E. A. Parker, Z. Rowe, A. Salvatore, M. Minkler, and J. Lopez. 2005. Community-based Participatory Research: Lessons Learned from the Centers for Children's Environmental Health and Disease Prevention Research. *Environmental Health Perspectives* 113 (10): 1463–1471.

Jones, Van. 2008. *The Green Collar Economy: How One Solution Can Fix Our Two Biggest Problems*. New York: Harper One.

Just Food. 2010. Food Justice. <http://www.justfood.org/food-justice> (accessed July 7, 2010).

Kingsolver, Barbara, with Steven L. Hopp and Camille Kingsolver. 2007. *Animal, Vegetable, Miracle: A Year of Food Life*. New York: Harper Collins.

LaDuke, Winona. 2004. *Indigenous People, Power and Politics: A Renewable Future for the Seventh Generation*. Minneapolis, MN: Honor the Earth.

Lipsitz, George. 1998. *The Possessive Investment in Whiteness: How White People Profit from Identity Politics*. Philadelphia: Temple University Press.

Magdoff, F., J. B. Foster, and F. Buttel, eds. 2000. *Hungry for Profit: The Agribusiness Threat to Farmers, Food, and the Environment*. New York: Monthly Review Press.

Massey, Douglass, and Nancy Denton. 1998. *American Apartheid: Segregation and the Making of the Underclass*. Cambridge, MA: Harvard University Press.

Mohai, P., and B. Bryant. 1992. Environmental Injustice: Weighing Race and Class as Factors in the Distribution of Environmental Hazards. *University of Colorado Law Review* 63 (4): 921–932.

Mohai, P., and R. K. Saha. 2003. Reassessing Race and Class Disparities in Environmental Justice Research Using Distance-Based Methods. Paper presented at the annual meeting of the American Sociological Association, Atlanta Hilton Hotel, Atlanta, GA, August 16. <http://www.allacademic.com/meta/p107607_index.html> (accessed May 26, 2009).

Moses, Marion. 1993. Farmworkers and Pesticides. In *Confronting Environmental Racism: Voices from the Grassroots*, ed. Robert Bullard, 161–178. Boston: South End Press.

Nestle, Marion. 2006. *What to Eat*. New York: Farrar, Straus & Giroux.

Nestle, Marion. 2002. *Food Politics: How the Food Industry Influences Nutrition and Health*. Berkeley, CA: UC Press.

Omi, Michael, and Howard Winant. 1994. *Racial Formation in the United States: From the 1960s to the 1990s*. London: Routledge.

Park, Lisa Sun-Hee, and David Pellow. 2004. Racial Formation, Environmental Racism, and the Emergence of Silicon Valley. *Ethnicities* 4 (3): 403–424.

Pastor, M., R. Morello-Frosch, and J. Sadd. 2006. Breathless: Air Quality, Schools, and Environmental Justice in California. *Policy Studies Journal: The Journal of the Policy Studies Organization* 34 (3): 337–362.

Pastor, M., J. Sadd, and J. Hipp. 2001. Which Came First? Toxic Facilities, Minority Move-in, and Environmental Justice. *Journal of Urban Affairs* 23 (1): 1–21.

Pellow, David N. 2002. *Garbage Wars: The Struggle for Environmental Justice in Chicago*. Cambridge, MA: MIT Press.

Peña, Devon G. 2002. Environmental Justice and Sustainable Agriculture: Linking Social and Ecological Sides of Sustainability. *Occasional Paper Series*, Second National People of Color Environmental Leadership Summit, Washington, DC, October 23–27. <http://www.ejrc.cau.edu/summit2/SustainableAg.pdf> (accessed July 1, 2010).

Petersen, D., M. Minkler, V. Vasquez Breckwich, and A. Baden. 2006. Community-based Participatory Research as a Tool for Policy Change: A Case Study of the Southern California Environmental Justice Collaborative. *Review of Policy Research* 23 (2): 339–354.

Pinderhughes, Raquel. 2003. Democratizing Environmental Assets: The Role of Urban Gardens and Farms in Reducing Poverty. In *Natural Assets: Democratizing Ownership of Nature*, ed. James Boyce and Barry Shelly, 299–312. New York: Russell Sage Publications.

Pollan, Michael. 2006. *The Omnivore's Dilemma: A Natural History of Four Meals*. New York: Penguin.

Pollan, Michael. 2007. Unhappy Meals. *The New York Times Magazine*, January 28. <http://www.michaelpollan.com/article.php?id=87> (accessed October 27, 2009).

Pollan, Michael. 2008. *In Defense of Food: An Eater's Manifesto*. New York: Penguin.

Popoendiek, Janet. 1999. *Sweet Charity?: Emergency Food and the End of Entitlement*. New York: Penguin.

Pothukuchi, Kami, and Jerome L. Kaufman. 2000. The Food System: A Stranger to the Planning Field. *Journal of the American Planning Association* 66 (2): 113–124.

Prentice, Jessica. 2007. The Birth of Locavore. Oxford University Press. <http://blog.oup.com/2007/11/prentice/> (accessed July 13, 2009).

Pulido, Laura. 1996. A Critical Review of the Methodology of Environmental Racism Research. *Antipode* 28 (2): 142–159.

Pulido, Laura. 1998. Ecological Legitimacy and Cultural Essentialism: Hispano Grazing in Northern New Mexico. In *Chicano Culture, Ecology, Politics: Subversive Kin*, ed. D. Pena., 121–140. Tucson: University of Arizona Press.

Pulido, Laura. 2000. Rethinking Environmental Racism: White Privilege and Urban Development in Southern California. *Annals of the Association of American Geographers* 90 (1): 12–40.

Schlosser, Eric. 2001. *Fast Food Nation*. New York: Penguin.

Shrader-Frechette, Kristen. 2002. *Environmental Justice: Creating Equality, Reclaiming Democracy*. New York: Oxford University Press.

Simmel, G. 1950. The Stranger. In *The Sociology of Georg Simmel*, trans. K. Wolff, 370–377. Glencoe, IL: The Free Press.

Strauss, Anselm. 1961. *Images of the American City*. Glencoe, IL: The Free Press.

Sze, Julie. 2006. *Noxious New York: The Racial Politics of Urban Health and Environmental Justice*. Cambridge, MA: MIT Press.

Taylor, Dorceta. 2000. The Rise of the Environmental Justice Paradigm. *American Behavioral Scientist* 43 (4): 508–590.

Terry, Bryant. 2009. *Vegan Soul Kitchen: Fresh, Healthy, Creative African-American Cuisine*. New York: Da Capo Press.

Tönnies, Ferdinand. [1887] 2005. *Gemeinschaft und Gesellschaft.* Darmstadt: Wissenschaftliche Buchgesellschaft.

Tuchman, Gail, and Harry Gene Levine. 1993. New York Jews and Chinese Food: The Social Construction of an Ethnic Pattern. *Journal of Contemporary Ethnography* 22 (3): 382–407.

United Church of Christ. 1987. *Toxic Wastes and Race in the United States: A National Report on the Racial and Socio-economic Characteristics with Hazardous Waste Sites.* New York: United Church of Christ.

United Church of Christ. 2007. *Toxic Waste and Race at Twenty.* New York: United Church of Christ.

Via Campesina. 2002. Food Sovereignty. Flyer distributed at the World Food Summit +5, Rome. <http://viacampesina.org> (accessed June 5, 2009).

Watson, James L. 1997. *Golden Arches East: McDonalds in East Asia.* Palo Alto, CA: Stanford University Press.

White, Monica Marie. 2010. Shouldering Responsibility for the Delivery of Human Rights: A Case Study of the D-Town Farmers of Detroit. *Race/Ethnicity: Multidisciplinary Global Perspectives* 3 (2): 189–211.

Wilkinson, Charles. 2005. *Blood Struggle: The Rise of Modern Indian Nations.* New York: W. W. Norton and Co.

Williams, Raymond. 1973. *The Country and the City.* New York: Oxford University Press.

Williams-Forson, Psyche A. 2006. *Building Houses out of Chicken Legs: Black Women, Food, and Power.* Chapel Hill: University of North Carolina Press.

Winne, Mark. 2008. *Closing the Food Gap: Resetting the Table in the Land of Plenty.* Boston: Beacon Press.

Winson, A. 1993. *The Intimate Commodity.* Toronto, Canada: Garamond Press.

Wit, D. 2004. *Black Hunger: Soul Food and America.* Minneapolis: University of Minnesota Press.

I

The Production of Unequal Access

2

A Continuing Legacy

Institutional Racism, Hunger, and Nutritional Justice on the Klamath

Kari Marie Norgaard, Ron Reed, and Carolina Van Horn

Karuk people have relied directly on the land and rivers of the Klamath Mountains for food since "time immemorial." So vast was the abundance of salmon, sturgeon, steelhead, lamprey, and forest food resources that the Karuk were among the wealthiest people in the region that would become known as California. These foods flourished in conjunction with sophisticated Karuk land management practices, including the regulation of the fisheries and the management of the forest through fire (Salter 2003; McEvoy 1986). Ceremonial practices including the First Salmon Ceremony regulated the timing of fishing to allow for escapement and thus continued prosperous runs. Forests were burned to stimulate production of food species, especially acorns and bulbs. Burning also influenced the local hydraulic cycles, increasing seasonal runoff into creeks. The diversity of available food resources provided a safety net should one species fail to produce a significant harvest in a given year. Thus while salmon were centrally important, other food resources were consumed fresh and preserved to provide throughout the seasons.

With the invasion of their lands by European Americans in the 1850s, the life circumstances of Karuk people changed considerably. Today Karuks are among the hungriest and poorest people in the state. Median income for Karuk families is $13,000, and 90 percent of tribal members live below the poverty line. Genocide and forced assimilation over the past century have damaged traditional knowledge and relationships with the land and led to changes in the people's tastes and desires. Yet despite dramatic events that took place during the Gold Rush, the testimony of elders about foods they ate until recently indicates that considerable changes have also occurred within the last generation, suggesting that contemporary circumstances, as well as historical ones, produce Karuk hunger. Even tribal members in their early thirties recount significant changes in the number of fish in their diet since childhood. Four dams on the Klamath River figure centrally in this fact. Since 1962, these dams have blocked access to 90 percent of the Spring Chinook salmon spawning habitat. When the Spring Chinook population plummeted in the 1970s, Karuk people attained the dubious honor of experiencing one of the most recent and dramatic diet shifts of any Native tribe in the United States.

Spring Chinook have been the single most important food source to decline, but there are at least twenty-five species of plants, animals, and fungi that form part of the traditional diet to which Karuk people are currently denied or have only limited access. Without salmon and tan oak acorns, Karuk people are currently denied access to foods that represented upward of 50 percent of their traditional diet (see figure 2.1). With the destruction of the once abundant riverine food sources, a significant percentage of tribal members rely on commodity or store-bought foods in lieu of salmon and other traditional foods. Food insecurity within the Karuk Tribe is evidenced by the fact that a survey conducted by the tribe in 2005 found that 42 percent of respondents living in the Klamath River area received some kind of food assistance.[1] One in five

Figure 2.1
Grinding acorns
Photo courtesy of the Karuk Tribe of California

respondents use food from food assistance programs on a daily basis. The percentage of families living in poverty in Karuk ancestral territory is nearly three times that of the United States as a whole. This dramatic reversal in food access is the direct result of the systematic, state-sponsored disruptions of long-standing traditional Karuk relationships with the land. Indeed poverty, hunger, and a wide range of cultural struggles experienced throughout Indian Country[2] today are the result of similar histories.

In this chapter we describe the processes through which Karuk people became hungry. This story is important on its own terms. And understanding why and how this group of people who had survived for tens of thousands of years off the land became hungry is also important for any understanding of food or environmental justice. We begin with a review of current literature on racism and environmental justice. We then use the ongoing struggle of the Karuk Tribe of California to maintain access to their traditional foods to illustrate how the production of hunger has been the result of a series of a "racial projects" through which traditional Karuk management practices have been damaged, wealth has been transferred to non-Indian hands, and the environment has been degraded. While a "materialist" basis for food and wealth in the natural world is acknowledged by Native scholars, the importance of land for the accumulation of wealth, and its absence for the production of hunger, has remained outside social scientific conceptions of institutionalized racism, environmental justice, racial formation, or food studies. We therefore aim to situate the production of Karuk hunger within a more integrated theory of environmental history, racism, and racial formation. Here we describe three racial projects significant for the production of hunger in today's Karuk community. These projects are outright genocide, lack of recognition of land occupancy and title, and forced assimilation. We indicate throughout how each of these state actions has disrupted Karuk cultural management practices, and in so doing produced hunger alongside ecological damage. Lest readers fall into the myth that the production of hunger took place in the past, we emphasize that lack of recognition of land title and forced assimilation are very much ongoing today. Current actions by the state of California and multiple bodies in the federal government, such as the failure to recognize Karuk fishing rights, Karuk land tenure, and Karuk traditional management practices, as well as the regulation of water resources by the Bureau of Reclamation and California Northwest Regional Water Quality Water Board, and the licensing of dams by the Federal Energy Regulatory Commission, can and should be understood as current racial projects that are very much behind the production of today's hunger. Forced assimilation happens as the aforementioned actions of the state deny Karuk people access to the land and food resources needed to sustain culture and livelihood. Forced assimilation happens even more overtly when, for example, game wardens arrest Karuk Tribe members for fishing according to tribal custom rather than state regulation.

Institutional Racism, Racial Formation, and Racial Projects

Our proposition that hunger in the Karuk community today is a product of denied access to traditional foods rests on the lens of a racialized environmental history. Early theories of racial inequality, including the work of W. E. B. DuBois ([1903] 2007), Manning Marabel (1983), and Walter Rodney (1974), explicitly include the importance of land as a source of wealth (and its absence as a source of poverty). Yet contemporary race scholarship has generally failed to incorporate the environment or environmental history in racial analyses. In tightening these connections we build on Omi and Winant's important work on racial projects, racial formation, and institutional racism (1994). Omi and Winant assert that racism and the racial categories of today can only be understood through attention to historical process they call *racial formation*. Racial formation occurs as the codification of economic and political conflict produces racial categories. Similarly, institutional racism indicates that racial disadvantage is built into the social structure. Howard Winant defines institutional racism as "the routinized outcome of practices that create or reproduce hierarchical social structure based on essentialized racial categories" (2004, 126). Yet despite the emphasis on history, attention to the importance of land as a source of wealth is surprisingly absent within scholarship on institutional racism. Instead, in contemporary theory, institutional racism has been understood as a function of disproportionate access to social resources such as educational opportunity or other forms of social, economic, and cultural capital (e.g., Stretesky and Hogan 1998), leaving aside the role of access to environmental resources in the reproduction of poverty and wealth. There are important exceptions. For example, in their study of racial formation in Silicon Valley, Park and Pellow apply the framework of racial formation to show how institutional racism is a "complex set of practices supported by the linked exploitation of people and natural resources" (2004, 403). They argue that "racial formation in the United

States has always been characterized by an underlying link between ecological and racial domination" (408), and emphasize that attention to ecological degradation enhances our understanding of race and racism in important ways: "If we follow racial formation theory and we agree that racism has shaped the very geography of American life across a number of sociohistorical periods, then we must admit that we cannot fully understand that social geography without also acknowledging that the exploitation of people of color and of natural resources have gone hand in hand" (421). In so doing, their work is a crucial and powerful piece connecting environmental and race theories.

Environmental Justice

The present situation in which Karuk people face hunger resulting from denied access to their traditional food fits within the framework of what is known as "environmental justice." For the past several decades it has been recognized that poor people and people of color are most likely to pay the price of various forms of environmental degradation ranging from toxic exposure in communities from landfill sites to workplace exposure (e.g., Bullard 1993). Within environmental justice scholarship, early work emphasized the need for the wider understanding of environmentalism that attention to race engendered. But as Park and Pellow also note, much of the environmental justice field has developed around the inclusion of race as a variable, focusing on descriptions of the unequal experience of people of color, but failing to incorporate powerful race theories such as racial formation or institutional racism with existing environmental theory.

Even in cases where institutional racism is employed (e.g., Bullard 1993), most discussions have focused on historical dynamics of housing segregation and the enforcement of health and safety violations. These instances reflect disparate access to social resources such as legal council and political representation, but stop short of taking into account a larger view of institutional racism in the production of wealth and poverty through disruption of relationships with land, or the importance of maintaining relationships with land as a means of carrying out culture. These latter features of institutional racism are central to understanding the impoverishment and genocide of Native people. Similarly, most discussion of environmental racism faced by Native Americans has focused on the very significant issues of mining and exposure through waste trading, landfills, and waste incinerators (Gedicks 1994). These circumstances are more similar to the conditions faced by African Americans fighting toxic exposure in urban settings, which led to the emergence of an environmental justice framing. Institutional racism with respect to Native people in the literature is most often discussed as the absence of economic infrastructure, unemployment, and inadequate education and health care, all forms of institutional racism that parallel the political circumstances and history of other urban-based racial minorities. Yet Native environmental justice calls us to move beyond the urban and spatial frames that have been so important to the field of environmental justice. Thus, we aim to show how institutionalized racism manifests not only as a disproportionate burden of exposure to environmental hazards, but also in denied access to decision making and control over resources. We aim to illustrate how the production of hunger has been simultaneous with the degradation of culture and the land. We draw upon and develop a lens of racialized environmental history to see what disrupted the Karuk people's ability to consume subsistence food.

The Production of Food Insecurity: A Racialized Environmental History

The diets of all peoples and cultures change over time. This fact can be seen as "natural." For the Karuk people, however, diet has shifted dramatically in the course of recent generations through what can only be understood as very "unnatural" conditions. While extensive cultural disruption from contact for many California tribes occurred up to five hundred years ago with the establishment of the Mission system, tribes in the northern and more remote part of the state experienced little contact with settlers until the Gold Rush (Norton 1979). As a result, these tribes have retained much more of their culture, population base, and traditional food use. The Karuk Tribe of California is today the second largest tribe in the state and is host to a large percentage of the total basket weavers, native language speakers, and cultural practitioners to be found statewide. Despite their relative intactness when compared to other American Indian tribes in California, the impacts of past activities from the Gold Rush to resource extraction and genocide on the lives, culture, and lifeways of Karuk people are enormous (Raphael and House 2007; Norton 1979). We next review the three racial projects carried out by the state that are significant for the production of hunger in today's Karuk community: outright genocide, lack of recognition of land occupancy and title, and forced assimilation. Each of these actions damaged the ecosystem and disrupted Karuk cultural management, and in so doing denied Karuk people access to food. Each set of actions was part of the process of racial formation: in each circumstance, the state's economic, political, and military actions were legitimated via the judicial system and justified by racialized rhetoric.

Genocide and Relocation

If environmental racism is the unequal burden of ecological hazards imposed on people of color and their surroundings, then the European conquest was the continental embodiment of this process.
—Park and Pellow 2004, 410

Although there was some prior interchange between Karuk and non-Indian people, violent dislocation began with the entry of miners to the Klamath region the Gold Rush of 1850 and 1851. During this period of explicit genocide, the outright killing of about two thirds of Karuk people, relocation of villages, and attempts to move people onto reservations all interfered with everyday food management and gathering activities (Lowry 1999; Norton 1979; Raphael and House 2007). Western scientists and social scientists alike follow in the tradition of claiming that prior to European contact our continent was an untouched wilderness. Yet in fact Native people actively managed salmon, acorns, and hundreds of other food and cultural use species. The abundance of these species was a product of this management in which high-quality seeds were selected, the production of bulbs enhanced through harvest, oak populations reinforced through fire, and fish populations carefully managed. Most non-Indians can identify ecological degradation in the form of severe manipulations of the rivers from hydraulic placer mining, or manipulations of the forest from the imposition of new fire regimes. What seems quite beyond comprehension, especially for non-Indians, is the ecological damage occurring from the disruption of Native cultural management. If the disruption to food management still seems insignificant in the face of genocide and relocation, recall that access to food is key to both immediate and long-term survival, which is why controlling access to food resources has long been such a favorite military strategy. Consider, as well, that while authors in this volume write of inequalities in the "production" and "consumption" of "food," Karuk people speak of the foods they eat as relations. They speak of a long-standing and sacred responsibility to tend to their relations in the forest and in the

rivers through ceremonies, prayers, songs, formulas, and specific practices they call "management." Rather than doing something *to* the land,
ecological systems prosper because humans and nature work together.
Working together is part of a pact across species, a pact in which both
sides have a sacred responsibility to fulfill. Traditional foods and what
the Karuk call "cultural use species" flourish as a result of human activities, and in return, they offer themselves to be consumed.

It seems impossible for non-Indians to fully grasp the meaning or
importance of this complete contrast to the non-Indian perspectives of
"food production" and "food consumption." Instead, the significance of
American Indian relationships with the natural world are at best lost in
overglamorized and essentialized characterizations of Noble Savages,[3]
and at worst, entirely invisible. To comprehend and acknowledge Native
relationships with food would require non-Indians to recognize not only
the depth of the human scale of Native American genocide, but also the
fact that this genocide has been an assault on a spiritual order that
nourished and governed an entire field of ecological relationships.

The disruption of Karuk cultural food management was carried out
by the first three governors of California, each of whom created state-
sponsored programs promoting the killing of Indians (Hurtado 1988).
Statements by these men illustrate both the racist ideologies of the time
and the role of the state of California in the racial project of genocide.
For example, in a message to the state legislature on January 7, 1851,
Governor Burnett said that "a war of extermination will continue to be
waged between the races until the Indian race becomes extinct" (Ibid.,
135). Racial ideologies are evident as justifications for state violence in
an 1852 letter by California Governor John Bigler, asking for assistance
from the federal government in protecting white settlers in northern
California from Indians: "The acts of these Savages are sometimes signal-
ized by a ferocity worthy of the cannibals of the South Sea. They seem
to cherish an instinctive hatred towards the white race, and this is a
principle of their nature which neither time nor vicissitude can impair.
This principle of hatred is hereditary . . . Whites and Indians cannot live
in close proximity in peace" (Heizer 1974, 189).

Largely due to state-sponsored Indian extermination, the Karuk population went from about 2,700 people pre-contact to about 800 people
some time between 1880 and 1910 (McEvoy 1986). Note that just as
Omi and Winant describe, the racial project of Karuk genocide was
achieved through both ideological justifications and legal mandates.
White notions of Indians allowed settlers to enter the region and extract

whatever resources they desired, while the Marshall Doctrine explicitly legitimated the perspective that Indian lands were available for the exploitation of whites without need for compensation. Despite the racism of the time, there were attempts from some corners to address the violence. Even in these critical voices, however, it was concluded, as in this editorial, that, "the fate of the Indian is fixed. He must be annihilated by the advance of the white man. . . . But the work should not have been commenced at so early a day by the deadly rifle" (Heizer 1974, 36).

In 1851 the U.S. government negotiated a treaty with the Karuk Tribe (Hurtado 1988). However, white landowners found the treaties unappealing as they gave Indians land, flour, pack animals, dairy cattle, and beef cattle, which would likely mean Native people would work their own ranches instead of providing cheap labor. "Treaties that conflicted with agriculture and mining interests had little hope of finding support in California's state government" which "did everything possible to thwart them" (Ibid., 139–140). On July 8, 1852, due to pressure from the governor of California, Congress refused to ratify this and other California treaties of that time. As a result, eighteen California tribes, including the Karuk Tribe, which agreed to treaty terms in good faith, were left without any of the protections, land, or rights they reserved in their treaties (Hurtado 1988).

Meanwhile, in 1851 and 1852, the state of California spent $1 million per year to exterminate Native people (Chatterjee 1998). Beginning in 1856, the governor issued a bounty of $0.25 per Indian scalp, increasing it to $5.00 per Indian scalp in 1860, and reimbursed bounty hunters for the cost of ammunition and other supplies. Then, in 1864 the Hoopa Valley Indian Reservation was established and all Karuk people were ordered to leave their ancestral lands along the mid-Klamath and lower Salmon rivers and relocate to the reservation. Many people did so. Others fled to the high country or escaped and returned. Yet due to this overt displacement and absence of a collective land base, many Karuk people continue to live on the Hoopa reservation, in cities on the coast, and spread across California and Oregon. This dispersal of people had significance for their access to food, the types of food they ate (and eat today), and their ability to participate in cultural activities to tend their food resources.

Both the human and environmental impacts of the Gold Rush and early settlement are impossible to grasp. As Karuk people were killed and forced to relocate, Karuk practices of tending the land to ensure food productivity were replaced by technologies such as hydraulic placer mining, which were enormously environmentally destructive. Forests and hillsides were washed away as highly pressurized water flushed an estimated twelve billion tons of mud and soil into California rivers statewide (Merchant 1998). These actions have obvious and lasting impacts on traditional Karuk riverine foods such as salmon, steelhead, sturgeon, and lamprey.

Lack of Recognition of Land Occupancy and Title

The period of overt genocide has now ended and a significant number of Karuk people have returned to their ancestral territory and continue to carry out traditional management. Upon their return however, they encounter another racial project that underlies today's hunger, the failure of the state to recognize their land occupancy and title. Access to land is central for the management and harvesting of food. Karuk people recognize over a million acres of biologically diverse mountains and rivers as their ancestral territory. Today Karuk-owned lands consist of only 793 acres, just 0.0007 percent of ancestral territory (Quinn 2007). Instead, 98 percent of the lands that were once occupied and managed by the Karuk are now officially under the management of the U.S. Forest Service (Ibid.).

The divergent, racialized European and Karuk conceptions of land, appropriate land use, and land "ownership" underlie and in turn becomes a vehicle for the lack of recognition of Karuk land title. "Prior to the infusion of Europeans into the Upper Klamath River in 1850, ownership of land by individuals was not recognized. But the tribes, and individual people did own rights to hunt, fish, gather and manage particular portions of the surrounding landscape" (Quinn 2007). As a result of these different conceptions of land "ownership," Karuks on the whole lost lands under racialized federal acts, while on an individual level some members within the Karuk Tribe later sold parcels into non-Indian hands for low prices.

By the time Karuk people were legally allowed to leave the Hoopa reservation and return to their ancestral territory, the U.S. Forest Service had already claimed it. But the state of California's refusal to recognize Karuk land title began with the aforementioned failure of the U.S. Congress to ratify the 1851 treaty. Then in 1887, the passage of the Dawes Act or General Allotment Act provided that small parcels of land were allotted to Karuk families, and simultaneously gave the federal government power to evict Indians from their land. An equally important aspect of the law enabled whites to cheaply acquire "surplus" lands that had not been allotted to Indians (Deloria 1970). The Dawes Act is widely recognized for its attempt to establish the European system of private ownership on Indian lands. Here, non-Indian conceptions of landownership are codified into laws which together with racialized rhetoric and ideology become the vehicle for the transfer of land from Indian into white hands. The Dawes Act was designed to break up tribal land and divide it among individuals: "It was hoped that initiating Indians to the concept of private landownership would aid in integrating them into white society" (Delaney 1981, 2). Because the Karuk people did not have a reservation, and were then living on lands claimed by the U.S. Forest Service, the 1910 amendment of the Dawes Act to include forest lands was particularly significant (Delaney 1981). Through this racial project, resources were diverted from Indian to non-Indian hands and land management practices shifted from activities geared toward food production to those that would achieve profits under capitalism (timber and farming).

Then in the 1950s, with the widening of State Highway 96, the Bureau of Indian Affairs transferred land to the state of California. In the process many Indian parcels were decreased further in size to accommodate the modern two-lane highway and mandatory right of way. By 2007, thirty-five of the original ninety parcels remained in the ownership of Karuk families. Today, because very little land within Karuk ancestral territory is in private ownership, land that does come onto the market is too expensive for most Indian families or even for the Karuk Tribe to purchase.

The state's failure to recognize the legitimacy of the Karuk aboriginal occupancy and land title is an enormous feature underlying present day hunger. Land management practices from burning to the collection of mushrooms are officially the jurisdiction of the U.S. Forest Service. Similarly, hunting and gathering regulations are set by the state of California according to "white man's" rather than tribal law. Furthermore, regulations regarding deer, elk, and other food species are written with recreational hunting in mind, not subsistence. Because the Karuk do not have a reservation, they hunt on federal forest lands, but these lands are not managed and regulated with the goal of providing subsistence foods. According to Jesse Goodwin, in Karuk tradition, "the only time that we consider not hunting the deer is . . . during mating season and early spring when they are dropping their babies. . . . We give them a chance to grow up, but any time in between there was fair game for getting out food." Within Karuk culture, hunting is part of management and respects the needs of the herds to ensure they are healthy, but is flexible enough to allow for taking deer at various times of year when it is needed. Management, however, also included making sure there was sufficient habitat for deer to flourish, in part through burning, rather than simply focusing on limiting how many deer could be killed.

In contrast, state fish and game regulations focus only on how many deer can be killed and when. In order for a Karuk Tribe member to get deer legally, he or she first must buy tags and a license, requiring proof of meeting California's hunter education requirements, all of which take time and cost money. As set by the California Department of Fish and Game, in 2008 hunting licenses for state residents over the age of sixteen cost $38.85, and the first deer tag cost $26.00 and the second cost $32.30 (California Department of Fish and Game Hunting Digest 2008, 8). There is no option to obtain a third tag. Yet the hunting season for the zones in Karuk ancestral territory lasts just over a month. And it is nearly impossible to make two deer last an entire year—especially when shared with extended family, including elders who can no longer hunt, and when the venison is being served at ceremonies. Karuk tribal member and cultural practitioner David Arwood notes, "Our way of life has been taken away from us. We can no longer gather the food that we [once] gathered. We have pretty much lost the ability to gather those foods and to manage the land the way our ancestors managed the land." If a Karuk person hunts "out of season" or gets a deer without purchasing a tag from the state it is considered poaching. Getting caught for poaching has a variety of consequences depending on the circumstances and if it is a repeated offense. Karuk tribal member Jesse Goodwin explains that "usually, they just take our gun rights away from us, try to see if there's any way of us never being able to do it again, and then after that they send you to jail." Mushroom regulations too are a source of tension. David Arwood relates how "there were two tribal members right up here and they had them sprawled on the ground with a gun on the back of their head because they didn't cut their mushrooms in half."

The lack of recognition of land title is coupled with a lack of recognition of fishing rights. During the 1970s the federal government stepped up enforcement and forcibly denied Karuk people the right to continue their traditional fishing practices (Norton 1979) by arresting them and even incarcerating them. Karuk fishing rights have yet to be acknowledged by the U.S. government, though now tribal members may fish at one "ceremonial fishery." As tribal member Jesse Coon explains: "We can fish at the falls. Dipnet and that, you know, that's the only place we can fish really. But we're not able to go out and go hunting anymore, without

Figure 2.2
Ron Reed dipnet fishing on the Klamath River
Photo courtesy of the Karuk Tribe of California

getting in trouble for it or something, you know, so—now we have to go to the store to buy our food, and get different kind of foods that aren't sustainable for our bodies, like food that was made here for our people, you know? So a lot of it has changed that way, you know" (see figure 2.2).

Access to food and notions of how land should be used may be contested, but the state holds the ability to assert its version. Traditional Karuk Fisherman Mike Polmateer describes his experience fishing at his family's long-established site:

I fish at my family's hole up here at Dillon Creek every single day during the winter, and I'm checked for my license no less than six times per year, by the same game warden, by the same two game wardens over and over and over, trying to catch me keeping fish. They sit up here on a point with binoculars watching me catch fish, and they watch me return them to the water. Because I'm—I'm afraid. . . . There's consequences to be suffered. . . . If you send your child out in to the world right now not knowing there's consequences to be suffered, they're going to end up like many many natives, not only in this country but in other countries, in the penal system. What I'm seeing now is this penal system is—they're raising our young kids now. They're going in at 18, 19, 20 years old, not coming out until they're 27, 28, 30 years old.

Land is also important in providing a home, which in turn facilitates the return of tribal members who have dispersed outside the ancestral territory. Land and having a home create the proximity needed for day-to-day social communication through which language can be used and culture carried out. Without recognized land title many Karuk people are dispersed, making it more difficult to maintain ceremonies, continue language use, maintain and strengthen cultural identity, or carry out other vital cultural practices. While some tribal members do travel to participate in ceremonies on the ancestral territory, many aspects of cultural practice, especially those related to food, cannot be continued in these distant locations. In addition to the cultural impacts from dispersing people, the absence of recognized land title makes for poverty, as Karuk people cannot use the land for subsistence or other income, and must instead pay rent to inhabit lands "owned" by others. Viewed in light of this information, present-day hunger is clearly a result of the state's failure to recognize land title. While events such as the failure to sign the treaty and the transfer of lands to the U.S. Forest Service happened over a century ago, the continuing consequences of such events are played out every day through the ongoing legal and criminal enforcement of racialized notions of how the land should be used and for whom. We argue that only by considering this *racialized environmental history* can one understand the hunger of and racism faced by Karuk people today. The management of Karuk cultural resources by non-Indian agencies, and the fact that Karuk cultural management is mostly illegal, is also part of the next racial project we describe underlying today's hunger: that of forced assimilation.

Forced Assimilation

Explicit forced assimilation of Native people into the dominant culture occurred through boarding schools and other institutional processes. Like youth from tribes throughout Canada and the United States, Karuk children were separated from families at young ages and taken to Bureau of Indian Affairs boarding schools in Oregon and California for the specific purpose of assimilation. Boarding schools for Indian children ages six to eighteen were mandated from the end of the 1880s up through the mid 1900s. They were prevented from speaking their native language and practicing their native customs and forced to eat a diet of "Western" foods. The result was that Karuk children were separated from families, communities, culture, and traditional foods, often for many years. They were unable to learn fishing, gathering, management practices, and cultural ceremonies. "One thing I do know that changed with a lot of the salmon too was all of the kids got shipped off the river to the boarding schools," said Carrie Davis, Karuk tribal member. "My father took initiative and he learned the fishing part of his culture. His best friend didn't really catch the fishing part as much as he knows language and a lot of the ceremonial stuff. My dad never danced in a ceremony. Four years ago was the first time he'd ever danced, because he was beat for even trying to be Indian." Karuk people still struggle today to recover economically, socially, politically, and mentally from the devastation of these policies.

Forced assimilation is ongoing today, although its vehicles may be less overt than in boarding schools. Instead as we discussed in the previous section, forced assimilation occurs because a significant proportion of Karuk cultural food management and production practices are illegal. Forced assimilation also happens when Karuk food sources are so depleted that tribal people must eat government commodity foods instead (see figure 2.3). While there is no policy *designed* to change how Karuk people view and use the land parallel to the ways that boarding schools explicitly enforced "white" behaviors onto Indian people, forced

Figure 2.3
Commodity canned foods
Photo courtesy of Kari Marie Norgaard

assimilation takes place at a variety of levels from explicit use of force, threat, and fear of force, to a range of reasons that keep Karuk people from participating in cultural practices. Again, the production of hunger is a present-day example of environmental justice intimately interwoven with racialized *environmental* history. The assimilation in question is assimilation to non-Indian understandings, values, and uses of the natural world. We therefore expand upon the significance of these disruptions of Karuk management for hunger here in our discussion of forced assimilation.

Whereas long-standing cultural traditions existed for regulating and sharing fish and other resources both within the Karuk Tribe and between neighboring tribes, the entry of non-Indian groups into the region led to conflict and dramatic resource depletion (McEvoy 1986). As noted earlier, cultural management practices used to enhance food resources from burning to fishing have been made illegal by federal, state, and other agencies. For example, Europeans did not understand the role of fire in the forest ecosystem. Since the Gold Rush period, Karuk people have been forcibly prevented from setting fires needed to manage the forest, prolong spring runoff, and create proper growing conditions for acorns and other foods (Margolin 1993; Anderson 2005). For many years following white settlement in their territory Karuk people were simply shot for engaging in cultural practices such as setting fires . Non-Indian fishing regulations, such as those developed and enforced through California Department of Fish and Game, have often failed to take into account the Karuk as original inhabitants, their inalienable right to subsistence harvesting, and the sustainable nature of Karuk harvests. As a result they have attempted to balance the subsistence needs of Karuk people with recreational desires of non-Indians from outside the area. Karuk tribal member Vera Davis notes the imbalance and injustice of this view:

Now I don't think that no one has a right to tell us when we can do it when you have people who pay hundreds of dollars to come in, kill the venison and get the horns. I don't think that is fair because this is our livelihood. . . . We had supplies from the river the year round. We hadn't been told that we couldn't get our fish any time of the year. That was put there for us by the Creator, and when we were hungry we went to the river and got our fish. Vera Davis. (qtd. in Salter 2003, 32)

Even more dramatic is the outright refusal of recognition of the Karuk fishery. In the 2005 Karuk Health and Fish Consumption Survey individuals were asked whether members of their household had been

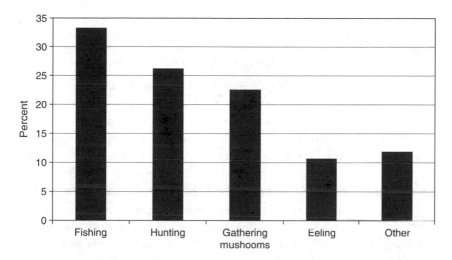

Figure 2.4
Percentage of tribal members who report harassment or questioning while gathering food by these methods. *Source*: Norgaard 2005

questioned or harassed by game wardens while fishing for a number of aquatic food species. As indicated in figure 2.4, 32 percent reported that they had experienced harassment while fishing.

To be fined or have a family member imprisoned imposes a significant economic burden on families. This is a risk that many are unwilling or unable to take. Of those reporting harassment, 36 percent reported that they had decreased their subsistence or ceremonial activities as a result of such contacts.

State regulations affect not only fishing and burning, but also hunting, mushroom gathering, and gathering of basketry materials. Tribal Vice-Chairman and Ceremonial Leader Leaf Hillman describes this situation: "The act of harvesting a deer or elk to be consumed by those in attendance at a tribal ceremony was once considered an honorable, almost heroic act. Great admiration, respect and celebration accompanied these acts and those who performed them. Now these acts (if they are to be done at all) must be done in great secrecy, and often in violation of Karuk custom, in order to avoid serious consequences." Tribal member Mike Polmateer explains the reality of growing up under these circumstances:

When I hunted with my uncles, for the longest time I never knew you hunted during the day. We always went and got our meat at night. And it was always about where's the game warden, you know, where's the cops, and you know, things like that. So that's one of the things that stuck in my mind as a young kid. We were always watching for headlights, you know, always trying to hide from the law, out doing what we were supposed to do, which was provide for our families. We weren't out selling meat. We weren't out selling hides.

In the 2005 Karuk Health and Fish Consumption Survey tribal members were also asked whether members of their household had been questioned or harassed by game wardens while gathering a variety of other cultural and subsistence items. Twelve percent reported such contacts while gathering basketry materials, and over 40 percent indicated harassment while gathering firewood. Twenty percent of survey respondents reported that they had decreased their subsistence or ceremonial activities as a result of such contacts. Denied access to traditional management at the hands of non-Native agencies has significant health, cultural, and spiritual impacts, including denied access to healthy foods (see Jackson 2005; Norgaard 2005). Forced assimilation through the imposition of non-Karuk management and denied access to traditional foods is the dominant racial project through which genocide, ecological degradation, and hunger are perpetuated in the present day. Yet Karuk lifeways continue to be practiced both overtly when tribe members can get away with it and covertly when they cannot. From a Karuk perspective, continuance of these traditional lifeways and practices is essential not only for food, but also for the maintenance of cultural and tribal identity, pride, self-respect, and above all, basic human dignity.

Forced assimilation reaches its most insidious form when the food species that Karuk people would like to fish for are simply not there. We began this chapter noting the importance of salmon as an abundant traditional food. The Klamath River was once the third-largest salmon producing river in the West. Yet as of 2009, the wild salmon populations of the Klamath River have been reduced to roughly 4 percent of their previous productivity. Traditional Karuk fish consumption is estimated at the extraordinary figure of 450 pounds per person per year (Hewes 1973). In contrast, today the Karuk people consume fewer than 5 pounds of salmon per person per year (Norgaard 2005). Now so few fish exist that even ceremonial salmon consumption is limited (see figure 2.5). Commercial canneries set up at the mouth of the Klamath severely impacted salmon runs during the 1920s. Then the building and operation of dams on the Klamath River, beginning in 1916 with Copco I, further decreased fish populations.

The construction of Iron Gate dam in 1962 appears to be significant, as over half of the respondents to the Karuk Health and Fish

Figure 2.5
Ron Reed and Merv George Jr. at Scottish Power Shareholders Meeting in Edinburgh,
Scotland, June 2004
Photo courtesy of Kari Marie Norgaard

Consumption Survey report that Spring Chinook became an insignificant source of food for their families during the 1960s and 1970s, although some families continued to gather significant food into the 1980s and 1990s. As coauthor Ron Reed notes, forced assimilation happens when you need something to feed your family:

A healthy riverine system has a profound effect on the people on the river. I have six children. If every one of those kids went down and fished and caught a good healthy limit like it was back in the 80s, you could pretty much fill a freezer and have nice good fish all the way through the year. But now, without a healthy riverine system the economy down here on the lower river is pretty much devastated. All the fishing community is devastated by the unhealthy riverine system. Instead of having healthy food to eat—fish—we are relegated to eating commodity foods that the government gives out. That's our subsidy: high starch foods, things that aren't so healthy that the Karuk people are pretty much forced to eat. (qtd. in Norgaard 2005, 18)

Fisheries scientists identify the five dams on the mainstem Klamath that are now owned by the corporation PacifiCorps as a major obstacle to fisheries health. As this book goes to press there is good news on the horizon. Tribes, environmentalists, commercial fishermen, farmers, and the dam owners have come to a settlement agreement on removal of the dams. If this takes place it will be the largest dam removal effort in the world.

Conclusion

We hope that this chapter has achieved a number of goals. First, we hope the story it tells helps to squarely situate food access as an issue of environmental justice. Early environmental justice work brought to light the crucial connections of racism and toxic exposure. As scholars continue to theorize the experiences of different racial groups, and activists define the connections between a wider range of social and environmental problems, our conception of environmental justice grows ever richer. Our story about how the Karuk people became hungry also contains important lessons about the long history of environmental injustice, a history that goes back much further than the commonly told history of environmental problems such as exposure to toxins—most of which were developed during World Wars I and II.

Second, while many environmental scholars and movement activists now integrate race as a key dimension of environmental problems, less attention has been paid to the incorporation of important racial theories (e.g., racial formation) with theory on the environment. Even environmental justice literature, while emphasizing the linked domination of the environment and people of color, has neither included the longer view of environmental history through which such relationships are visible, nor adequately theorized racial formation, which would allow for understanding of their significance. We hope that we have made clear the imperative of these links, and among other things demonstrated why racial formation and environmental history must inform our conceptions of food justice.

Third, we hope that race scholars will further integrate the role of environmental degradation as an interacting factor in the production of racism. This understanding that racial formation and environmental exploitation are intertwined has important lessons not only for hunger and food justice studies, but also for sociological understanding of the role of land as a source of wealth (and its absence for the reproduction of poverty and racism). In making a case that theory on institutional racism must incorporate the environment, we hope we have also illustrated why theory on racial formation and environmental history must be integrated more generally.

Finally, we have emphasized here how the destruction of the land becomes a vehicle for racism and hunger, but the reverse is also true. Traditional ecological knowledge and management have made the ecology of the Klamath River what it is today. Thus racism and cultural genocide produce further environmental decline. As they gain political and economic standing, Native American tribes including the Karuk have become increasingly involved in natural resource management. Yet tribes are disadvantaged in these settings due to both a lack of broader social understanding of their unique cultural perspectives, and a lack of acknowledgment of the violent history perpetuated against them—much less the continuing effects of this history. It has been our aim to enhance broader public understanding of both this history, and the importance of Karuk management for ecological health on the Klamath. May the Karuk people, the Klamath River, and all who live there continue to flourish.

Acknowledgments

The authors wish to thank all who were interviewed for this project for sharing their experiences and time. We thank our families for their support.

Notes

1. The 2005 Karuk Health and Fish Consumption Survey contained sixty-one questions designed to evaluate the range of economic, health, and cultural impacts for tribal members resulting from the decline in quality of the Klamath River system. Open- and closed-ended questions on the consumption and harvesting of traditional foods were developed in response to interview data. Personal and family history information on medical conditions was included, as well as information on age of death of family members. The survey was distributed to adult tribal members within the ancestral territory. The survey had a response rate of 38 percent, a total of ninety questionnaires. This is a relatively high response rate for this rural, impoverished community; still, we are unable to know the views of those who did not respond. Given community demographics, we speculate that many of those Karuk Tribe members who did not respond were more traditional, and had less income than those who did respond.

2. The term *Indian Country* is widely used by Native people in the United States to refer to lands that are legally owned and controlled by tribes, as well as metaphorically to refer to the fact that Native people create and occupy cultural spaces within the dominant culture of the United States (e.g., a major Native newspaper is *Indian Country Today*).

3. The phrase *Noble Savage* comes from a characterization of Native Americans by some European colonists in which Native people were idealized for positive qualities, yet were simultaneously viewed as inferior for being "closer to the earth." The term idealizes Native people, but is deeply racist. Native American agricultural technologies and social achievements were a source of wonder for the Europeans, but rather than recognize these as the result of sophisticated cultural accomplishments that had been learned over time, Europeans assumed that Native people were primitive and their achievements "natural." The *Noble Savage* concept emerges in conjunction with Romantic critiques of the harshness of civilization. The first use of this term is widely credited to the philosopher Jean-Jacques Rousseau, in 1755.

References

Anderson, M. Kat. 2005. *Tending the Wild: Native American Knowledge and the Management of California's Natural Resources*. Berkeley: University of California Press.

Bullard, Robert D., ed. 1993. *Confronting Environmental Racism: Voices from the Grassroots*. Boston: South End Press.

California Department of Fish and Game Hunting Digest. 2008. <http://www.dfg.ca.gov/publications/digest> (accessed January 22, 2011).

Chatterjee, Pratap. 1998. The Gold Rush Legacy: Greed, Pollution and Genocide. *Earth Island Journal* 13 (2): 199.

Delaney, Lisa. 1981. *Klamath National Forest Karuk Allotment Situation Assessment: Past, Present and Future*. Unpublished U.S. Forest Service document.

Deloria, Vine. 1970. *Custer Died for Your Sins*. New York: Avon Books.

DuBois, W. E. B. [1903] 2007. *The Souls of Black Folk*. Reprint. London: Oxford University Press.

Gedicks, A. 1994. *The New Resource Wars: Native and Environmental Struggles against Multinational Corporations*. Boston: South End Press.

Heizer, Robert F., ed. 1974. *The Destruction of California Indians: A Collection of Documents from the Period 1847 to 1865 in Which Are Described Some of the Things That Happened to Some of the Indians of California*. Santa Barbara, CA: Peregrine Smith.

Hewes, G. W. 1973. Indian Fisheries Productivity in Pre-Contact Times in the Pacific Salmon Area. *Northwest Anthropological Research Notes* 7 (3): 133–155.

Hurtado, Albert L. 1988. *Indian Survival on the California Frontier*. New Haven, CT: Yale University Press.

Jackson, Jennifer. 2005. *Nutritional Analysis of Traditional and Present Foods of the Karuk People and Development of Public Outreach Materials*. Happy Camp, CA: Karuk Tribe of California.

Lowry, Chag, ed. 1999. *Northwest Indigenous Gold Rush History: The Indian Survivors of California's Holocaust*. Arcata, CA: Indian Teacher and Educational Personnel Program.

Marabel, Manning. 1983. *How Capitalism Underdeveloped Black America: Problems in Race, Political Economy and Society*. Boston: South End Press.

Margolin, Malcolm. 1993. *The Way We Lived: California Indian Stories, Songs & Reminiscences*. Berkeley, CA: Heyday Books.

McEvoy, Arthur E. 1986. *The Fisherman's Problem: Ecology and Law in the California Fisheries 1850–1980*. New York: Cambridge University Press.

Merchant, Carolyn. 1998. *Green Versus Gold: Sources in California's Environmental History*. Washington, DC: Island Press.

Norgaard, Kari Marie. 2005. *The Effect of Altered Diet on the Health of the Karuk People*. Happy Camp, CA: Karuk Tribe of California.

Norton, Jack. 1979. *Genocide in Northwestern California: When Our Worlds Cried*. San Francisco: Indian Historian Press.

Omi, Michael, and Howard Winant. 1994. *Racial Formation in the United States: From the 1960s to the 1980s*. New York: Routledge and Kegan Paul.

Park, Lisa Sun-Hee, and David Pellow. 2004. Racial Formation, Environmental Racism, and the Emergence of Silicon Valley. *Ethnicities* 4 (3): 403–424.

Quinn, Scott. 2007. *Karuk Tribe of California Aboriginal Territory Acreage Assessment*. Orleans, CA: Karuk Department of Natural Resources.

Raphael, Ray, and Freeman House. 2007. *Two Peoples, One Place*. Vol. 1, Humboldt History. Eureka, CA: Humboldt County Historical Society.

Rodney, Walter. 1974. *How Europe Underdeveloped Africa*. Washington, DC: Howard University Press.

Salter, John F. 2003. White Paper on Behalf of the Karuk Tribe of California: A Context Statement Concerning the Effect of Iron Gate Dam on Traditional Resource Uses and Cultural Patterns of the Karuk People within the Klamath River Corridor. <http://www.mkwc.org/publications/fisheries/Karuk%20White%20Paper.pdf> (accessed January 21, 2011).

Stretesky, Paul, and Michael Hogan. 1998. Environmental Justice: An Analysis of Superfund Sites in Florida. *Social Problems* 45 (2): 268–287.

Winant, Howard. 2004. *The New Politics of Race: Globalism, Difference, Justice*. Minneapolis: University of Minnesota Press.

3

From the Past to the Present

Agricultural Development and Black Farmers in the American South

John J. Green, Eleanor M. Green, and Anna M. Kleiner

Action to achieve social justice in agrifood systems should be informed by research from a variety of perspectives. Yet, in scholarly literature concerning changes in American agrifood systems, minority producers and their communities have received limited attention. There are exceptions, of course, with the growing body of work on farmworkers and recent attention to new immigrant farmers being noteworthy. Unfortunately, the world of African American/black farmers has long gone understudied, although a small but vocal group of authors has made major contributions to our understanding of patterns of change occurring among these producers and their attempts to construct alternative organizations and institutions.

Some noted works on agrifood systems provide helpful insights into the ever-changing scale, technological advancement, capital intensiveness, and structure of agricultural production, distribution, and consumption (see: Buttel, Larson, and Gillespie 1990; Heffernan 2000; Heffernan, Hendrickson, and Gronski 1999; Hendrickson et al. 2001; Lobao and Meyer 2001). Conventional agricultural markets have become increasingly cut off by the growing concentration of control over the agrifood system by corporate firms (Heffernan 2000). Marketing problems encountered by family farmers consist of the system privileging large-scale producers, insufficient information on market outlets and prices, and the ongoing cycle of market price disasters (Green 2001; Green and Picciano 2002). Many of these characteristics leave producers with limited resources vulnerable to risks (Dismukes, Harwood, and Bentley 1997; Green 2001).

Beyond specific community and regional studies, much of the literature is general in nature and focuses on theoretical insights concerning what appear to be national and international trends. These works provide helpful conceptual tools for interpreting change across time and exploring how microlevel phenomena fit into these broader trajectories. At the same time, such studies often miss variations in the experiences of different groups in terms of their regional location, production types, and sociodemographic characteristics, especially gender, race, and ethnicity. Many theoretical and empirical works leave out express attention to black farmers, an ironic omission given their presence in American agriculture since the time of slavery. When they are mentioned, it is often in the context of the legacy of slavery, sharecropping, and the Great Migration northward. After doing a cursory read of the mainstream literature, a student of agrifood systems may arrive at the faulty conclusion that there are no black farmers left in the United States.

In this chapter, we contribute to the growing literature on black farmers on the basis that investigating the challenges they face and the successes they achieve is important in its own right, in addition to what it tells us about the dominant agrifood system and countervailing social movement responses to it. Interpreted through a "rural livelihoods" theoretical framework, we use the case of black farmers in the southern United States, especially those in Mississippi, to explore the dynamic set of circumstances faced by these producers and how they have managed to persist in the face of great odds. Information is drawn from the broad range of published literature and the research projects conducted by the authors over the course of the past decade, including archival sources, interviews, focus groups, and secondary data analysis (Green 2001, 2002; Green and Kleiner 2009a; Green et al. 2003; Kleiner and Green 2008). We pay special attention to the stresses placed on black farmers' livelihood systems through structural inequality, limited access to land, and indifference and hostility from the institutions that are supposed to assist producers, mainly the U.S. Department of Agriculture. We then focus on the grassroots organizations black farmers and their allies have developed to pursue more sustainable livelihoods.

"Rural Livelihoods" Theoretical Framework

Occupying an important place in international development studies, theoretical work on livelihood systems draws from an eclectic blend of theoretical frameworks (for an overview, see Green and Kleiner 2009a). The term *livelihood* refers to the strategies and tactics used by individuals, their households, and their communities in working to make a living. In doing so, they must negotiate socioeconomic relations and political terrain (Bebbington 1999; De Haan 2000; Ellis 1998; Farrington et al. 1999; Hall and Midgley 2004).

Dynamic political, economic, and ecological arrangements sometimes pose threats to livelihood systems while opening opportunities at other times. Of particular interest are the situations where access to resources or exclusion from them influence people's ability to handle changes in the broader environments in which they are embedded (De Haan and Zoomers 2005; De Haan 2000; Bebbington 1999). Analysts often seek to determine whether people with specific resource sets are more vulnerable or resilient when faced with change.

Drawing from work as diverse as social movement theory, networks, and studies of social capital, many livelihood theorists have directed their attention to the ways in which different resources are accessed or constrained and how they influence people's capabilities for taking action. Bebbington (1999) focuses on five forms of capital important to livelihood systems: natural, human, cultural, social, and produced. These are similar to the forms of assets/capital discussed in contemporary community development literature (e.g., Green and Haines 2008; Flora, Flora, and Fey 2004). In some circumstances people must trade certain resources to access others. It is argued that situations where people have been able to substantially improve their livelihoods resulted from having the capability to access diverse resources, take advantage of opportunities, and enhance the ways in which resources are used to develop more of a resource or gain another form of resource. To the extent that resources can be invested and used for expansion of capabilities, they are considered forms of capital (Bebbington 1999). In reviewing the literature and reflecting on his own research, Bebbington (1999, 2028) argues, "Where rural people have not been able to improve their livelihoods, the principal reasons seem to derive from failure or inability to: defend their existing assets; identify and secure opportunities to turn assets into livelihoods; or protect existing ways of turning assets into livelihoods."

As critical scholars interested in the political economy of agrifood systems have long noted, the power that groups have over their own livelihood systems results, at least partially, from their position in the socioeconomic structure. Many studies of agricultural production focus attention on farm scale in terms of acreage or sales or both, but scholars including Heffernan (1972) and Rodefeld (1974) note the importance of other dimensions such as ownership and decision making. Thus, it is not just the scale of the farm enterprise, but also control of labor and the means of production that are important factors. The often-used conceptual understanding of "family farming" is based on the assumption of the family owning land and equipment, providing labor, and making decisions about their business (Lobao and Meyer 2001). From this perspective, one can argue that the vulnerability or resiliency of a farmer's livelihood system is influenced by the structural arrangements in which he or she is embedded. Important issues include ownership, who makes decisions, and the scale of production. In each of these dimensions, black farmers have traditionally been at a disadvantage.

Organizations and institutions mediate people's access to the assets necessary for coping with change by opening and closing pathways for improvement (De Haan and Zoomers 2005), especially through the spheres of the market, the state, and civil society (Bebbington 1999; Meert 2000). Power relationships, processes, and policies ranging from the local to the global levels influence which pathways are accessible to groups versus being constrained. Restricted access to particular resources results in greater vulnerability. People are often excluded from accessing resources on the basis of their location, nationality, class, gender, race, and ethnicity.

In the context of this discussion, race/ethnicity is of concern because it is an ascribed social status, with associated social, economic and political rewards and penalties. Being black has important symbolic meaning in American society, and this is amplified in the agrifood sector. From their beginning days in the United States, black farmers were defined institutionally as being less worthy than white farmers, first through slavery, then via tenancy and sharecropping. Even today, black farmers are viewed with indifference at best and contempt at worse by the mainstream agricultural establishment. As will be shown in the following sections of this chapter, this has resulted in discrimination and limited access to resources for some groups relative to others. Black farmers in the southern United States have faced challenges and blocked opportunities, thereby threatening their livelihood systems. In some cases, they responded by constructing alternative organizations and placing pressure on the state for change.

Agriculture, Black Farmers, and Livelihood Systems

The twentieth century brought vast changes in agriculture that severely limited farmers' power in decision making and their ability to survive on the land in the face of traditional and new agricultural risks. These changes included an increase in large-scale, technically advanced, capital-intensive, and structurally complex modes of production, processing, and distribution (Heffernan 2000). Corporate concentration occurred in both the supply and sales sectors through horizontal and vertical integration (Heffernan 2000; Heffernan, Hendrickson, and Gronski 1999; Hendrickson et al. 2001). This coincided with the increasing presence of multinational and transnational corporations and efforts to reduce domestic agricultural protections under free trade agreements and global governing bodies (i.e., the World Trade Organization).

These changes have impacted the whole of farm production, and some analysts have noted increased polarization and the development of a dual system of agriculture (Lobao and Meyer 2001) with many small-scale farmers dependent on off-farm income on one side and large-scale industrialized corporate farms on the other side (Albrecht 1997; Lobao and Meyer 2001). In the middle, the traditional, idealized family farm is increasingly at risk. Recognizing these general patterns, it is important to consider that the people involved in agriculture are not from a homogenous group. There is heterogeneity between farmers and farms along several different dimensions, including scale (Albrecht 1997, 1992; Thomas et al. 1996; Wimberley 1987) and structural organization (Heffernan 1972; Rodefeld 1974, 1978), goods produced (Albrecht 1998), production practices employed, and the farmers' positions in terms of industry entry and exit (Jackson-Smith 1999). There is also variation among socioeconomic, gender, and racial/ethnic groups. Failure to fully acknowledge this diversity has led to limited government assistance for small, limited-resource and minority farmers who tend to face disproportionate challenges relative to their larger, capital-intensive and white counterparts (USDA Civil Rights Action Team 1997; U.S. Government Accountability Office 2001; USDA National Commission on Small Farms 1998). Therefore, it is important for us to better understand the challenges and needs not only of farmers, but of family farmers in general. We must also consider the complexities faced by specific subgroups in the farming population, such as black producers.

A starting point in this endeavor is to recognize regional differentiation based on historical agrifood development patterns. In his analysis of the origins of three general production systems in the United States, Pfeffer (1983) maintains that variability in the structure of agricultural enterprises across the United States is largely a result of labor management constraints in the context of natural, social, economic, and political conditions. He compares the development of corporate, sharecropping, and family farming systems. Arguably, the South's unique history, especially with regard to slavery, the Civil War and its aftermath, the large population of black residents, the institutional mechanisms of plantation society, and maintenance of white privilege resulted in a particular development trajectory. This development history led Beckford (1999) to place the American South in a broader category of countries with plantation legacies that led to underdevelopment.

Between the time of European settlement and the end of the Civil War in 1865, blacks provided much of the labor in southern agriculture as slaves, especially in cotton production (Fite 1984; Genovese 1976). Their labor was used to build a system of elite white wealth and power. Although all farms did not have slaves, and all slave-holding farms were not large in scale, the social and legal environment allowing for white ownership of black labor provided the avenue by which large plantation systems were organized with owners (often called "planters") practicing almost complete control over labor. Using the physical labor of slaves, crop production was increased even with few technological developments beyond the cotton gin, a processing invention that substantially increased production capacity.

Following the Civil War and emancipation of slaves, planters and their managers made attempts to engage the so-called Freedmen into wage labor arrangements. For the most part, this was not successful, because the wage system was not a part of the culture of work in the region's dominant agricultural system, and it would not provide the basic necessities of life such as food and shelter. For former slaves, there were few incentives to remain in agriculture other than there being limited off-farm employment opportunities. Planters soon found a mix of structural arrangements for agriculture that was more productive while also allowing them to keep blacks and poor whites subjugated through vulnerable land tenure arrangements, perpetual indebtedness, and coercive violence. Control over labor switched to sharecropping, tenant farming, and the crop-lien system (see Fite 1984 and Marable 1979 for instructive reviews of this history).

Tenancy systems consisted of the farmer working the land and paying rent to the landowner, typically as a set price or as a portion of crop sales. These arrangements were often problematic for the actual working farmers, because they were at the mercy of the landowner in terms of decision making. If the landowner did not like what a tenant was doing, he could simply rent the land to a different producer in the future. Sharecropping systems typically involved the farmer working the land and obtaining supplies or access to land or both from the landowner in exchange for a share of the actual crop or crop sales. Sharecroppers were often forced to plant specific crops as dictated by the landowner, and they could only obtain seed and supplies from approved sources, such as a plantation store where they would be charged excessive prices and interest rates. The landowners and their partners were generally the ones weighing the crops, marketing crops, keeping records in their own favor, and distributing funds (Marable 1979). As a result, many sharecroppers stayed in perpetual debt to the landowner. Some debt was owed to lenders beyond the landowner. The crop-lien system entailed farmers accessing much needed credit by mortgaging their crops for supplies (Marshall and Godwin 1971), often from banks.

As Fite (1984) argues, tenancy, sharecropping, and the crop lien were more than mere economic arrangements. They provided the basis of social control by elite whites over both black and poor white producers. The vulnerability of these arrangements, and the extreme pressure the wealthy used to enforce their power, led Rogers to reference the "interlocking systems of sharecropping and segregation" (2006, 24).

Despite these challenges, black landownership grew in the Reconstruction period and continued in many places up through its peak in the early 1900s (Marable 1979). In fact, early movement of blacks into the Mississippi Delta region was in the hope of landownership, and many achieved this goal. The land was wooded and swampy and many whites were not interested in settling in an area with those conditions. The railroad, in a progressive move for the time, looked to the black population for settlement of the area and offered decent terms and prices. This is how many of the families farming in the area today originally gained their land. As an early farmer in the Mississippi Delta proclaimed, "I came to Mound Bayou in 1887 and purchased from the Louisville, New Orleans and Texas Railroad Company forty acres of land on the terms fixed by them for such purchases. This was usually seven dollars per acre, one dollar per acre down, the balance in five equal payments" (Jackson 1937, 31). It was the period when some notable black communities were established, such as Mound Bayou, which was settled by former slave families from Joseph Davis's Hurricane Plantation (Hermann 1999).

However, landownership dwindled as more lands were consolidated into white-owned plantations (Rogers 2006). Thoughts of Reconstruction and Freedmen gave way to a new racial order. Institutionalized segregation of the races and the so-called "black codes" and "Jim Crow"

Table 3.1
Number of farm operators by race and tenure, 1920

	National		Mississippi	
Farm tenure	White	Black	White	Black
Owner	67%	24%	61%	14%
	(3,691,868)	(218,612)	(68,131)	(23,130)
Manager	1%	<1%	1%	<1%
	(66,223)	(2,026)	(797)	(192)
Tenant	32%	76%	38%	86%
	(1,740,363)	(705,070)	(41,954)	(137,679)
Total	(5,498,454)	(925,708)	(110,882)	(161,001)

Source: United States Department of Commerce 1922. Table constructed by John Green.

laws limited political and economic opportunities and served as mechanisms for continuation of white privilege in the South (Woodward 1966).

The year 1920 is noted as the height of the number of black farms in America. At that time, there were disparities between whites and blacks in owner-operator status compared with tenancy, both nationally and even more so in the state of Mississippi. As demonstrated in table 3.1, more than two-thirds of white farmers were owner-operators, compared with less than one-quarter of black farmers at the national level. The gap was even greater in Mississippi, where 86 percent of black farmers were tenants.

For those black farmers who were able to obtain and control land, it proved a valuable resource for their livelihood strategies and communities. One black farmer interviewed by an author of this chapter for an oral history project spoke of how important this land was, stating, "It is our livelihood. It is our life. Definitely the land is more valuable than money to me and my family" (Burkett 2007). As noted by many analysts of black rural communities, landownership provided economic security, independence, and community stability (Dittmer 1995; Gilbert, Sharp, and Felin 2002). This assessment was borne out in the Civil Rights Movement when black farmers and black landowners in general had the greater level of autonomy and resources necessary to help support rural organizing efforts (Dittmer 1995).

Over the course of the twentieth century there was a widespread loss of farms and an exodus of people and families from agriculture across the United States (Lobao and Meyer 2001). Along with a rapid decline in the number of farms and farmers, there was a parallel increase in the average size of farm operations between the 1920s and 1970s. This included periods encompassing the bottoming out of the global cotton market, the Great Depression, and the Dust Bowl era, as well as rapid mechanization and chemicalization of agricultural production following World War II. Between the 1970s and early 1980s, the number of farms remained fairly constant until the severe farm crisis of the 1980s. By this time, much of agricultural production was dependent on credit for capital-intensive systems, and producers were heavily indebted. In past crises, farm families would work more and exploit themselves to meet their economic needs. However, by the 1980s there was so much financial capital involved that no level of belt tightening or extra work could save some families from foreclosure (Heffernan 2008). The number of large-scale and corporate farms increased.

Although agricultural crises have proven problematic for family farmers and their communities across racial and ethnic groups, blacks have exited agricultural production at rates disproportionately faster and more severe than those of whites. Wood and Gilbert (2000) estimate a 98 percent loss of black farm operations between 1900 and 1997 compared with a nearly 66 percent loss among white operations.

This begs two questions. What challenges to their livelihood systems have black farmers faced? Which of these are general to family farmers—especially those that are characterized as limited resource—and which are unique to this minority group defined through their ascribed racial status? Drawing from a range of studies (USDA Civil Rights Action Team 1997; Gilbert, Sharp, and Felin 2002; Green 2001, 2002; Grim 1995; Jones 1994; Kleiner and Green 2008; Lobao and Meyer 2001; Mitchell 2000; Schor 1996; Wood and Gilbert 2000), the following challenges appear to be particularly troubling.

• General decline in the ability of small and medium-sized family farmers to make a living from the land in the face of competition from large-scale, industrial, and corporate production units.
• Mechanization of production, requiring less labor and additional capital and technological investment.
• Vulnerable tenure status in terms of tenancy arrangements.
• Property disputes in situations where landowners did not leave a will, often resulting in forced sale of heir property.
• Corporate consolidation of commercial markets with demands for low-cost high-volume production.
• Restricted access to timely and appropriate credit because of discrimination by lenders, both commercial and governmental entities.

• Limited knowledge of, participation in, and access to government agricultural programs, and the existence of few programs specifically designed to assist small-scale and limited-resource producers.

The last point deserves considerable attention. In situations where people face particular challenges to their livelihood systems, they often turn to their government for protection and assistance. However, black farmers have been underserved by governmental bodies. The state has not provided an adequate or timely response for protection of livelihoods for black producers (USDA Civil Rights Action Team 1997; Green et al. 2003; Grim 1996; Jones 1994; USDA National Commission on Small Farms 1998), and the USDA Civil Rights Action Team (1997) reported both historic and contemporary discrimination against the needs and interests of black farmers. Some of these entailed aggregate personal discrimination, while others were more institutional and structural forms of discrimination.

The characteristics of agricultural programs have changed over the course of several decades, but there continue to be initiatives to provide access to credit, price supports, and insurance for production of traditional commodities including cotton, corn, and soybeans. A variety of barriers block pathways for many black farmers, preventing them from fully participating in and receiving the benefit of these programs. Limited-resource and minority farmers, including black producers, cite preferential attention given to large-scale producers, personal and institutional discrimination, and insufficient outreach efforts (Green 2001, 2002). Furthermore, most of the prominent government agricultural programs were designed to provide the greatest benefits to those farmers with the highest level of commercial production rather than those in the greatest need of assistance (Jones 1994). These farmers also face challenges managing risk and the financial intricacies of agricultural production and marketing, while little technical assistance is available to them to address these aspects of operating a farm. Overall, both anecdotal evidence and empirical research suggest that there is inequality in access to and outcomes from existing agricultural programs (USDA Civil Rights Action Team 1997; U.S. Government Accountability Office 2001; Green 2001; Jones 1994).

Many of these criticisms have been voiced not only through protest and advocacy, but also through legal challenges such as the *Pigford v. Glickman* (1999) class action case of black farmers who believed that the U.S. Department of Agriculture had been discriminatory in its credit programs and other programs. A settlement was reached in the case, with payouts to some claimants as well as other forms of relief. Additionally, changes were to be made in the operation of USDA programs and offices. Although progress has been made in efforts to push civil rights in the USDA (some of the successes of which are outlined in the following section), many producers have had to focus more of their attention on grassroots development efforts to survive.

Responses from the Grassroots

The short social history of black farmers in the American South provided in this chapter demonstrates the challenges they have faced over multiple generations. This rather negative history can be counteracted, however, by directing attention to the progressive initiatives black farmers have made in their efforts to survive and achieve greater livelihood security. Much of this mobilizing and organizing work took place around development of self-help associations, cooperatives, and interaction with civil rights initiatives. These efforts continue to be a part of contemporary social movement responses to the dominant global industrialized agri-food system.

Collectively referred to as community-based cooperatives (CBCs), these groups share similarities with traditional producer and consumer cooperatives in terms of member ownership, democratic control, limited return on investment, and patronage refunds (Green 2002). Different from their larger-scale, commodity-specific counterparts in mainstream agriculture, CBCs focus more on local geographic boundaries, generally have a small membership base, and express a social justice as well as an economic agenda. Many of these organizations are legally chartered as cooperatives, while others are nonprofits. Some of them are more informal self-help groups (Green and Kleiner 2009b).

Some self-help associations and "poor people's cooperatives" originated in the late 1800s as small-scale farmers and sharecroppers faced competition with their larger-scale counterparts in combination with the high prices they had to pay for inputs, the low commodity prices they received for their goods, and the crop-lien system (Marshall and Godwin 1971). As already discussed in a previous section of this chapter, tenancy and sharecropping arrangements resulted in major power differences between landowners/managers and those who actually worked the soil.

The Colored Farmers Alliance and Cooperative Union sought to address the mutual interests of small farmers, sharecroppers, and hired laborers in a collective manner (Marshall and Godwin 1971). Beginning in Arkansas and then spreading regionally, the Southern Tenant Farmers' Union (STFU) was established in response to laborers and sharecroppers being pushed off the land by owners trying to take advantage of government programs intending to raise prices by paying farms to curb production levels. STFU, cooperative organizers, and some progressive policy makers advocated for rules to provide payment to workers, expand services, and create the U.S. Department of Agriculture's Farm Security Administration in 1937. In contrast to most other agricultural agencies, this agency sought to help displaced workers by creating resettlement communities and developing cooperative businesses (Ibid.). Some of the cooperative organizations started during this time helped set the stage for future mobilization efforts (Green 2002).

Along with the Civil Rights Movement of the late 1950s through early 1970s came another push to develop community-based cooperatives, in the form of "New Poor People's Cooperatives" (Marshall and Godwin 1971; Ulmer 1969). As voting rights were achieved and public accommodations more widely accepted, some community organizers turned their attention toward the next step in a broader struggle—attaining economic justice through access to markets and jobs. Agricultural and consumer cooperatives were organized across the South. These cooperatives often included residents of rural and urban communities, producers, consumers, churches, and civil rights groups.

Cooperatives pooled resources to purchase supplies and equipment, combined their produce to market to wholesale buyers, established local markets, and formed credit unions. Recognizing the need for collective action, several combined forces to create umbrella organizations to serve producers and communities across the southern region. Lead among them was the Federation of Southern Cooperatives (which later became the Federation of Southern Cooperatives/Land Assistance Fund). Through capacity building, advocacy, and providing direct services, these organizations filled many of the gaps left by the market and state in meeting people's livelihood needs.

Scholars have investigated the successes and challenges of CBCs (Green 2002; Green and Kleiner 2009a; Marshall and Godwin 1971; Reynolds 2003; Ulmer 1969). One of the primary purposes of CBCs was to create self-help organizations for poor, rural, and minority communities to mediate the impact of the broader socioeconomic forces and pursue more secure livelihoods for their members. Because dominant institutions in the market and state have often failed to address black farmers' needs, CBCs engage in social movements to organize alternatives.

The successes achieved by these groups have gone beyond grassroots development to influence national-level policies and programs. Most well known was the role that organized black farmers played in gaining attention to past discrimination and the need for justice in services offered to black farmers, such as the *Pigford v. Glickman* (1999) case. Through their mobilization, organizing, and advocacy efforts, as well as their ability to demonstrate field-based successes, CBCs and their associations also had input in constructing more progressive alternatives. Efforts included participation in commissions to study and make recommendations on the plight and future survival of underserved producers through the USDA National Commission on Small Farms (1998) and the USDA Civil Rights Action Team (1997).

Even though the last several federal farm bills have extended many of the programs supporting capital-intensive industrial agriculture, there were advances made in the direction of access and equity in government programs. A list of accomplishments from the last decade includes the appointment of an assistant secretary for civil rights position in the USDA; development and implementation of procedures for nominating and electing Farm Services Agency Committee members and counting ballots; public access to county-level USDA program participation rates; and creation of a volunteer minority farm registry. Collaborating with numerous other organizations from a broad range of backgrounds (e.g., sustainable agriculture, community food security), they also promoted the establishment and funding of programs of great benefit to black farmers, including the Outreach and Technical Assistance Program for Socially Disadvantaged Farmers and Ranchers Program (section 2501 of the 1990 Farm Bill), among numerous others. They advocated for dialogue between community-based organizations and the USDA through the latter's Annual Partners Meeting. These groups were organized locally, regionally through groups such as the Federation of Southern Cooperatives/Land Assistance Fund, and nationally through the Rural Coalition.

Recognizing the successes of CBCs, it is important to note the challenges. Traditionally underserved producers continue to face challenges with establishing alternative markets, accessing insurance, and realizing benefits from programs that tend to support a small number of the largest, highly mechanized, and capital-intensive farm operations.

Working to identify and document these challenges and possible responses, the Rural Coalition conducted focus groups across the nation, including places in the southern United States, to inform the 2002 Farm

Bill (Green 2001, 2002). Kleiner and Green (2008) followed this work in partnership with the Mississippi Association of Cooperatives and Heifer International, conducting focus groups with members of organizations in Mississippi and Louisiana. Through these venues, participants discussed existing and desired market outlets and the opportunities and challenges associated with each one. They explored the meaning of sustainable agriculture, what resources exist, and the challenges faced in this approach to production. These studies highlighted common barriers to pursuing alternative markets, such as land insecurity, lack of affordable credit, and limited access to profitable markets. Many of their participants called for greater collaboration between individual producers and community-based organizations to overcome these challenges.

Discussion

Reviewing the position and experiences of specific groups in development processes provides the basis for a more nuanced understanding of social transitions and the complexities involved in terms of who wins and who loses under particular structural arrangements. Given black farmers' important albeit often unrecognized role in American agrifood systems, directing attention to them is a rewarding enterprise. Shining light on some of the extreme injustices and their costs to individuals, families, and communities allows students of inequality to more clearly articulate structural constraints on livelihood systems and the sustainability of a people. At the same time, exploring the history of black farmers also provides lessons on mobilizing and organizing to construct alternative institutions to meet people's needs in a more socially just manner.

As a growing body of literature in the academic and popular presses emphasizes, the dominant agrifood system is unsustainable and unjust. Many of the developments in recent decades have resulted in problems for farmers, their families, rural communities, and the environment. The narrative of black farmers shows greater details of the transitions that have taken place, and may help keep us from being overly romantic about the past. The story of black farmers is the story of justice pursued but never fully realized. For black farmers, it is not a matter of going back to the days of yesteryear, for how can a system based on slavery and sharecropping be considered idyllic? Instead it is a matter of continuously striving to achieve justice in the future. Until black farmers and agricultural workers of all races, ethnicities, and genders are treated equitably, a more just and sustainable agrifood system will be too far out of reach.

References

Albrecht, Don E. 1992. The Correlates of Farm Concentration in American Agriculture. *Rural Sociology* 57 (4): 512–520.

Albrecht, Don E. 1997. The Changing Structure of U.S. Agriculture: Dualism Out, Industrialism In. *Rural Sociology* 62 (4): 474–490.

Albrecht, Don E. 1998. Agricultural Concentration: An Analysis by Commodity. *Southern Rural Sociology* 14 (1): 18–40.

Bebbington, Anthony. 1999. Capitals and Capabilities: A Framework for Analyzing Peasant Viability, Rural Livelihoods and Poverty. *World Development* 27 (12): 2021–2044.

Beckford, George L. 1999. *Persistent Poverty: Underdevelopment in Plantation Economies of the Third World*. 2nd ed. Kingston, Jamaica: University of the West Indies Press.

Burkett, Ben. 2007. Interviewed by Eleanor M. Green, January 23. Mississippi Black Farmer Oral History Collection, Delta State University Archives, Cleveland, MS.

Buttel, Frederick, Olaf F. Larson, and Gilbert W. Gillespie Jr. 1990. *The Sociology of Agriculture*. New York: Greenwood Press.

De Haan, Leo J. 2000. Globalization, Localization, and Sustainable Livelihood. *Sociologia Ruralis* 40 (3): 339–365.

De Haan, Leo J., and Annelies Zoomers. 2005. Exploring the Frontier of Livelihoods Research. *Development and Change* 36 (1): 27–47.

Dismukes, Robert, Joy L. Harwood, and Susan E. Bentley. 1997. *Characteristics and Risk Management Needs of Limited-resource and Socially Disadvantaged Farmers*. Washington, DC: United States Department of Agriculture.

Dittmer, John. 1995. *Local People: The Struggle for Civil Rights in Mississippi*. Urbana: University of Illinois Press.

Ellis, Frank. 1998. Household Strategies and Rural Livelihood Diversification. *Journal of Development Studies* 35 (1): 1–38.

Farrington, John D., Diana Carney, Caroline Ashley, and Cathryn Turton. 1999. Sustainable Livelihoods in Practice: Early Applications of Concepts in Rural Areas. [London: Overseas Development Institute.] *Natural Resource Perspectives* 42 (June): 1–15.

Fite, Gilbert C. 1984. *Cotton Fields No More: Southern Agriculture, 1865–1980*. Lexington: University Press of Kentucky.

Flora, Cornelia, Jan Flora, and Susan Fey. 2004. *Rural Communities: Legacy and Change*. Boulder, CO: Westview Press.

Genovese, Eugene D. 1976. *Roll Jordan, Roll: The World the Slaves Made*. New York: Vintage Books.

Gilbert, Jess, Gwen Sharp, and M. Sindy Felin. 2002. The Loss and Persistence of Black-owned Farms and Farmland: A Review of the Research Literature and Its Implications. *Southern Rural Sociology* 18 (2): 1–30.

Green, Gary P., and Anna Haines. 2008. *Asset Building and Community Development*. Thousand Oaks, CA: Sage Publications.

Green, John J. 2001. Developing Programs from the Grassroots: Assessing the Needs and Interests of Limited Resource Farmers. Working Paper Number 1, Program Responsiveness Series. Missouri Action Research Connection, University of Missouri-Columbia.

Green, John J. 2002. Community-Based Cooperatives and Networks: Participatory Social Movement Assessment of Four Organizations. PhD diss., University of Missouri-Columbia.

Green, John J., Tomeka Harbin, Christopher Pope, Lorette Picciano, and Heather Fenney. 2003. A Multi-Community Assessment of the Risk Management Needs of Small, Limited Resource and Minority Farmers. Working paper for the Rural Coalition and U.S. Department of Agriculture, Risk Management Agency.

Green, John J., and Anna M. Kleiner. 2009a. Exploring Global Agrifood Politics and the Position of Limited Resource Producers in the United States. In *The Politics of Globalization*, ed. Samir Dasgupta, 313–333. London: Sage Publications.

Green, John J., and Anna M. Kleiner. 2009b. Escaping the Bondage of the Dominant Agrifood System: Community-based Cooperative Strategies. *Southern Rural Sociology* 24 (2): 149–168.

Green, John J., and Lorette Picciano. 2002. Amplifying the Voices of Community-based Organizations through Action Research: Obtaining Input from the Grassroots on Agricultural Policies and Programs. Paper presented at the Annual Meeting of the Community Development Society, Cleveland, MS, July 22.

Grim, Valerie. 1995. The Politics of Inclusion: Black Farmers and the Quest for Agribusiness Participation, 1945–1990s. *Agricultural History* 69 (2): 257–271.

Grim, Valerie. 1996. Black Participation in the Farmers Home Administration and Agricultural Stabilization and Conservation Service, 1964–1990. *Agricultural History* 70 (2): 321–336.

Hall, Anthony, and James Midgley. 2004. *Social Policy for Development*. Thousand Oaks, CA: Sage Publications.

Heffernan, William D. 1972. Sociological Dimensions of Agricultural Structures in the United States. *Sociologia Ruralis* 12 (3–4): 481–499.

Heffernan, William D. 2000. Concentration of Ownership and Control in Agriculture. In *Hungry for Profit: The Agribusiness Threat to Farmers, Food, and the Environment*, ed. Fred Magdoff, John Bellamy Foster, and Frederick H. Buttel, 61–75. New York: Monthly Review Press.

Heffernan, William D. 2008. Interviewed by Anna M. Kleiner and John J. Green, December 27.

Heffernan, William D., Mary Hendrickson, and Robert Gronski. 1999. *Consolidation in the Food and Agriculture System*. Washington, DC: National Farmers Union.

Hendrickson, Mary, William D. Heffernan, Phil Howard, and Judy B. Heffernan. 2001. *Consolidation in Food Retailing and Dairy: Implications for Farmers and Consumers in a Global Food System*. Washington, DC: National Farmers Union.

Hermann, Janet Sharp. 1999. *The Pursuit of a Dream*. Oxford: University Press of Mississippi.

Jackson, Maurice Elizabeth. 1937. Mound Bayou: A Study in Social Development. Master's thesis, University of Alabama.

Jackson-Smith, Douglas. 1999. Understanding the Microdynamics of Farm Structural Change: Entry, Exit, and Restructuring among Wisconsin Family Farmers in the 1980s. *Rural Sociology* 64 (1): 66–91.

Jones, Hezekiah S. 1994. Federal Agricultural Policies: Do Black Farm Operators Benefit? *Review of Black Political Economy* 22 (4): 25–50.

Kleiner, Anna M., and John J. Green. 2008. Expanding the Marketing Opportunities and Sustainable Production Potential for Minority and Limited Resource Agricultural Producers in Louisiana and Mississippi. *Southern Rural Sociology* 23 (1): 149–169.

Lobao, Linda, and Katherine Meyer. 2001. The Great Agricultural Transition: Crisis, Change, and Social Consequences of Twentieth Century U.S. Farming. *Annual Review of Sociology* 27:103–124.

Marable, Manning. 1979. The Land Question in Historical Perspective: The Economics of Poverty in the Blackbelt South, 1865–1920. In *The Black Rural Landowner—Endangered Species: Social, Political, and Economic Implications*, ed. Leo McGee and Robert Boone, 3–24. Westport, CT: Greenwood Press.

Marshall, Ray, and Lamond Godwin. 1971. *Cooperatives and Rural Poverty in the South*. Baltimore, MD: The Johns Hopkins University Press.

Meert, Henk. 2000. Rural Community Life and the Importance of Reciprocal Survival Strategies. *Sociologia Ruralis* 40 (3): 319–338.

Mitchell, Thomas W. 2000. From Reconstruction to Deconstruction: Undermining Black Landownership, Political Independence, and Community through Partition Sales of Tenancies in Common. Research Paper No. 132, Land Tenure Center, University of Wisconsin-Madison, Madison.

Pfeffer, Max J. 1983. Social Origins of Three Systems of Farm Production in the United States. *Rural Sociology* 48 (4): 540–562.

Pigford v. Glickman. 1999. 185 F.R.D. 82 (D.D.C.).

Reynolds, Bruce. 2003. *Black Farmers in America, 1865–2000: The Pursuit of Independent Farming and the Role of Cooperatives*. Washington, DC: United States Department of Agriculture.

Rodefeld, Richard D. 1974. The Changing Organizational and Occupational Structure of Farming and the Implications for Farm Workforce Individuals, Families, and Communities. PhD diss., University of Wisconsin-Madison.

Rodefeld, Richard D. 1978. Trends in U.S. Farm Organizational Structure and Type. In *Changes in Rural America*, ed. Richard D. Rodefeld, Jan Flora, Donald

Voth, Isao Fujimoto, and James Converse, 158–177. St. Louis: C. V. Mosby Company.

Rogers, Kim Lacy. 2006. *Life and Death in the Delta: African American Narratives of Violence, Resilience, and Social Change*. New York: Palgrave Macmillan.

Schor, Joel. 1996. Black Farmers/Farms: The Search for Equity. *Agriculture and Human Values* 13 (3):48–63.

Thomas, John K., Frank M. Howell, Ge Wang, and Don E. Albrecht. 1996. Visualizing Trends in the Structure of U.S. Agriculture: 1982–1992. *Rural Sociology* 61 (2): 349–374.

Ulmer, Al. 1969. *Cooperatives and Poor People in the South*. Southern Regional Council, Atlanta, GA.

United States Department of Agriculture (USDA) Civil Rights Action Team. 1997. *Civil Rights at the U.S. Department of Agriculture*. Washington, DC: U.S. Government Printing Office.

United States Department of Agriculture (USDA) National Commission on Small Farms. 1998. A Time to Act: A Report of the USDA National Commission on Small Farms. Washington, DC: U.S. Government Printing Office.

United States Department of Commerce, Bureau of the Census. 1922. *Agriculture: General Report and Analytical Tables*. Washington, DC: U.S. Government Printing Office. http://www.agcensus.usda.gov/Publications/Historical_Publications/1920/1920_General_Reports_and_Analysis.pdf (accessed January 15, 2011).

U.S. Government Accountability Office. 2001. *Farm Programs: Information on Recipients of Federal Payments*. Washington, DC: United States Government Accountability Office.

Wimberley, Ronald C. 1987. Dimensions of U.S. Agristructure: 1969–1982. *Rural Sociology* 52 (4): 445–461.

Wood, Spencer D., and Jess Gilbert. 2000. Re-entering African American Farmers: Recent Trends and a Policy Rationale. *Review of Black Political Economy* 27 (4): 43–64.

Woodward, C. Vann. 1966. *The Strange Career of Jim Crow*. 2nd ed. New York: Oxford University Press.

4

Race and Regulation

Asian Immigrants in California Agriculture

Laura-Anne Minkoff-Zern, Nancy Peluso, Jennifer Sowerwine, and Christy Getz

Yang[1] planted strawberries and a collection of "Asian" vegetables on his small, two-acre plot in Fresno County in California's Central Valley. He and his wife handled the bulk of the labor, from time to time hiring in people to help. In the spring, as the strawberries began to fruit, family members from Fresno, Oakland, Richmond, and elsewhere showed up to help with the harvest—taking home small quantities of the fruit for their children and other family members in exchange for helping the Yangs get their strawberries to market. That is, they did until the day a Labor Standards Enforcement agent came to the roadside farm and asked Yang's mother about the terms of her work. As the older woman did not have a thorough enough command of English to answer, the agent found Yang and asked to see proof of her worker's compensation insurance, which unsurprisingly, he did not have. Reciprocal labor had been part of his family's practices since they had farmed in the mountains of Laos, long before they were forced to leave in the wake of the American war. But new laws protecting family workers in California were about to change such practices. Facing a fine of several thousand dollars for the "offence" of hiring his mother and other extended family members as unpaid and uninsured labor to pick strawberries, it looked as if Yang would no longer be able to afford farming on his tiny rented plot of land. While the war[2] had not stopped him from farming, U.S. labor laws might, even though these, in principle, are meant to protect and help workers. Yang and other Hmong farmers clearly did not fit the picture of the "average" farmer in California.

How do state agricultural policies affect both access to material resources and the construction of racial identities? In this chapter, we trace the effects of such policies on three Asian immigrant groups.

Working-class Chinese were the first to be barred from legal immigration through the Chinese Exclusion Act of 1882. The designation of immigrants from Asia as nonwhite and noncitizens set the stage for later enclosure laws in which Japanese immigrants, who had become well established in California as both farmworkers and small farm owners, were barred from owning land. Although for a time Japanese farmers succeeded in resisting such state practices, their gains were erased through the sentencing of both immigrant and American-born Japanese to internment camps during World War II. More recently, Hmong refugees, bringing agricultural experience from their native Southeast Asia, have begun to practice small-scale farming in California's San Joaquin Valley. However, U.S. labor regulations[3] have defined many Hmong agricultural practices as illegal, contributing to some images of these refugees among regulators as "problematic." These three examples demonstrate that as the state regulates agricultural resources, it creates racial categories that separate lawful members of society from "alien" outsiders. These designations then legitimate and are reinforced by everyday experiences of racial "othering" or racial exclusion. We refer to the combined effects of these processes as *agricultural racial formations*.

The state historically created racial hierarchies through the restriction of landownership among those designated as "non-white,"[4] and therefore ineligible for citizenship.[5] Japanese immigrants succeeded in circumventing the alien land laws in the early twentieth century, however, their internment during World War II served as an explicit, state-sanctioned enclosure, stripping Japanese of property that could be expropriated by neighboring whites. In most cases, such land was never returned. Our research reveals that racial hierarchies still exist in complex and indirect ways, though often unintentionally. With regard to the Hmong, the application of labor laws protecting workers, which were primarily intended to regulate much larger farms, have dissuaded many Hmong from participating in agriculture, as they carry penalties disproportionate to Hmong farming incomes. At the extreme, they can be seen to serve as institutionalized enclosure laws, operating through seemingly neutral, and indeed even progressive legislation. Despite the protections afforded by their refugee status, Hmong and other Southeast Asian refugee farmers are obstructed from participating in agriculture due to California agricultural laws that continue to assume their agricultural subjects are white.

The Law and Asian American Farmers

What is the role of law and policy in creating racialized agricultural formations with political and economic consequences? In this section, we draw on theories relating the law, accumulation, and racial formation, in order to explore the role of the state in creating racial distinctions (and hierarchies) in California agriculture. Law and policy play an important role in the processes that constitute "dispossession by extra-economic means" (Glassman 2006). Civil rights laws in the United States have rendered it difficult, if not impossible, to implement new laws explicitly dispossessing people from land and other resources, based on racialized identities (Civil Rights Acts of 1866, 1964, and 1991). However, law and policies meant to facilitate accumulation in the agricultural sector, and those meant to protect the rights of workers, have had indirect effects on the production of racial identity and racialized dispossessions.

We see racial formation as "the sociological process by which racial categories are created, inhabited, transformed, and destroyed" (Omi and Winant 1994, 55–56). As key aspects of "projects . . . [that] . . . reorganize and redistribute resources along racial lines" (56), agricultural laws and policies construct racial difference through assumptions about what constitutes normal or appropriate agricultural practice and the effects of these assumptions. Though it is beyond the scope of this chapter to comprehensively explore the relationships between law, policy, and racial difference, we examine some of the ways regulations to protect workers have made it difficult for Hmong farmers in California, thus creating racialized agricultural formations. We contrast these indirect forms of exclusion with the overt and well-documented experiences of Chinese immigrant farmers due to the federal exclusionary acts and with those of Japanese farmers who lost their land due to California's alien land laws.

In the late nineteenth and early twentieth centuries, Chinese and Japanese farmers and farmworkers experienced dispossession in ways that resemble the practices of enclosure described by Karl Marx, except that where class was the primary form of differentiation in the cases Marx described, race has served that purpose in the United States (Marx [1867] 1977, 876).[6] In this regard, our chapter contributes to recent debates on Marx's original idea of primitive accumulation, in which it has been argued that dispossession is an ongoing process (Glassman 2006; Harvey

2004; De Angelis 2004; McCarthy 2004).[7] In California, farmworkers were denied access to land and labor opportunities in order to enable accumulation by Americans of European ancestry. In the case of the Chinese, enclosures were made possible by the Chinese Exclusion Act of 1882, prohibiting further entry of Chinese workers into the country and dampening their presence in the California agricultural labor and land market (Almaguer 1994; Chan 1986; McWilliams 2000; Omi and Winant 1994). Japanese farmers were expropriated first through the Alien Laws of 1913 to 1927 and then more violently through the Internment Acts of 1942. Being defined as Chinese or Japanese thus facilitated exploitative treatment of Chinese and Japanese laborers and farmers. Racism and the production of white privilege—not to mention citizenship and landownership—were explicit motives in the development of these U.S. laws (Almaguer 1994).

In contrast to the overtly racist motives behind legislated dispossession of Chinese and Japanese farmers from agriculture in earlier times, contemporary dispossession of the Hmong is more subtle. In the case of the Hmong, we argue that California labor regulations are dispossessing farmers of their ability to engage in agricultural livelihoods. Such regulations, and their definitions of "family," render long-standing cultural traditions of family labor reciprocity financially onerous, if not impossible, for the majority of undercapitalized Hmong farmers in the state.

The following sections trace the origins of racialized dispossessions of so-called Asian American groups, enabled by laws and regulations with racialized intents or effects. We argue that dispossession is racialized when laws, policies, or practices specifically affect particular racial groups, whether purposely or unintentionally. To locate such dispossessions historically, we describe the experiences of Chinese and Japanese immigrant farmers and Hmong refugee farmers at different moments in California history.

Chinese Agriculture and the Exclusion Act of 1882

As the first federal discriminatory immigration law in United States history, the Chinese Exclusion Act of 1882, turned a page on the perception of race in America, defining immigration as a privilege generally correlated with whiteness. The law reflected and legitimated existing racist ideologies and institutionalized them, in an attempt to deny poor Chinese immigrants access to citizenship, employment, and landownership in the United States (Gyory 1998, 2; Calavita 2006; Lee 2003; Lieb 1996).

In agriculture, the law limited the availability of Chinese immigrants as laborers, changing the ethnic makeup of the farmworker population. By restricting Chinese immigration, it minimized competition for white workers seeking agricultural jobs. Although the law was implemented nationally, its effects on the agricultural sector were most felt in California, which by then had the fastest growing agricultural economy in the country. The Chinese had first arrived in California in large numbers around 1852, as part of the Gold Rush. As mining declined, the main occupations of Chinese in California shifted from self-employed miners to wage workers, many of them in agriculture (Chan 1986). In the early 1870s, the onset of economic depression, the completion of the transcontinental railroad, and the declining number of jobs in California had exacerbated anti-Chinese sentiment among white laborers. In 1882, when the law was passed, seven out of eight farmworkers in California were Chinese (National Farm Worker Ministry 2009).

In addition to discriminating based on race and nationality, the law also had a class component: while the working class laborers were disallowed, merchants, diplomats, students, and temporary travelers from China all were permitted to enter the country. Nonetheless, many Chinese immigrants evaded the law, finding ways to enter the country and practice agriculture. They worked as both farmers and farmworkers in the United States, primarily in California, during the exclusionary period. There were in fact many loopholes in the laws, based on class and gender exemptions, which allowed for falsification of paperwork and identities (Lee 2003). By using false documentation, taking advantage of loopholes in the law, and maintaining a presence in California agriculture, Chinese workers in agriculture proved resistant to the racist law (Chan 1986; Lee 2003).

Though managing to continue in agriculture, the Chinese nevertheless suffered exclusion in other ways. The law was extended and amended in 1892, 1902, and 1904, proving a powerful agent in shaping the identities of multiple generations of Chinese immigrants as excluded from the categories of whiteness and citizenship. It was finally repealed in 1943, when, in the context of World War II, it hampered U.S. foreign relations with China (Calavita 2007; Gyory 1998).

The law had lasting repercussions on society in its determination of racial categories and subsequent racialized agricultural formations. It excluded the Chinese from having a legitimate presence in the United States and California and from legal participation in agriculture. Further, the law underwrote and institutionalized individual acts of racism, which had been frequent since Chinese immigration to California (Calavita 2006 and 2007; Gyory 1998; Lee 2003).

Placing limitations on the immigration and citizenship of certain races and nationalities helped define the parameters of U.S. citizenship and determined the constitution of racial categories in the United States more broadly. Regarding such restrictive legislation, Lee argues: "The Chinese exclusion laws and other government legislation excluding all other Asian immigrants reflected and maintained an exclusionary and racialized identity that marked Asians, African Americans, Latinos, and Native Americans as outsiders" (2003, 6–7).

Much of the negative racial stereotyping and general anti-Asian sentiment that was augmented by the Chinese Exclusion Act was also applied to Japanese immigrants. Many white political and labor leaders warned that Japanese immigrants would take the place of the Chinese in creating a threat to white farm labor (Lee 2003). In 1907 President Roosevelt passed an Executive Order restricting Japanese immigration to the U.S. mainland from Hawaii. Japan also agreed to stop issuing travel papers to Japanese laborers, according to a "gentleman's agreement" with the United States the same year (Calavita 2007; Lee 2003). In the next section of this chapter, we discuss the ways that the explicitly racist alien land laws specifically limited Japanese American landownership and agricultural capacity.

Japanese Farmers and the Alien Land Laws of 1913–1927

The first substantial wave of Japanese immigrants to California arrived in the 1890s. This first generation, consisting primarily of young unmarried males, filled some of the labor gaps left by Chinese transitioning out of the farming sector in the early 1900s. Japanese association records show that by 1910 two-thirds of Japanese laborers employed in California were working in agriculture. This was the peak of Japanese employment as agricultural wage workers, as labor-intensive crop cultivation had expanded in many regions of California during the preceding decade (Higgs 1978, Matsumoto 1993).

In order to secure opportunities as agricultural laborers, many Japanese workers underbid Chinese, Mexican, and white laborers for wages. Once established in an area, they would demand higher wages and contract renegotiation, striking or initiating work slowdowns in the process. This willingness and ability to organize and protest, and thereby improve their wages and economic status, differentiated them from the Chinese

Japanese escaped.
+ demanded
better use

workers who preceded them. Organizing both agitated their employers and threatened white growers, corporate and small alike (Almaguer 1994).

From 1900 to 1920, Japanese immigrants' primary role in agriculture shifted from farm labor to farm tenant and, less frequently, farm owner-operator. By 1920, Japanese farmers were producing 30 to 35 percent of all trucked crops in California (Higgs 1978). By 1925, almost 50 percent of the Japanese population in California was active in small farming enterprises (Almaguer 1994).

Although wage discrimination may have been overcome as more Japanese transitioned to become tenants and owners, there is evidence of further discrimination against Japanese farmers in the land rental market. Japanese land renters generally paid higher rents relative to land values than did white land tenants. As some Japanese gained access to land as tenants and owners, many also hired Japanese workers, paying them higher wages than the white farmers had paid (Higgs 1978).

These changes in Japanese access to land and labor caused many white farmers to justify their own discrimination against Japanese, as they had viewed the control of land, workers, and capital as white privileges. As Almaguer (1994, 186) notes, "Japanese farmers became the first group of racialized immigrants to challenge the precarious position of Anglo small family farmers in the state."

Legal discrimination targeting Japanese farmers began in 1913 with the passing of the state of California's first Alien Land Law. The law precluded them from owning land and limited land leases to three years. Section 1 of the Alien Land Law of 1913 states: "All aliens eligible to citizenship under the laws of the United States may acquire, possess, enjoy, transmit, and inherit real property, or any interest therein, in this state, in the same manner and to the same extent as citizens of the United States, except as otherwise provided by the laws of this state" (Alien Land Law of 1913, 206).

Section 2 of the law stipulated that all other aliens not included above could "lease land in this state for agricultural purposes for a term not exceeding three years" (Guskin and Wilson 2007).

The objective of the Alien Land Law, as described by the law's coauthor, State Attorney General U.S. Webb, was explicitly "to limit [Asian] presence by curtailing the privileges which they may enjoy here; for they will not come in large numbers and long abide with us if they may not acquire land" (Higgs 1978: 215, brackets in original).

Although the law did not state any particular nationality or ethnic group, it affected mainly Japanese immigrants, who were at that time the largest Asian immigrant group legally ineligible for citizenship,[8] as Chinese immigration had already been limited via the exclusion laws. The law applied not only to noncitizen individuals, but also to corporations when the majority of the shareholders consisted of aliens ineligible for citizenship. The law was in effect until 1956, when it was repealed (Higgs 1978).

From 1790 to 1952 racialized limits had been legislated concerning who could become a naturalized citizen of the United States. The Naturalization Act of 1790 limited eligibility to "free white persons." In the Naturalization Act of 1875, persons of African descent were allowed to become naturalized as citizens. All Asian immigrants, classified as neither white nor black, continued to be "ineligible" for naturalization (Guskin and Wilson 2007).

Like Chinese before them, Japanese farmers found many ways to evade the law's restrictions. The most common tactic was to purchase land in the names of their U.S.-born children, who were legal U.S. citizens. Since the land bought under the names of minors had to be under the guardianship of an adult, some sympathetic white and Hawaiian-born Japanese citizens (who had immigrated generations earlier) were named as trustees. A second way Japanese farmers evaded the land law was through the creation of "dummy corporations." These were companies that were formed in order to purchase land, where the majority of members or stockholders were citizens but owners in name only, allowing noncitizen Japanese to manage the land (Higgs 1978, Matsumoto 1993).

The Alien Land Law was amended in 1920, in an attempt to clarify prior ambiguities and increase its restrictiveness. Noncitizen Japanese farmers had been finding loopholes and succeeding at farming. After the amendment, noncitizen Japanese were denied rights to lease as well as own agricultural land. Further, they could not be members of any corporation with entitlement to land, and could not act as guardians for minors owning or leasing land. Again, the amendments did not prevent Japanese farmers from finding ways around the restrictions. As long as white landowners could gain financially by leasing to the Japanese, some were willing to do so, even if it meant breaking the law. After 1920, it was common practice for noncitizen farmers to be named in writing as "managers" of white-owned farms, while actually leasing the land from a white farmer.

The 1923 amendment to the Alien Land Law took the restrictions even further, not allowing those Japanese ineligible for citizenship to "acquire, possess, enjoy, use, cultivate, occupy, and transfer real property" (Higgs 1978, 220–221). Racialized enclosure could not have been more blatant.

As shown by the actual number of Japanese farm operators in 1940, which was approximately the same as the number in 1920, many Japanese immigrant farmers managed to resist the restrictions on their agricultural activities. Yet their successful resistance to racist discrimination via exclusionary land laws was ultimately thwarted by other, more ominous racist laws during World War II. From 1942 to 1945, the Japanese living on the West Coast of the United States were transported to internment camps, in the course losing their farms, their homes, and the businesses they had established in the United States. After internment, most Japanese were unable to return to their homes and reestablish their previously cohesive communities (Higgs 1978; Matsumoto 1993).

In 1923 several U.S. Supreme Court cases challenged the Alien Land Law and its amendments as unconstitutional. It was defended by Justice Pierce Butler as directly affecting the "strength and safety" of California agriculture. Using state security as a defense, Butler held that the law was "not arbitrary and unreasonable in its discrimination" (Higgs 1978, 219). Yet "how the operation of a few thousand small vegetable and fruit farms by the Japanese threatened the strength and safety of California was not explained by the learned Justice" (Ibid., 219). The alien land laws were racially motivated, without question, and the rhetoric of security used as an excuse to defend unjust legal decisions.

Excluding racialized immigrant groups by dispossessing them of their rights to control land is one way that the white population of California ensured that capitalist farming would remain white. As Almaguer (1994, 186) poignantly writes, "These laws represented yet another attempt at social closure by the white population in the state." Using the legal system to define the racial makeup of landownership is one way that the law has been used to align race and class, resigning people of color to the unpropertied classes of America.

The Hmong 1975–2009: Protecting Workers, Challenging Family Farmers

Hmong refugees are some of the most recent immigrants of Asian descent in California. In 1975 they were deemed political refugees by the United

Nations as a result of the United States' Secret War[10] in Laos. Prior to their arrival in the United States, most Hmong were engaged in subsistence agriculture, many in remote areas of their own countries. Most arrived in the United States with minimal English abilities or other skills for employment.[11] Upon settling in the United States, many of the Hmong reached back to their agricultural roots by leasing land and initiating very small-scale or what we call "microfarming."

In contrast to the Chinese and Japanese, who eventually attained the status of "model minority" by the 1960s, Laotian and Cambodian refugees became the new subjects of overt discrimination. They were largely deemed "social failures" because of their agrarian backgrounds, war experiences, poor performance in schools, poverty, and dependence on government welfare (Ong 2003).

Cambodian and Laotian refugees started to be associated with a lower class of immigrants than Chinese and Japanese immigrants before them, which constituted an "ideological blackening" (Ong 2003, 86). Because of the effects of their tragic history on families and the lack of culturally appropriate social services support, many of the youth have turned to gang violence, contributing further to their racial stereotyping. At the same time, evasion of government authority and institutions by the Hmong, stemming from centuries of government mistrust, inadvertently contributes to the stereotyping. As a result, assumptions made about the Hmong (and other Laotian and Cambodian immigrants), in reference to their agricultural practices, histories, capabilities, and "remoteness" have acted to construct racialized identities that have contributed to new racialized agricultural formations.

While the prior, legalized forms of dispossession discussed earlier were overtly racist, contemporary laws and institutions are not explicitly so. Instead, the implementation of certain regulations, when mapped onto the cultural and socioeconomic landscape of small-scale farming in California, has had subtle but profound racialized *effects*. In this section, we discuss how asymmetries between Hmong cultural norms, rules, and practices and those in the United States, as manifested in laws and other institutions, result in various forms of racialized dispossession of the means of agricultural production.[12]

Small-scale family farms have not been characteristic of California agriculture historically; rather, medium-sized, large-scale, and corporate agribusinesses (although often family owned) have been more typical (Walker 2004). However, extremely small-scale, often immigrant, farm families comprise a growing movement of new farmers (Molinar and

Yang 2000; Walker 2004). Their farms are typically so small that the vast majority of them are not even captured by the U.S. Census of Agriculture. Yet they are subject to many of the same agricultural regulations as are their corporate counterparts with vastly different historical circumstances.

Since their arrival in the United States, many Hmong have engaged in agriculture on a truly small scale. Early on, families grew herbs and a few vegetables for their own consumption. At that time, only a few better-off and more powerful Hmong families succeeded in commercial farming. In 1992, the average Southeast Asian farm in Fresno County was 3.25 acres, and only several hundred had been identified by The University of California's Cooperative Extension programs (Molinar and Yang 2000). By 2008, extension advisors Richard Molinar and Michael Yang had identified 1,500 Hmong, Mien, Laotian, and Cambodian farms. In 2008–2009, a team led by UC Berkeley conducted in-depth interviews with sixty Southeast Asian farmers in Fresno. The average farm size at that time was 8.8 acres, ranging from 0.5 to 60 acres, with nearly half being 5 acres in size or less. Average gross revenues from these farming operations ranged between $5,000 and $50,000 annually.[13] Although farming is not typically the chosen profession of Hmong young people, Sowerwine, Getz, and Peluso found that many older Hmong and some younger adults wish to continue in agriculture (see also Nguyen 2005).[14]

In spite of a variety of challenges, some 1,500 to 2,000 Hmong and other Southeast Asian refugee farmers have managed to establish very small farms in Fresno and Sacramento, and possibly several thousand more farm throughout other California counties (Molinar 2007). Hmong farmers today are most visible in the production of strawberries and "Asian" vegetables.[15] Producing for the commercial U.S. market was extremely different from subsistence or market-garden vegetable production in Laos. Hmong refugees left everything behind and, except for some U.S. government assistance, had few financial resources, no connections to established farmers or extension resources (especially in their early years here), and extremely limited communication skills and marketing connections (Fass 1986; Sowerwine, Getz, and Peluso 2009). Moreover, as mentioned, Hmong farming practices diverge significantly from those of small and medium-sized farming practices in the United States. In particular, most of the Southeast Asian–owned farms are significantly smaller and less capitalized than the average owner-operated small farm, particularly in California. The United States Department of Agriculture defines a small farm as having less than $250,000 in annual gross sales, whereas Southeast Asian farmers earn on average $25,000 in a good year. Second, the vast majority of Southeast Asian farmers lease rather than own land, thus disincentivizing the transition to more profitable organic or perennial crops. Third, the cultural institution of labor reciprocity, in which families share labor or in kind services with extended family members, is central to the economic viability of their farms. Fourth, Southeast Asian farmers lack the language and social networks to tap into government and other institutional resources available to them. Finally, there is a kind of self-selected invisibility of the Hmong, a cultural legacy of the aforementioned government mistrust, which manifests in numerous ways contributing to the racialization of their particular agricultural formation.

The exact number of Hmong farmers is extremely difficult to estimate because many fail to register with the county agriculture commissioner, the most reliable source of agricultural data. Though they work on the land, Hmong farmers remain difficult to find and contact because many work elsewhere as well. The spatial arrangements of their microproduction, in the midst of the biggest vegetable and fruit-producing region in the country, do not facilitate their being found. In Fresno, for example, Hmong farms are rarely, if ever, adjacent to their homes; most live in Fresno city or other nearby areas. Some drive up to forty-five minutes to reach their farm plots. Hmong farmers typically do not own the land they farm. Commonly, one better-off farmer leases a larger plot of land, and then engages in various informal subleasing arrangements to other farmers, thereby making it extremely difficult to locate those sublessees. In addition, most Southeast Asian farmers only have short-term leases on their farm lands. Because of their short-term leases, and because the land they cultivate typically lies in the urban-agricultural interface, which is frequently targeted for conversion to suburban development, some farmers cultivate different plots every year or few years. As a result, each time they move, they become invisible by another degree.

Language is another huge barrier to accounting for the number of Hmong farmers and their practices. While some Hmong have learned rudimentary English, few government agencies or their employees have language capacity in Hmong or Lao. Thus it is not surprising that Hmong are underrepresented in state government statistics and underserved by Cooperative Extension services, even when extension officers are sympathetic to their needs, as we have found in Fresno and Sacramento counties. In addition, their physical and practical "remoteness," driven in part by their mistrust of any government officials or agents, contributes to a mystique of Hmong farmer identity, making it easier for mainstream/industrial farmers, marketers, and regulators to stereotype their racialized identities and practices (Sowerwine, Getz, and Peluso 2009).

The "remoteness" of the Hmong is exacerbated by their prevalent perceptions of hostile regulatory climates in regard to their most common status as microscale farmers. This perception has added to the mistrust of government officials who must enforce regulation, and renders the process of establishing trust-based extension relationships with them even more difficult. The perception of hostility has been borne out in practice: Hmong farmers have been cited for numerous labor and Occupational Health and Safety Act (OSHA) violations. These citations have required poor farmers to pay onerous fines for having minors (often nieces and nephews) working in their fields, not buying workers' compensation insurance for extended family members, providing only one portable toilet for thirty employees, not having an illness and injury prevention plan, not training employees about heat-related illness, and not providing single-use cups for their workers' use. In addition to reflecting completely different cultural norms and expectations of family labor, many of these OSHA and U.S. Department of Labor regulations are difficult to find and understand for even the most literate immigrant farmers.

While the Hmong have been cited for diverse infractions, we focus specifically here on the regulation of labor practices. At certain points in the crop cycle, strawberries and Southeast Asian vegetables need high levels of labor inputs, particularly around harvest time. For small Hmong growers, this labor typically has been provided by extended family members, sometimes in exchange for working on each other's small plots, just as they would have helped each other farming in Laos. Although in Laos, small-scale market farming, in contrast to subsistence farming, changed the expectations of family laborers. With the coming of market farming, hourly or daily wages were paid in addition to the traditional allocation of shares of the harvest.

In California, however, workers' compensation law regulates this long-standing cultural practice of unpaid labor sharing among extended family members and by doing so increases the farmers' costs. All workers on a farm, paid or unpaid, family or not, must be covered by workers' compensation insurance, which can cost a small Hmong farmer/owner

$445 on average a year. For many whose farms are barely viable, any additional cost is a sizable burden, especially when bad weather results in crop failure. Most Hmong are either unaware or unsure of how to comply with workers compensation insurance requirements due to language and institutional barriers; as such, they are vulnerable to violations. Moreover, many would simply not be able to farm if they had to bear those costs for workers who came as part of a family harvest event or just to help out. Rather than protect workers from exploitation by growers as it is meant to do, this otherwise progressive legislation has had devastating effects on Hmong microfarmers' capacities to continue farming. It also adds to their racialized construction as "difficult" and "remote." For Hmong, however, part of the difficulty of regulatory compliance comes in the definition of "family" and the ways titles and leases are written.

In 2004 a multiagency "sweep" was carried out in Fresno County, seeking to identify and prosecute growers who did not carry the appropriate workers' compensation insurance for workers on their farms. Pushed by labor unions, and purportedly aimed at larger growers, several small Southeast Asian growers were hard hit; many felt that they were unfairly targeted. An advocate for these Hmong farmers stated that agents conducting the sweeps failed to have translators on hand, making it difficult for the agents to communicate with the farmers. He questioned whether proper information had been gathered and whether or how information on the violations and requirements for action were transmitted to the accused. In one instance, a farmer that was cited actually farmed an adjacent plot to the one in question, yet he happened to be on-site at the time of the inspection and, because he was unable to communicate with the inspector, was wrongly cited. Large numbers of Asians working in a field become easy targets. The law states that a farmer is required to carry workers' compensation insurance coverage for each and every person who is not named on the title of the farm. This applies to both nuclear and extended family members, to those who work for pay or exchange labor, and to permanent and temporary laborers. Because the farm owners did not carry the insurance for those differentiated by the law as extended family members (whose names would not be listed on a title or lease agreement), they were cited between $14,500 and $25,000 (Fresno Bee 2007; Sowerwine, Getz, and Peluso 2009). As the Fresno Southeast Asian community became aware of the sweep, historical anxieties around government authority reemerged and nearly fifty Hmong farmers stopped farming for fear of being caught (Fresno Bee 2007). Today, farmers remain fearful and resentful of the law. Nevertheless, to fulfill their needs for labor during high-input times, some farmers who are trying to stay in business have been forced to hire contract workers. Others count on their luck to not be discovered.

When the workers' compensation law is implemented and mapped onto the socioeconomic and cultural landscapes of microfarming in California, arguably a niche landscape dominated by Southeast Asian farmers, the Hmong appear as particularly vulnerable to its effects. They pay fines grossly disproportionate to their incomes, to such an extent that many have either been forced out of farming altogether or continue to farm in fear of the authorities.

Legislative attempts[16] to amend workers' compensation insurance requirements to exempt small farms have been introduced into the California legislature. Senate Bill 452, introduced in 2007 and framed as protecting small Hmong farmers in particular, is an attempt to exempt any grower with less than $100,000 a year in taxable income from buying workers' compensation insurance for unpaid relatives working on the farm. Had it passed, immediate and extended family, including spouses, children, parents, grandparents, aunts, uncles, and first and second cousins, would have been exempted if they worked without wages on a "small" family-owned farm. All of the sixty farmers interviewed in our 2008–2009 survey had annual revenues between $5,000 and $50,000, well within the limit. The labor unions and their allies, however, opposed this, because so many growers that hire labor fall under this $100,000 ceiling and the impacts would extend far beyond the Hmong and Southeast Asian grower community. Because of the strong resistance, the senator sponsoring the bill dropped it (Sowerwine, Getz, and Peluso 2009).

Many Hmong growers are additionally affected by the insurance costs because they are unable to get (and often unwilling to take) bank credit. Few own homes, equipment, or land to use as collateral; few have credit histories that would enable them to acquire working capital to expand their farming activities enough to make the insurance fees affordable. Ironically, lack of capital precludes many Southeast Asian growers from taking commercial advantage of their low-input farming techniques. Many growers use few or no pesticides, but since they do not own their own land, it is difficult for them to transition to organic certification. The extra costs and time that would be involved in certification and record keeping are additionally prohibitive.

The culture of regulation in California agriculture, and the underlying assumptions about accessing capital, farming practices, ethics, rights, and responsibilities, has been articulated in such a way as to marginalize and discriminate against Hmong farmers, particularly the smallest ones. The circumstances of their arrival and status as refugees had, in fact, led us to expect otherwise. On the contrary, regulations intended to protect workers are in fact beginning to have the unintended effects of pushing Hmong out of farming (Molinar 2007).[17] Before examining the emergence of this race-class agricultural formation and its relations to the law, we briefly explain how refugee status differentiates Hmong from other immigrants.

Unlike the Chinese or Japanese immigrant farmers previously discussed, the Hmong immigrated to the United States as documented refugees, affording them many benefits that Chinese and Japanese immigrants were denied. In 1965 refugees were recognized as a special category of immigrant in the United States, with the understanding that they would not be returning to their home country (Ong 2003). In the case of Southeast Asian refugees, most had to leave their home countries because of U.S. intervention (and failure) in the region. Over more than twenty years, waves of Southeast Asian refugees were transported to the United States directly from Laos or from refugee camps predominantly in Thailand, Hong Kong, and Indonesia. Through various volunteer resettlement agencies, the U.S. government settled them in California, Minnesota, Wisconsin, and other states. In order to distinguish legally between refugees and immigrants, the United States passed the Refugee Act in 1980.

Because of their status as refugees or resident-aliens, Hmong immigrants have different political and citizenship status than that of early Chinese and Japanese agricultural workers. They receive a higher rate of government assistance than previous immigrants, have the ability to work legally as soon as they enter the country, and are eligible for permanent resident status one year after immigration (Ong 2003). They are legally entitled to negotiate financial assistance in the United States, being eligible for welfare, loans, and other government support services (Lieb 1996).

However, this does not mean they actually seek or gain access to these services, again, because language and cultural barriers are difficult to overcome. Ultimately, refugee status, though apparently a more privileged status than that of other immigrants, has only partially assisted Hmong wanting to farm in California. The intention of refugee assistance is not to help them achieve upward class mobility or to become landowners, which might allow some to return to agriculture full time, as they had done in Laos. Instead, it is intended to remove them from the welfare system as soon as possible, allowing them to fill lower-tier positions in labor markets (Ong 2003).

Although Hmong growers have not been entirely excluded from participation in California farming, some have decided to move or return to Minnesota (where they may have been originally placed upon immigration) to farm or to work factory jobs (Fresno Bee 2007). Clearly, labor regulations have combined with cultural incongruities and contributed to new racialized agricultural formations on the California landscape. This has happened as the traditional reciprocal labor practices of small Hmong farmers and their extended family members have simultaneously come under fire and created an administrative impression of Hmong as irresponsible, if not outlaw, farmers. In effect, the practice of family farming Hmong-style has been criminalized—the ironic racialized effect of a workers' compensation law with progressive intentions.

Conclusion

While the contemporary production of racialized agricultural formations in the United States, and California in particular, is nowhere near as baldly racist as during the times of the exclusionary and alien land laws, the effects of law and policy on different communities practicing agriculture can have racialized effects. The unique agricultural history of California, one in which large-scale, corporate (albeit family owned) agribusiness has dominated the structure of state agriculture, has produced a Polanyian-style response from organized labor in the state to legally protect workers through requiring insurance provision (Polanyi 1944). Yet this push to protect both documented immigrant and unprotected family workers has inadvertently contributed to the construction of Hmong family farm labor as a "problem" and thereby made Hmong farmers a "problem population." The difficulties experienced by Hmong in accessing markets and production opportunities through the usual mechanisms sought by even small U.S. growers, such as credit and extension services, further complicate their potential for transcending the stereotypes we have called a new racialized agricultural formation. The dilemma is perhaps a uniquely twenty-first century one: displaced by U.S. interventions in their home country in the late twentieth century, Hmong are still being dispossessed of even the smallest gains they have achieved in their adoptive country. The terms are only subtler and more closely related to some relatively unexplored, racialized, dimensions of an old agrarian question about the effects of capitalism on agriculture.

Notes

1. Yang is a pseudonym here.

2. Most Hmong came to the United States from mainland Southeast Asia. Many came to the United States as refugees seeking political asylum after the Vietnam War, in which they fought against communist forces in the region.

3. In critiquing such laws, we do not mean to undermine the important progress made by labor unions and others working for farmworker rights, or to deny that family members may treat their own relatives unjustly. Rather, we argue that such laws, when applied without regards to cultural histories, actually work to discriminate against those who are economically and socially marginalized.

4. Who was actually included in the definition of *white person* has been a topic of legal contestation and deliberated on a case-by-case basis. Legal debates from 1878 to 1944 designated specific ethnic and national groups as either included or excluded from the legally defined category of "white" in the United States. Persons of certain nationalities, such as Japanese, Filipinos, and Syrians, were designated as both white and nonwhite in separate cases (Lopez 2006). It was not until the McCarran-Walter Act of 1952 that whiteness as a precondition for citizenship was formally abolished (Calavita 2007; Lopez 2006).

5. Congress made the first declaration that naturalization for citizenship be restricted to white persons in 1790. All persons deemed as nonwhite or other were considered "aliens ineligible to citizenship."

6. In volume 1 of *Capital (Das Kapital)*, Karl Marx ([1867] 1977) explains his theory of primitive accumulation as the separation of the workers from the means of production. He states that there are two processes in this transformation: the social means of subsistence and production that are turned into capital (accumulation) and the creation of the wage-laborer class (via loss of the means of production and subsistence). The "classic" form of primitive accumulation that he refers to is specifically the removal of agriculturalists from their land or the expropriation of "the peasant from the soil" (876), which he depicts in the enclosure of common grazing land in the English countryside in the fifteenth and sixteenth century.

7. Harvey (2004) differentiates between his notion of "accumulation by dispossession" and Marx's historical notion of primitive accumulation in that accumulation by dispossession is ongoing and is more complex than a one-time expropriation of peasants from their land.

8. See note 5.

9. The exclusion order was officially rescinded on January 2, 1945, seven months before the official end of the war.

10. The United States officially denied involvement in the war in Laos. On May 15, 1997, the United States reversed its policy, acknowledging its role and dedi-

cating the Laos Memorial at the Arlington National Cemetery in honor of the Hmong and other combat veterans from the "secret war." It wasn't until this time that Hmong vets received their due benefits.

11. Some women embroidered traditional textiles (Fass 1986), yet others, including former military and militiamen, had difficulty finding employment, due to language barriers.

12. Our research explores disjuncts in the realm of agriculture similar to the "collision of cultures" between the Hmong and Western health care systems that Anne Fadiman deftly describes in her eye-opening book about the Hmong, *The Spirit Catches You and You Fall Down.*

13. Sowerwine, Getz, and Peluso comprise the UC Berkeley team that developed and implemented the survey in 2008–2009. Hmong language collaborators from The Fresno Economic Opportunities Commission, UC Cooperative Extension, and Fresno State were trained in interviewing techniques and carried out the majority of the surveys. Some of the interviewees were sampled randomly, from lists provided by UCCE Fresno and from pesticide permits on file with the Agricultural Commissioner's Office. Others were sampled using the snowball method.

14. All "interviews" referred to in this section were conducted by Sowerwine, Getz, and Peluso in 2008–2009 as part of their ongoing research on Hmong farmers in Fresno and Sacramento counties.

15. Most of the strawberry farmers in Sacramento are actually Mien; however, many of the earliest farmers, producing for the processing industry in the Central Valley, especially around Fresno, were Hmong.

16. Most recently, in February 2009, California Senate Bill No. 677 was introduced, which would lower the ceiling to $10,000 and add a restriction prohibiting growers that hire contract laborers (Fresno Bee 2007). No action has been taken yet, as the bill is still in committee (California Senate Bill No. 677 2009).

17. Richard Molinar is a small-farm advisor in Fresno County, with a long history of assisting Hmong farmers. Yet even for him, accurate estimates of the number who have tried and given up on farming are extremely difficult to make.

References

Alien Land Law of 1913. 1913. Statues of California 206.

Almaguer, Tomas. 1994. *Racial Fault Lines: The Historical Origins of White Supremacy in California.* Berkeley: The University of California Press.

Calavita, Kitty. 2006. Collisions at the Intersection of Gender, Race, and Class: Enforcing the Chinese Exclusion Laws. *Law & Society Review* 40: 249–281.

Calavita, Kitty. 2007. Immigration, Law, Race, and Identity. *Annual Review of Law and Social Science* 3: 1–20.

California Senate Bill No. 677. 2009. 2009–10 S., Regular Session (Ca.)

Chan, Sucheng. 1986. *This Bittersweet Soil: The Chinese in California Agriculture, 1860–1910.* Berkeley: The University of California Press.

Civil Rights Act of 1866. 1866. Public Law, 14 Stat. 27–30.

Civil Rights Act of 1964. 1964. Public Law, 78, Stat. 241.

Civil Right Act of 1991. 1991. Public Law, 105, Stat. 1071.

De Angelis, Massimo. 2004. Separating the Doing and the Deed: Capital and the Continuous Character of Enclosures. *Historical Materialism* 12 (2): 57–87.

Fass, Simon. 1986. Innovations in the Struggle for Self-Reliance: The Hmong Experience in the United States. *International Migration Review* 20 (2): 351–380.

Fresno Bee. 2007. Bill Gives Ag Families a Break on Job Insurance. *Fresno Bee.* March 19.

Glassman, J. 2006. Primitive Accumulation, Accumulation by Dispossession, Accumulation by "Extra-economic" Means. *Progress in Human Geography* 30 (5): 608–625.

Guskin, Jane, and David Wilson. 2007. *The Politics of Immigration: Questions and Answers.* New York: Monthly Review Press.

Gyory, Andrew. 1998. *Closing the Gate: Race, Politics, and the Chinese Exclusion Act.* Chapel Hill: The University of North Carolina Press.

Harvey, David. 2004. The "New" Imperialism: Accumulation by Dispossession. In *The New Imperial Challenge*, ed. L. Panitch and C. Leys, 62–87. London: Merlin Press.

Higgs, Robert. 1978. Landless by Law: Japanese Immigrants in California Agriculture to 1941. *Journal of Economic History* 38 (1): 205–225.

Lee, Erika. 2003. *At America's Gates: Chinese Immigration during the Exclusion Era, 1882–1943.* Chapel Hill: The University of North Carolina Press.

Lieb, Emily Belinda. 1996. The Hmong Migration to Fresno: From Laos to California's Central Valley. Master's thesis, California State University, Fresno.

Lopez, Ian Haney. 2006. *White by Law: The Legal Construction of Race.* 10th ed. New York: New York University Press.

Marx, Karl. [1867] 1977. *Capital.* Vol. 1. New York: Vintage Books.

Matsumoto, Valarie J. 1993. *Farming the Home Place: A Japanese American Community in California 1919–1982.* Ithaca: Cornell University Press.

McCarthy, James. 2004. Privatizing Conditions of Production: Trade Agreements as Neoliberal Environmental Governance. *Geoforum* 35 (3) (May 2004): 327–341.

McWilliams, Carey. 2000. *Factories in the Field: The Story of Migratory Farm Labor in California.* Berkeley: University of California Press.

Molinar, Richard. 2007. Personal communication.

Molinar, Richard, and Michael Yang. 2000. Family Farms in Fresno, California. Cooperative Extension Work in Agriculture and Home Economics. U.S. Department of Agriculture, University of California and Fresno County Cooperating. <http://www.sfc.ucdavis.edu/research/molinar.html> (accessed August 11, 2008).

National Farm Worker Ministry. 1999. Farm Worker Conditions. <http://www
.nfwm.org/category/map/learn-more/farm-worker-conditions> (accessed June
20, 2009).

Nguyen, Daisy. 2005. Children of Ethnic Hmong Leaving Family Farms. *The
San Diego Union-Tribune*. December 28.

Omi, Michael, and Howard Winant. 1994. *Racial Formation in the United
States: From the 1960s to the 1990s*. 2nd ed. New York: Routledge.

Ong, Aihwa. 2003. *Buddha Is Hiding: Refugees, Citizenship, the New America*.
Berkeley: University of California Press.

Polanyi, Karl. 1944. *The Great Transformation: The Political and Economic
Origins of Our Time*. Beacon Press.

Sowerwine, Jennifer, Christy Getz, and Nancy Peluso. 2009. Unpublished
interviews.

Walker, Richard. 2004. *The Conquest of Bread: 100 Years of Agribusiness in
California*. Berkeley: The University of California Press.

II

Consumption Denied

5

From Industrial Garden to Food Desert

Demarcated Devaluation in the Flatlands of Oakland, California

Nathan McClintock

A dilapidated liquor store stands at the corner of 17th and Center in West Oakland. With its plastic sign cracked and yellowed, its paint pockmarked and peeling away in long lesions from the store's warped clapboard siding, it could be a clichéd metaphor for the decay of America's "inner cities" during the postindustrial era (figure 5.1). But it is also representative of the disproportionate number of liquor stores in urban communities of color. Establishments such as these often serve as the sole food retailer in areas that planners and food justice activists have come to call "food deserts."[1]

A recent report to Congress by the USDA Economic Research Service defines *food desert* as an area "with limited access to affordable and nutritious food, particularly such an area composed of predominately lower income neighborhoods and communities" (USDA 2009). A number of articles and reports over the last few years have attempted to characterize and identify food deserts in the United States, Canada, Britain, and Australia. Most have concluded that in the United States, food deserts disproportionately impact people of color (Smoyer-Tomic, Spence, and Amrhein 2006; Beaulac, Kristjansson, and Cummins 2009). While many studies have drawn spatial or statistical correlations or both between race and the absence of supermarkets (Raja, Ma, and Yadav 2008; Lee and Lim 2009; Zenk et al. 2005), researchers have also found that small corner stores and ethnic grocers are abundant in these food deserts (Short, Guthman, and Raskin 2007; Raja, Ma, and Yadav 2008). Nevertheless, fresh and nutritious produce is rarely available at these small stores, and the type of food generally tends to be of poorer quality and less healthy, high in sugars and saturated fats (Cummins and MacIntyre 2002).

Food access in Oakland's food deserts falls under a similar rubric. The socioeconomic terrain demarcating poverty and affluence in this Bay

Figure 5.1
Corner store sign, Lower San Antonio (East Oakland). This kind of store serves as the primary food retail in Oakland's flatlands neighborhoods. Photo by the author, May 2010

Area city of about 391,000 (2010) roughly follows the contours of its physical geography of flatlands and hills (figure 5.2). Census data reveal that the vast majority of Oakland's people of color live in the flatlands (figure 5.3). Between a quarter and a third of people in the flatlands live below the poverty line; median income is 25 percent lower than the citywide average. The flatlands host the lowest percentage of home ownership and the lowest levels of educational attainment. Unemployment here is roughly twice the citywide rate. Crime and public health statistics overlap in a more or less identical fashion. In predominantly black flatlands neighborhoods, such as West Oakland and Central East Oakland, these statistics are even bleaker.

It is precisely in these flatlands neighborhoods that the city's food deserts can also be found. And it is here that food justice movements have taken root. Yet to better understand Oakland's food deserts and to recognize the emancipatory potential of the initiatives that have emerged as a solution, it is helpful first to understand the forces that have hewn the urban landscape into a crude mosaic of parks and pollution, privilege

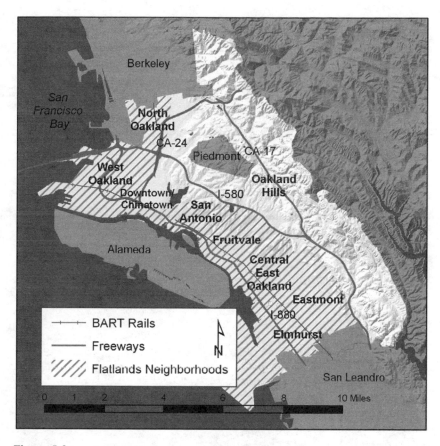

Figure 5.2
Oakland, California, and its major districts. Flatlands neighborhoods are shaded. Map by the author

and poverty, Whole Foods and whole food deserts. Few studies move beyond a geospatial or statistical inventory of food deserts to unearth these historical processes. In this chapter I focus on the structural role of capital (with an implied capital "C") in order to emphasize the extent to which capital defines the urban environment. Driving down MacArthur or International Boulevards "in the cuts" of the Oakland flatlands provides a glimpse into how capital's dynamic cycles—its ebbs and flows—have shaped both the built environment and the social relations woven through it, leaving an almost entirely treeless and worn landscape of used car dealerships, taco trucks, liquor stores, dilapidated storefronts, and the occasional chainlinked vacant lot.

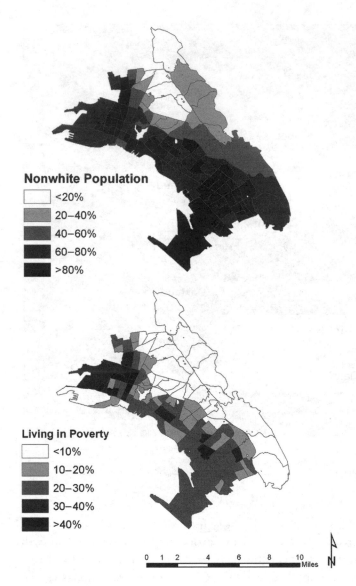

Figure 5.3
Oakland's racialized socioeconomic landscape. The Oakland hills are home to the majority
of the city's white (non-Latino) population, while the majority of people of color reside in
the flatlands (bottom). Poverty is also concentrated in the flatlands (bottom). Maps by the
author. *Source*: U.S. Census, 2000 (SF-3)

Understanding the historical and structural roots of this urban land-scape is fundamental to understanding the individual and collective agency that adapts to or resists its development. With this in mind, I tap existing histories of Oakland and urbanization in California, demo-graphic and economic data, and current "grey literature" (e.g., planning, economic, and public health reports) to broadly trace the historical geography of Oakland's flatlands during the periods of industrialization and deindustrialization, roughly from the turn of the century to the "neoliberal turn" of the 1980s. I draw on theoretical insights from the growing field of urban political ecology to shed light on the structural processes that have restricted access to healthy food for residents of the flatlands, arguing that a combination of industrial location, residential development, city planning, and racist mortgage lending unevenly devel-oped the city's landscape and concentrated the impacts of capital devalu-ation within the flatlands, a process I refer to as "demarcated devaluation" and which ultimately created the city's food deserts.

Root Structure: Devaluation of Urban Capital

To understand Oakland's food deserts, the related diet-related illnesses impacting flatlands residents, and the food justice initiatives that have arisen in response, an analysis of the historical processes that have unevenly shaped the city's socioecological landscape is a necessary first step. Environmental sociologists, political ecologists, and urban geogra-phers have described the material transformation, or "metabolism," of the biophysical environment and human populations by political eco-nomic processes such as capitalism (Foster 1999; Gandy 2003; Heynen, Kaika, and Swyngedouw 2006). David Harvey (2006) stresses the inter-connected nature of society and environment; understanding one cannot be done without understanding its relation to the other:

On the ecological side . . . we have to understand how the accumulation of capital works through ecosystemic processes, re-shaping them and disturbing them as it goes. Energy flows, shifts in material balances, environmental trans-formations (some of them irreversible) have to be brought thoroughly within the picture. But the social side cannot be evaded as somehow radically different from its ecological integument. . . . The circulation of money and capital have to be construed as ecological variables every bit as important as the circulation of air and water. (88)

Such an analysis necessarily takes place at multiple levels. In his analy-sis of urban hunger in Milwaukee, Nik Heynen (2006) underscores the importance of looking across scales to understand the connections between hunger and its causes. The physical experience of hunger, malnutrition, or the body's biochemical metabolic process cannot be treated as disconnected from the larger-scale processes determining the availability of food. Indeed, the chain of causality spans several levels of scale, from the individual to the household, from neighborhood to municipality, and from national to global.[2] Viewing socioecological change this way certainly complicates analysis (and demands a certain level of interdisciplinarity) but may ultimately offer a fuller, if not more nuanced, understanding of the links among ecology, public health, and social change.[3]

The web of social and political relations driving and shaping these changes is complex and multidimensional. Nevertheless, at the risk of being seen as an economic determinist, I want to focus on one process that is fundamental to the transformation of the urban landscape and creation of food deserts: the devaluation of certain types of capital. It undergirds the structural processes of uneven development and the social disruption that emerges in response. Nowhere is this process so readily apparent as in postindustrial cities such as Oakland. Cities are ground zero of humans' transformative power, where the influx of capital is visibly inscribed on the landscape in the form of buildings and infrastructure, as roads, bridges, power lines, rail lines, sewers. During historical moments of capital overaccumulation following economic booms, surplus capital is invested in this kind of fixed or immobile capital, transforming the urban environment.[4] During economic downturns, as capital retreats from urban industrial zones, the postindustrial city nevertheless retains its industrial character, albeit devalued, dilapidated, and scarred by pollution, often to such a great degree that it precludes future investment.[5] Rents fall, unemployment rises. Both labor and fixed capital are devalued. Harvey (2001) writes, "The geographical landscape which fixed and immobile capital comprises is both a crowning glory of past capital development and a prison which inhibits the further progress of accumulation" (247). These zones left fallow inside the city by capital's retreat belong to what Richard Walker (1978) has called "a lumpengeography of capital," or "a permanent reserve of stagnant places" awaiting new investment once land and labor values have been sufficiently devalued.[6]

From this perspective, the contemporary cityscape is a map of previous cycles of capital accumulation and devaluation, a palimpsest of building, decay, and renewal.[7] The walls of this prison of fixed capital are often clearly delineated by planning, policy, property taxes, and political boundaries. These buttresses and ramparts, whether or not they were crafted with intention, effectively *demarcate and quarantine devaluation* to prevent its impacts from bleeding over, both metaphorically and materially.[8] As environmental justice literature reveals, this process of demarcated devaluation has been highly racialized historically through zoning, redlining, and neighborhood covenants (Matsuoka 2003; Maantay 2002; Self 2003; Morello-Frosch 2002; Boone et al. 2009).

Human populations viscerally experience these ebbs and flows of capital. As countless cases in the era of deindustrialization illustrate, capital devaluation has historically been the harbinger of social upheaval in the form of migration, poverty, hunger, crime, and declining public health. Given the extent to which the urban landscape is shaped by capital and its crises of accumulation, urban social struggles against the socioeconomic upheaval that follows are interwoven with struggles for a more equitable environment. Perhaps less obvious to many mainstream environmentalists, struggles to protect or clean up the urban environment are equally as entwined within struggles for social justice; as Swyngedouw and Heynen (2003) point out, "processes of socioecological change are . . . never socially or ecologically neutral" (911). Understanding the food justice movement in Oakland and elsewhere therefore depends on understanding the structural forces, generally, and capital devaluation more specifically, that gave rise to the movement in the first place. Applying this analytical framework, I devote the remainder of this chapter to outlining Oakland's twentieth-century history of industrialization and deindustrialization, demarcated devaluation, and the consequent creation of the city's food deserts.

An Industrial Garden Grows

In reference to her childhood home of Oakland, Gertrude Stein famously wrote, "there is no there there." While these words have been used to belittle Oakland for the seventy years that have passed since their publication, they remain poignant when taken in their original context. Stein had returned to the city decades later and was unable to recognize the childhood home of her memories in the vast expanse of new housing sprawling eastward from downtown (Rhomberg 2004). The transformative power that had effaced the "there" of Stein's turn-of-the-century childhood home continued to reshape Oakland as industrial and residential capital flowed and ebbed throughout the rest of the twentieth century.

Advertising Oakland as a "city of homes," speculators from the mid-nineteenth century onward hoped to cash in on its proximity to San Francisco's bustling commercial center (Scott [1959] 1985). The promise of the seemingly paradoxical union of Arcadia and Utopia that was the aesthetic hallmark of California development—pastoral landscapes embodied within an ordered, neighborhood logic (McClung 2000)—fueled a vibrant housing sector in Oakland, drawing the wealthy merchant class to the Oakland hills and foothills. A booster for housing in Oakland's lower foothills in 1911 advertised "home sites from which [to] look down on the cities about the bay . . . far removed from the dirt and turmoil of the work-a-day world" (Scott [1959] 1985; Bagwell 1982).

At the same time, completion of the transcontinental railroad and construction of its terminus in Oakland in 1869 accelerated the expansion of industry from San Francisco to the East Bay; the arrival of iron works, canneries, cotton and lumber mills, breweries, and carriage factories fueled further industrial agglomeration around the rail terminals in West Oakland and the estuary waterfront at the southern edge of downtown (Bagwell 1982; Walker 2001). A 1910 promotional booklet published by the Oakland Chamber of Commerce features a world map with all shipping lines leading to "Oakland Opposite the Golden Gate, The Logical Port and Industrial Center of the Pacific Coast" (Scott [1959] 1985).[9]

Worker housing emerged primarily in West Oakland, between the downtown business district and the rail and shipping terminus. The displacement of San Francisco residents following the 1906 earthquake was a boon for Oakland, bringing in a new workforce and new demands for housing. With population and industry growing at a rapid pace and aided by the extension of horse-drawn and electric streetcar lines, Oakland expanded to the north and east, annexing previously autonomous communities such as Temescal, Brooklyn, Fruitvale, Melrose, and Elmhurst by the end of the first decade of the twentieth century (Groth 2004; Bagwell 1982; Scott [1959] 1985).

World War I saw a massive influx of military capital into Oakland. Automotive manufacturers such as the Durant Motor Company, Hall-Scott Motor Company, Chevrolet, and General Motors expanded considerably during these years, earning Oakland the moniker "Detroit of the West." Shipbuilding dominated the port, and employed upward of 40,000 in 1920. Drawn by the promise of jobs, new workers, many of them African Americans and immigrants, flooded in by the thousands.

Wartime industrialization and the boom that continued through the 1920s saw the expansion Oakland's residential development alongside the construction of new factories eastward into the orchards and pastures of the annexed townships (Ma 2000; Walker 2001; Bagwell 1982). Integrating the pragmatism of locating industry where land was available with the reformist planning vision of Ebenezer Howard and Lewis Mumford, planners and developers in Oakland (as in Southern California) embraced the paradigm of the "industrial garden": the dispersal of industry away from the mixed-use downtown core but closely tied to nearby, semiautonomous residential neighborhoods. In these industrial garden suburbs, factory workers would return home by bus or rail to a neighborhood of small, single-family homes, each with a yard or garden. Proponents pushed "garden living" in these quiet and tranquil respites far—but not too far—from the factory grind as a cure to the social and health risks already well documented in the mixed-use urban slums of the Northeast, Chicago, and to a lesser extent in the older downtown cores of San Francisco, Oakland, and Los Angeles (Self 2003; Hise 1997, 2001). Urban and rural modes of survival came together here, as workers clocked out and headed home to tend vegetables, chickens, and goats in their yards (Nicolaides 2001; Johnson 1993). As Mike Davis (1997) writes, the industrial garden was "a new kind of industrial society where Ford and Darwin, engineering and nature, were combined in a eugenic formula that eliminated the root causes of class conflict and inefficient production" (358); in essence, by keeping the worker happy, productivity could increase while nipping a restive labor movement at the bud.

During the New Deal the vast expanse of small homes that had cropped up as part of the industrial garden expanded rapidly. Beginning in 1934, a flood of highly subsidized, low-interest mortgage loans from the newly created Federal Housing Administration (FHA) fed the growing suburbs; East Oakland soon filled in with suburban developments of small, Mediterranean-style single-family homes. As in other California industrial centers, developers consolidated land purchase, subdivision, construction, and sales in order to maximize efficiency and minimize costs. Vast tracts of small houses, mostly prefabricated or built from kits with nearly identical floor plans, created an economy of scale that dovetailed nicely with the contemporary planning vision of neighborhood cohesion, mixed use, and garden cities to create quintessential industrial gardens. In order to expand homeownership, housing production had to be reorganized into a quasi-Fordist system of on-site assembly of prefab components to perfect the "minimum house": a small, single-family home constructed as cheaply as possible but comfortable and unique enough to satisfy the dream of home ownership (Hise 1997). The newly subdivided suburban landscape was rapidly filled in with these small, single-family homes erected virtually overnight.

However, the federally subsidized dream of homeownership in the industrial garden was not available to everyone. The social idealism of Ebenezer Howard's garden cities and Lewis Mumford's inclusive "ecotopian" regions undergirded the vision of many suburban planners. Nevertheless, the pragmatism of industrial location, the whims of individual developers, and the rising power of racist homeowners' organizations soon elided their utopian vision. People of color rarely qualified for FHA loans because these were to be applied only to newly constructed homes and, contrary to Howard's vision of universalist garden cities that welcomed and nourished all workers, new home developments in the suburban industrial gardens were racially exclusive. Until 1948 racial covenants established by developers and homeowners' associations prevented people of color from moving in and disturbing social divisions seen as "natural" (Hise 2001; Self 2003; Sugrue 2005). Even after the Supreme Court made racial covenants illegal via *Shelley v. Kraemer* in 1948, such obstacles remained in practice. Contractors were rarely able to secure loans for construction for nonwhites in a "Caucasians only" neighborhood and realtors feared "the wrath of white homeowners" (Sugrue 2005).

The racialized demarcation of urban space taking place between the wars was not new in California. For decades the labor movement in California had already laid the groundwork for the formation of a virulent form of white class-consciousness via their aggressive exclusion of Asian, Latino, and African American workers (Daniels [1959] 1977; Saxton 1971; McWilliams [1949] 1999). Easy access for whites to low-cost, single-family homes in close proximity to East Oakland's factories simply fueled racist and exclusionary sentiments by creating a sense of bootstrap entitlement, where hard work alone was seen as the key to material success. Homeownership thus helped heterogeneous European and Euro-American populations of workers consolidate as a spatially and racially homogenized labor force of "whites," geographically distinct from the radicalism of recent European immigrants and African Americans in West and North Oakland and along the estuary.[10] Suburbanization of industry and housing was thus a way to escape from the working class and "to attract a better brand of labor, removed from the 'bad moral atmosphere' of the inner city, and promising the stability of homeownership for the 'better class' of workers" (Walker 1981, 400).

As new workers flooded into Oakland during World War II, housing was scarce. Trying to defuse tensions between blacks and southern white migrants, the Oakland Housing Authority located black-only housing projects in West Oakland and corresponding projects for whites in East Oakland. Most of these housing projects were located in industrial areas on landfill and adjacent to railroads. The black population of Oakland grew nearly sixfold in Alameda County between 1940 and 1950, but African Americans were rarely allowed to rent outside of West Oakland due to racial covenants and similar barriers to renting in the new industrial gardens. Ramshackle dwellings in West Oakland were converted and sub-divided to accommodate the new migrants. In the postwar years the razing of temporary wartime migrant housing in the East Bay only increased the housing squeeze. In 1940, 15 percent of West Oakland's housing units were overcrowded; the percentage doubled a decade later (Johnson 1993).

The practice of bank redlining also stopped the flow of mortgage and property investment capital into parts of the city where people of color resided. Working with banks and local realtors, the Home Owner's Loan Corporation (HOLC) and its parent organization, the Federal Home Loan Bank Board, developed Residential Security Maps and Surveys that divided cities into ranked sections. Most African American neighbor-hoods were ranked "D–Fourth Grade" for "hazardous" and colored red on the maps. Homes in these areas rarely qualified for loans. Meanwhile white neighborhoods were ranked higher if they had racial covenants that offered "protection from adverse influences" such as "infiltration of inharmonious racial or nationality groups" (Sugrue 2005; Maantay 2002).[11] While discriminatory lending existed before the creation of these maps, they helped to reify the delineation between rich and poor, between whites and people of color.[12] Even after redlining was prohibited under the 1968 Fair Housing Act, it continued in a self-reproducing, de facto manner due to a complex of factors, from zoning and housing prices to the spatialized legacy of denied loan applications (Kantor and Nyusten 1982), as well as the relocation of home insurance agencies to the suburbs (Squires, Velez, and Taueber 1991).

A 1937 HOLC Residential Security Map of Oakland (figure 5.4) and associated report reveals the spatial logic of redlining. The area reports for most flatlands neighborhoods warned potential investors of "detri-mental influences," notably the "infiltration" of "lower grades" such as "Negros," "Orientals," "shopkeepers," "lower classes," "relief

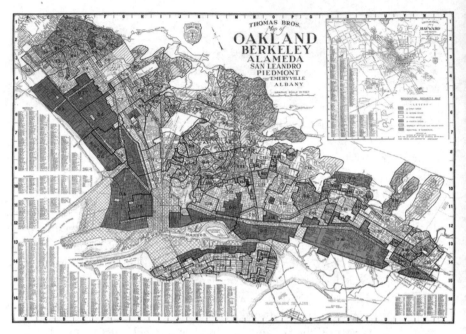

Figure 5.4
The 1937 Home Owner's Loan Corporation Residential Security Map for Oakland. "Red-lined" neighborhoods (Class D) appear here as dark grey. Along with the adjacent Class C areas (which appear yellow in the original map), these delineations continue to define Oakland's flatlands neighborhoods. Source: T-RACES: a Testbed for the Redlining Archives of California's Exclusionary Spaces. Available at <http://salt.unc.edu/T-RACES> (accessed June 9, 2010)

families," and "foreign born." On Oakland's north–south axis, neighborhoods west of Grove Street (now Martin Luther King Jr. Way) all appear as Grade D. This redline separated blacks from whites, effectively ghettoizing North and West Oakland.[13] East 14th Street (now International Blvd.) served as the east–west redline in East Oakland through the 1950s, limiting blacks to a few blocks adjacent to the industrial zones. Oakland's Asian population was effectively quarantined, as well, from the late nineteenth century until 1920. Chinatown, south of downtown and west of Lake Merritt, received a D rating (HOLC Area D-11) due to the "predominance of Orientals," an "indication of future slum condition" (HOLC 1937). By the 1930s, some Asians were able to move to blue-collar neighborhoods along San Pablo Avenue in West Oakland and into the San Antonio district, precisely the "infiltration" that the HOLC Area Reports used to redline a neighborhood.

Like the Chinese, the presence of any "low class foreign born" laborers was enough for HOLC to paint a neighborhood red. The Area Report for the Jingletown neighborhood (Area No. D-15), home to a largely Portuguese millworker population, was also classed as "hazardous" due to "Detrimental influences: Odors from industries; heterogeneous mixtures of old two-story homes and old one-story cottages (latter predominating). Predominance of foreign inhabitants, infiltration of Negroes and Orientals. . . . This area lies below east [sic] Fourteenth Street (below the tracks) and is poorly regarded; semi-slum area. There are only a few Negroes and Orientals, but the low class foreign element is large" (HOLC 1937).

By the late 1930s, large swaths of the flatlands, the first of the industrial garden neighborhoods constructed during the interwar boom years, had already been rated "Yellow" for "C–Third Grade" or "Declining," the result of "decreasing desirability" due to aging homes and "infiltration" by "lower grade elements." One such area was the Fruitvale district, where a large Mexican population had developed much earlier to work in the adjacent canneries and orchards (Ma 2000; Self 2003). By the late 1930s, the ten- to twenty-year-old worker cottages in HOLC Areas C-19, C-20, and C-26 were no longer "highly regarded by mortgage institution officials" due to the "threat of infiltration by lower grades," "proximity to areas infiltered [sic] by Negroes," and the growing population of "foreign born" and "Latin races" who already comprised up to 20 percent of the district at the time (HOLC 1937). The mere arrival of blacks, however, seemed to be enough to tip the risk scale from yellow to red. A large part of the adjacent San Antonio district (Area D-10) received a D grade: "This area is similar to C-19 in appearance but infiltration of Negroes necessitates hazardous rating" (Ibid.).

Redlining and yellowlining, along with racial covenants and federal housing subsidies, stewarded and demarcated a highly racialized urban landscape of prosperity and neglect for much of Oakland's industrial boom years and after. The Oakland hills and most of East Oakland's industrial garden suburbs remained predominantly white and affluent, while West Oakland, Chinatown, and the slightly more dilapidated East Oakland neighborhoods adjacent to E. 14th Street (San Antonio and Fruitvale) were left high and dry as investment waned. Like West Oakland's housing stock, labor—human capital—was also devalued as an influx of postwar migrants saturated the labor market, joining the ranks of the unemployed.[14] As Massey and Denton (1993) argue, segregation bred "hypersegregation," the emergence of "ghetto culture," and the decline (and flight) of the black middle class, cleaving an even greater economic rift between West Oakland and the East Oakland garden suburbs, migrants and old timers, blacks and whites, industrial growth and senescence.

Demarcated Desertification

If industrial relocation and FHA-funded residential development were the source of capital flows that irrigated East Oakland's industrial garden from the 1920s to the 1940s, homeowners associations, zoning, and redlining were the dikes that initially prevented this capital from flowing back toward West Oakland, and then effectively quarantined its devaluation to the few areas where people of color were allowed to live. New capital continued to flow in. Between January 1945 and December 1947 roughly $300 million was spent on the expansion of new industrial plants in the Bay Area (Whitaker 1992). Within the city itself, however, devalued fixed capital—a landscape of aging housing stock and obsolete factories (exemplified in figure 5.5)—left little room for new industry to take root.

Figure 5.5
Abandoned ironworks, Elmhurst (East Oakland), one of more than a hundred factories that stopped production between the 1950s and 1980s. Photo by the author, February 2008

Table 5.1
Decline of manufacturing in Oakland and increase in Alameda County

Year	Manufacturers Oakland	Rest of Alameda Co.ª	Workers Total	Rest of Alameda Co.ª	Value added by manufacture $ (millions)	Share of Alameda Co. total (%)
1939	549	344	15,935	10,911	67.7	55
1947	701	485	25,601	28,437	207.6	51
1958	824	727	24,305	25,751	377.1	47
1967	748	956	19,100	36,200	417.1	32
1977	692	1,365	16,300	42,200	739.1	34
1987	717	1,735	11,800	35,500	1,095.7	16

Data source: U.S. Census Bureau, United States Census of Manufactures 1947, 1958, 1967, 1977, and 1987.
ªCalculated by subtracting Oakland data from Alameda Co. data.

A highly coordinated growth machine of industry, developers, boosters, and white laborers driven by the promise of homeownership and jobs diverted this latest flow of capital to the greenfields of the newly incorporated industrial suburbs—San Leandro, Hayward, Fremont, San Lorenzo, Newark, Union City, Milpitas—that flanked the East Bay between Oakland and San Jose. Vast tracts of agricultural land were incorporated into these pro-business municipalities, zoned as industrial, and sold for prices below industrial land prices in Oakland. National companies such as General Motors and Caterpillar built branch plants on these fertile greenfields, and defense contracts showered the new industrial suburbs with federal capital, ensuring rapid growth. As the data in table 5.1 illustrate, manufacturing nearly doubled in Alameda County (outside of Oakland) between 1948 and 1967. Here at the urban edge of the new suburbs, industry was given a tabula rasa. In essence, these new suburban municipalities provided a more favorable business climate, spatially removed from the pressure cooker of the urban center's working class and the grip of recalcitrant city politicians (Walker 1981; Self 2003). In the words of the Bay Area Council, which helped drive industrial suburbanization, suburban employees were "more loyal, more cooperative, more productive workers than those in big cities" (cited in Johnson 1993, 212). The implicit (and at times explicit) message to future investors was that this suburban workforce was largely white.

Just as in East Oakland during the interwar years, industry and housing in the new suburbs went hand in hand, part of a concerted planning effort to disperse industry and the suburban residential developments that followed in its stead. These industrial shifts and the prosperity of the postwar era further fertilized the American dream of homeownership. Large-scale housing developments in the urban periphery and the expansion of automobile ownership cultivated suburban development and white flight, draining urban areas of their tax base. Just as the industrial garden of East Oakland was watered with a strong mix of industrial and residential capital during the World War I and 1920s boom years, and with capital available through FHA loans in the 1930s and 1940s, the new industrial garden suburbs grew rapidly in the post-World War II era as a result of this same combination of industrial capital and federal housing subsidies. As Oakland deindustrialized and new factories sprouted in the suburbs, working-class white Oaklanders followed, lured by homeownership and proximity to jobs, just as they had done in the previous wave of interwar and wartime suburbanization. Between 1949 and 1951 only 600 units among the 75,000 constructed in the Bay Area were open to blacks (Johnson 1993). Upwardly mobile whites left the East Oakland flats to join the downtown ruling elite in their Oakland foothills and hillside neighborhoods, taking their cash with them.[15] In Elmhurst, for example, white residents made up 82 percent of the neighborhood's population in 1960 and median income was $6,154, only about 2 percent lower than the citywide median income; a decade later whites made up only slightly more than a third, while on the other side of the city boundary in San Leandro, people of color were excluded. Median income in Elmhurst dropped to 10 percent lower than that of the city (Whitaker 1992).

As capital was channeled into the industrial suburbs, it began to dry up inside the city's boundaries, leaving the once-verdant urban economy parched of tax revenue. By the mid 1960s, the number of manufacturers within Oakland had begun its steady decline. Between this downward trajectory and the steady growth of manufacturing in the new industrial suburbs, Oakland's share of Alameda County's industrial productivity dropped from more than half to less than a third in the four decades following World War II (cf. table 5.1).[16] More than 130 factories shut their doors and nearly 10,000 manufacturing jobs were lost by 1977. Unemployment skyrocketed as a result. The unemployment rate in 1964 was 11 percent but for blacks was almost twice that high. Business ownership was absentee for the most part; by 1978, only 25 percent of businesses in East Oakland were locally owned (Henze, Kirshner, and Lillow 1979).

This trend continued in the 1980s as jobs shifted from the traditional manufacturing and warehousing sectors to a service-based industry. The Bay Area on the whole benefited from a boom during this period, with a 15 percent growth in jobs between 1981 and 1986. Oakland, however, reaped little in the way of this regional bounty; employment grew only by 1.5 percent during these same years. The flatlands bore the brunt of job loss during this period. West Oakland and Fruitvale lost 8 to 10 percent of jobs. In the Elmhurst and San Antonio districts, employment decreased by roughly a third (Landis and Guhathakurta 1989).

As East Oakland's industrial garden withered, housing became available to upwardly mobile people of color for the first time. The Oakland border with San Leandro truly became a color line. Just as East Oakland's industrial garden communities had excluded people of color via racial covenants, new housing developments in places like San Leandro and San Lorenzo excluded people of color using racial covenants and informal "gentlemen's agreements" between realtors and homeowners' associations. Creating a class alliance with developers, increasingly conservative white homeowners in the new suburbs helped to exert political pressure to further confine devaluation to the Oakland flatlands. Proposition 14, a 1964 ballot initiative sponsored the California Real Estate Association and supported by 65 percent of voters statewide, essentially overturned the federal Fair Housing Act, passed the year before. In 1978 this same alliance was able to pass the infamous Proposition 13, which severely limited cities' ability to raise property taxes. The resulting decrease in property taxes took a toll on Oakland's already impoverished flatlands, as inflow of revenue was squeezed by more than $14 million, leading to facilities closures and cuts to public services (Rhomberg 2004; Self 2003).

As earlier in the century, Oakland's demographic shifts in the era of deindustrialization were not simply black and white, but multihued. Changing immigration policies in 1965 allowed a greater influx of Latinos into Oakland, primarily into the already heavily Mexican Fruitvale district. Many of the new arrivals worked in low-end service jobs in the industrial suburbs to the south (Hondagneau-Sotelo 1994). By the late 1970s and early 1980s, the impoverished flatlands became a major center of refugee resettlement for Salvadoran, Guatemalan, Khmer, Lao, Hmong, Khmu, Mien, and Vietnamese fleeing the Cold War's bloody battlegrounds in Central America and Southeast Asia. Resettlement programs in poor areas of East Oakland kept the majority of these immigrants poor, adding to an already large and devalued pool of cheap labor for the postindustrial economy (Ong 2003). Social networks provided entry into formal market niches and a vibrant, yet self-exploiting, informal economy, much of it centered in Chinatown, San Antonio, and Fruitvale (Marech 2002).

As the former industrial garden dried up, some new capital (in the form of federal urban redevelopment and freeway construction) did flow into the economically parched urban landscape, yet the promised jobs and opportunities never emerged. To the contrary, urban redevelopment ultimately displaced thousands of residents from their homes. Several of the most "blighted" areas were razed under the aegis of urban renewal. Thousands were displaced and forced to relocate. Single-family homes and duplexes were subdivided to accommodate those displaced, adding an additional strain on the dilapidated housing stock. Redlining prevented or dissuaded any new investment for housing repair. Housing in the East Oakland flatlands eventually became dilapidated, as well, due in part to a large number of absentee landlords who were homeowners who had followed the industrial garden to the suburbs, or speculators who bought their devalued property at fire sale prices. By 1978 more than two-thirds of East Oakland's single-family homes and apartments with more than five units were owned by absentee landlords (Henze, Kirshner, and Lillow 1979). Rents grew for increasingly decrepit housing, driving up vacancy rates to the point where the City of Oakland declared a "state of emergency" in April 1974 in response to the high number of vacant and abandoned housing units in East Oakland. These 1,200 empty units were seen as a result of the "blighting influence" of E. 14th Street, the major artery running the length of East Oakland. More than half of the structures assessed in the 1972 Elmhurst Redevelopment Project were categorized as containing "building deficiencies."[17] By the late 1980s, almost a third of vacant houses in the flatlands were considered in "poor" condition by the City of Oakland's Office of Community Development (Whitaker 1992).

As this chapter demonstrates, the devaluation of capital in Oakland was contained in the flatlands via racist policy and practice. The construction of major transportation corridors through the flatlands also helped to materially reinforce these existing spatial and socioeconomic divisions in Oakland, as in other postindustrial American cities, physically demarcating the boundaries between investment and abandonment, rich and poor, whites and people of color. Plans for the Nimitz, MacArthur, and Grove-Shafter Freeways were approved in 1958 by the all-white Oakland city council (Self 2003). The Grove Shafter (California Route 24/Interstate 980), which was placed immediately adjacent to the old Grove Street redline, effectively severed West Oakland from downtown. The MacArthur (Interstate 580) divided the flatlands from the hills. The Nimitz (Interstate 880), which parallels the MacArthur, was sited through the city's industrial corridor along the city's southwestern edge, roughly separating the majority of factories and warehouses and access to the estuary from the flatlands residential areas. Other construction projects were sited in devalued flatlands neighborhoods where land values were low and the political power of the community marginal. The Cypress Freeway was constructed right through the middle of West Oakland, razing hundreds of homes and displacing thousands of residents.[18] The Bay Area Rapid Transit (BART) system, which began in 1964, had a similar impact on the flatlands. In most of the flatlands, the BART tracks were placed above ground to reduce costs. Construction of the BART line between downtown and the trans-Bay tunnel destroyed 7th Street in West Oakland, the cultural and economic center of Oakland's African American community, and displaced several hundred families, many of whom moved to East Oakland where they were faced with rents two to three times as high as what they paid in West Oakland (Whitaker 1992). Small businesses (including grocers) also felt the impact of redevelopment as their clientele was displaced.

The port and its rail lines, the freeways, the Bay Bridge, and the BART were constructed to link Oakland to the region and to position it as a major transportation hub for the economically vibrant Bay Area. But as Self (2003) argues, capital and people flowed above West Oakland on freeway overpasses and BART tracks, channeled to San Francisco's enduring commercial center and Oakland's growing industrial suburbs. These conduits of capital served as physical boundaries of devaluation of existing fixed capital in the flatlands, material structures demarcating what zoning and redlining succeeded in doing invisibly on paper. Not only did the benefits of the freeway and BART system—the hallmarks of urban modernity—bypass the flatlands, their construction was marked by dispossession and displacement of Oakland's flatlands residents.

Retail in the Red

As capital devaluation became more and more contained in the flatlands, the city's retail landscape changed dramatically. A depressed flatlands economy made it difficult to retain major retailers, including supermarkets. For example, when the new Eastmont Mall, built on the site of the former East Oakland GM factory, held its grand opening in November 1970, it beckoned customers with the promise of unlimited parking and two major department stores, a four-plex movie theater, and food court. By the 1980s, however, falling purchasing power and an increase in drug dealing and related violent crime around the mall led to a major decline in retail sales. During the 1990s both department stores closed, as did the mall's Safeway supermarket. With the mall's anchor stores gone, business occupancy dropped to only 30 percent (Oakland Tribune 2007). By 1987 only four department stores continued to operate within the city limits (Rhomberg 2004).

This pattern of capital flight and devaluation transformed food access during the era of deindustrialization in the Oakland flatlands and in U.S. "inner cities" on the whole. Across the country, food retail had been gradually changing since first the arrival of chain grocery stores prior to World War I and by chain supermarkets in the 1930s. After World War II, supermarkets (both chain and independent) dominated the lion's share of food retail. Driven by the entry of women into the workforce, a growing demand for one-stop shopping, automobile culture, and a massive influx of new processed foods derived from subsidized commodities, supermarkets became more and more popular. Shopping centers, a new model of retail often "anchored" by a supermarket, sprouted up in the new white suburbs across America. By 1960 more than two-thirds of groceries were purchased at supermarkets. Unable to compete with the economies of scale enjoyed by supermarkets, many small grocers went out of business. The power of corporate supermarket chains increased during this period as well. Chain supermarkets slowly drove the independent chains out of business, waging "price wars" to secure turf. By 1975 corporate food retailers controlled about two-thirds of the food retail market, draining capital from the local economy and funneling it off to corporate headquarters (Walker 2004; Eisenhauer 2001).

As food retail became concentrated in the aisles of major supermarkets, food access became increasingly dictated by supermarket location. By the 1970s nationwide economic "stagflation" caused supermarket retail to founder. Mergers and leveraged buyouts of competing chains hit less competitive, inner-city markets hard; between 1978 and 1984, Safeway alone closed more than 600 stores in these neighborhoods (Eisenhauer 2001). The boarded-up hulls of failed supermarkets littered the shoals of America's postindustrial cities; many remained shuttered, others converted to churches, and only some rigged anew as thrift or dollar stores for consumers with declining purchasing power. While the number of supermarkets in urban areas declined, however, the overall number of supermarkets increased. By the mid 1990s, in urban areas the poorest urban neighborhoods had roughly half the retail supermarket space than did the richest urban neighborhoods (Ibid.).

During the 1980s and 1990s superstores took over the helm of food retail, spatially concentrating food access in locations often only accessible by car. For working class people, falling wages and retail capital's retreat from postindustrial urban centers meant that cheap food availability was limited to big box stores and fast food joints (Walker 2004; Mamen 2007). A "junk food jungle" took root in the barren stretches of the fresh food desert throughout poor neighborhoods in postindustrial America, capitalizing on the niche left by the retreat of groceries and supermarkets and a demand for food that was easily accessible, convenient, and cheap, sending the incidence of diabetes and obesity skyrocketing (Goldstein et al. 2008). Liquor stores followed a similar successional logic. With the ebb of food retail capital, liquor stores began to serve as the primary source of food provisioning in America's inner cities, yet prices for their goods were often higher than those found at a supermarket, and fresh fruits and vegetables were unavailable.

As table 5.2 reveals, food retail in the Oakland flatlands paralleled these national trends. Between 1935 and 1987, the total number of grocery stores in Oakland dropped fivefold, from over 1,000 to about 200 while the average number of employees per store increased nearly tenfold. These shifts signal not only the arrival of supermarkets and consequent concentration of the food retail sector, but also the steep decline in service to the city's growing population, an overall decrease from 36 to 5 stores per 10,000 residents. The decline hit the flatlands even harder. In West Oakland, the number of grocery stores in West Oakland declined from 137 in 1960 to 22 in 1980, due largely to supermarket penetration (Fuller 2004), a drop from nearly 25 percent of all of the city's stores to just above 10 percent. By the 1990s, many of these same supermarkets that had pushed out the small grocers in the flatlands had also closed their doors in response to falling profits. The Safeway at Eastmont Mall, one of the mall's anchor stores, closed at this time. In a particularly ironic twist, two of the country's four leading supermarkets, Safeway and Lucky Stores, were headquartered in Oakland, yet access to quality food in the once bountiful industrial garden of Oakland's flatlands had evaporated as capital reinvested outside of the city lines. One can conclude from the data in table 5.3 that the rapid growth of

Table 5.2
Consolidation and decline of grocery stores in Oakland

Year	Number of grocery stores[a]	Number of paid employees	Employee-to-store ratio	Stores per 10,000 people[b]
1935	1,086	1,923	1.8	35.9
1948	828	1,783	2.2	21.5
1958	525	1,513	5.3	14.3
1967	394	2,065	10.8	10.9
1977	257	1,913	11.1	7.6
1987	201	2,349	11.7	5.4

Data source: U.S. Census Bureau, Census of Business 1935, 1948, 1958, and 1967; U.S. Census Bureau, Census of Retail Trade 1977 and 1988.
[a]For 1958 to 1987 retail data, Standard Industrial Classification (SIC) Code 541 was used. For 1935, "Grocery stores without meat" and "Combination stores (Grocery stores with meat)" were aggregated; for 1948, grocery stores with and without meat were aggregated. Grocery stores accounted for roughly two-thirds of "Food Stores" (SIC Code 54) for all years.
[b]Calculated using population data from the nearest census year (1940 to 1990).

Table 5.3
Decline of Oakland's share of food stores and sales in Alameda County

	Oakland's share of Alameda Co. totals		
Year	Population (%)[a]	Food stores (%)[b]	Sales (%)
1935	59	66	64
1948	52	63	57
1958	40	55	45
1967	34	50	38
1977	31	37	29
1987	29	34	24

Data source: U.S. Census Bureau, Census of Business 1935, 1948, 1958, and 1967; U.S. Census Bureau, Census of Retail Trade 1977 and 1988.
[a]Calculated using population data from the nearest census year (1940, 1950, 1960, 1970, 1980, and 1990).
[b]For 1958 to 1987 retail data, Standard Industrial Classification (SIC) Code 54 was used. For 1935, data for the category "Food Stores" was used; for 1948 data, "Food Group" was used.

the suburbs precipitated the decline of Oakland's share of food stores, but Oakland's sales nevertheless began to lag disproportionately due to the declining purchasing power of the city's population. By the late 1980s, a third of Alameda County's food stores were located in Oakland, but these accounted for only a quarter of the county's total food sales.

With the retreat of the supermarkets and closure of small-scale groceries, food retail in the flatlands has been largely left to liquor stores and corner stores that serve as de facto liquor stores. Statistics help to describe a landscape of food access not unlike that of many other food deserts. In 1935 there were more than eight grocery stores for every liquor store in Oakland; by 1977, there were fewer than two.[19] In the flatlands the number of liquor stores per person (three to six stores per 1,000 residents) was two to four times the city average in 2007. There are four times as many fast food restaurants and convenience stores as grocery stores and produce vendors in the East Bay (Spiker, Sorrelgreen, and Williams 2007). No supermarkets serve residents in West Oakland and recent plans for British supermarket giant Tesco to open a West Oakland store have fallen through. A recent survey by a food justice initiative found that in six flatlands neighborhoods, residents reported they have to leave their neighborhoods to find affordable, healthy food (HOPE Collaborative 2009). West Oaklanders have to cross into the redeveloped box store land of neighboring Emeryville to shop at Pak-n-Sav. Similarly, in East Oakland's Council District 6, no national grocery chain exists.[20] Most East Oaklanders find the best deals across the city border; one focus group participant noted, "Oakland dollars are going to San Leandro" (Ibid., 16). Another noted, "I wish we could have more fresh foods rather than junk food, candy, and soda that we're all used to eating because that is the only thing around" (Ibid.). Participants said they want more stores that sell healthier foods and better quality produce. Another study highlights residents' acute awareness of the difference not only in availability, but also of quality: "Yes, there's a difference in the stores in our area compared to the stores in Montclair or somewhere else [in the Oakland hills]. You know, the vegetables are great up there, everything is so beautiful. And you come down, I think we get ours last off the truck" (Treuhaft, Hamm, and Litjens 2009).

Conclusion

Across the street from the liquor store on 17th Street where we began, the verdure of an urban garden spills through a chainlink fence.

A colorful orange and yellow sign hanging on the gate advertises a community food security project, welcoming passersby into the cultivated chaos of garden vegetation. Flanking the entrance to the garden, a produce stand is stocked with a kaleidoscope of brightly colored peppers, persimmons, chard, and salad greens, sold at cost to the ethnically diverse crowd gathered around the display. When viewed as a metaphor, this actual urban streetscape seems almost contrived—a moral standoff between garden and liquor store, nutrition and intoxication, growth and senescence, stewardship and abandon. As symbols, these two spaces have come to represent opposing forces in the struggle for food justice in the food deserts of the flatlands and elsewhere. But on a material level, these two types of food outlets have very real impacts on urban livelihoods, provisioning low-income communities with quite different types of food—fresh organic produce or highly-processed packaged food—leading to very real differences in nutritional intake and wide-reaching effects on public health.

Over the last five years, several food justice organizations such as People's Grocery, City Slicker Farms, Village Bottom Farms, Phat Beets Produce, and Planting Justice have taken over vacant lots and underutilized park land in West and North Oakland to provide flatlands residents with fresh produce either via community supported agriculture (CSA), sliding-scale farm stands, or farmers markets.[21] In East Oakland, Oakland Food Connection reintegrates production and consumption by teaching a curriculum that includes urban gardening and cooking classes to educate children about food culture and nutrition. PUEBLO's Youth Harvest program works with at-risk teenagers to harvest fruit for distribution at senior centers in the Fruitvale and San Antonio neighborhoods. Several of the organizations help Oaklanders build gardens in their yards and provide mentorship from sowing to harvest. More than a hundred elementary, middle, and high schools in Oakland use gardens as classrooms to teach science, health, and nutrition. To do justice to the accomplishments of these organizations and individual activists is impossible within the scope of this chapter, as is a discussion of the radical roots and emancipatory vision of these food justice organizations.

This history of demarcated devaluation of the Oakland flatlands suggests that food deserts arise from an incredibly complex intersection of historical forces operating at multiple spatial and temporal scales. In this chapter I uncover only a few of the many sedimentary layers of the urban palimpsest, that of industrial, residential, and retail capital and some of the ways in which their ebbs and flows were spatially demarcated.

Further excavation and new geographies are still needed to more fully map the uneven terrain of food access in the flatlands. Other layers need to be uncovered: the role of food policies operating at multiple scales, from federal to local, farm subsidies to food stamps and free lunch; the politics of city contracts and bidding, development and redevelopment programs, planning and zoning; how current economic and demographic shifts in the flatlands may both fuel and fight the advances of food justice activists.

What becomes clear then is that the fight for food justice cannot be waged with urban gardens and produce stands alone. This hands-on, experiential, and participatory approach is powerful and effective, both through its ability to bring food to those in the surrounding neighborhood and to rally newcomers to the food justice movement. Yet it functions only at the microscale; even massive agglomerations of urban gardens are unlikely to meet more than 5 percent of the vegetable demands of a city such as Oakland (McClintock and Cooper 2009). The passion and vigor with which food justice activists break new ground in the urban fallows in Oakland and elsewhere must extend also to rethinking and rebuilding the entirety of the metropolitan and regional food system—production, processing, distribution, retail, and waste recovery—in both urban and peri-urban areas. Creative new economic incentives and land use protections will be needed to buffer a fledgling local food system from the continuous cycle of economic booms and busts and competitive pressures of the global food system. Perhaps most importantly, jobs paying a living wage must be fundamental to the design. Once capital flows are diverted and used to cultivate a more just food system in the urban fallows, keeping it bountiful will remain one of the great challenges.

Notes

1. The jury is still out on whether or not "food desert" is an appropriate metaphor. Food may well be available in these so-called food deserts, but it is generally of poor nutritional value. Fast food outlets may abound while fresh fruit or vegetables are nowhere in sight. Some opt for the term *health food deserts* or *fresh food deserts* while others reject the image of a bleak and parched urban landscape, opting for the lush and primordial *junk food jungles*. Others hope to throw out such sensationalist taxonomy altogether, with its potentially racialized subtext linking people of color to exotic and/or depraved environments. I use the term *food deserts* here simply in metaphorical contrast to Oakland's history as an "industrial garden."

2. Rather than envisioning these relations as a nested hierarchy, however, it is helpful to think of a complex web of interconnectivity. Global economic restructuring in the neoliberal era, as well as increasing access to technology and information, have undermined and reorganized the traditional hierarchical relationships.

3. Rather than stop at an explanation of how the biophysical environment, human bodies, or social relations are transformed by flows of capital, we should also address how these flows are then resisted, reconfigured, or redirected in response. This dialectic helps unravel the classic "structure versus agency" binary by instead emphasizing the creative and destructive tension between "actors" (biophysical and social, individual and collective) operating at the same or different spatiotemporal scales. Distinguishing structure from the agency of individual actors becomes simply a question of shifting the spatiotemporal grain and extent of analysis, in essence, zooming in to identify the actions of an individual actor and zooming out to see how these individual actions operate collectively on larger scales over time and space.

4. According to Harvey's analysis, when there is an overaccumulation of surplus capital or labor, it either seeks a spatial fix to find new spaces for investment (2001) or enters into a "second circuit" of capital, and is invested in this kind of "fixed capital" to avoid a crisis of devaluation of one or the other (1989).

5. In such cases capital actually undermines its own means of production by fouling its resource base; see James O'Connor (1998) on capitalism's "second contradiction."

6. In the urban morphology literature, the term *urban fallow* denotes derelict land and buildings, abandoned, obsolescent, and awaiting redevelopment, the final successional phase of a "burgage cycle" of urban development (Clark 2001). Viewing urban fallow as part of a broader lumpengeography of capital helps to locate these investment cycles within a larger spatial geography of capital.

7. Doreen Massey (1995) incorporates social relations into this palimpsest. Using a vivid geomorphological metaphor, she describes the series of "sedimentary layers" laid down by past cycles of investment. These layers embody not only physical fixed capital, but also the associated negotiations and struggles between capital and labor (and society more broadly).

8. As Harvey (2006) elaborates, this concentration of devaluation constitutes another form of capital accumulation by dispossession; by confining devaluation elsewhere, new sites can monopolize production.

9. Urban growth obviously does not arise of its own accord but is stewarded by a "growth machine," a coalition/class alliance of business owners, developers, media, and industrialists (Logan and Molotch 1987). In Oakland much of the growth in the earlier part of the century was due in large part to the efforts of the city's powerful growth machine, a class alliance that included Francis "Borax" Smith, owner of the Key System, mayors Frank Mott (in office 1905–1915) and John Davie (in office 1915–1931), and the city chamber of commerce. The dynamo at the center of it all was the conservative pro-business *Oakland Tribune* under the ownership of the Knowland family from 1915 to 1977. The

Knowlands' powerful control of media consolidated the growth machine's grip on city politics for much of the twentieth century. This growth machine resisted San Francisco's repeated efforts to incorporate Oakland into a regional metropolis. Rather than being periphery to San Francisco's core, Oakland's growth machine pushed on several occasions to become the core of an East Bay metropolis (Rhomberg 2004; Self 2003; Scott [1959] 1985).

10. This promise of homeownership, which in the Hoover years had risen to be the symbolic pinnacle of American citizenship, was central to the reformist planners' attempt to "Americanize" (read "deradicalize") recent European immigrants and subsume them into a growing class alliance of white, working-class homeowners (Hise 1997).

11. For an example of the actual documents used, see part II: Home Rating Instructions of the 1935 FHA's *Underwriting Manual: Underwriting and Valuation Procedure under Title II of the National Housing Act. Federal Housing Administration, Washington, DC*. Available at <http://salt.unc.edu/T-RACES/fha.html> (accessed August 10, 2010).

12. Some argue that redlining did not actually restrict lending, but that higher interest rates in redlined areas may have prevented investment by builders and buyers (Hillier 2003).

13. As Self (2003) describes, this boundary gradually moved farther east to Telegraph Ave., the major north–south artery connecting downtown Oakland to Berkeley.

14. The ranks of the unemployed become the rank-and-file of the "industrial reserve army" (Marx [1867] 1976; Harvey 2001), brought in when necessary to meet production demands or to lower wages when production costs rise, and cast back into the reserve when no longer needed.

15. Explanations of "white flight" from the black city center largely revolve around (a) white fear of an inundation of blacks into their neighborhoods, (b) the American dream of homeownership fueled by postwar prosperity, and (c) the expansion of automobile ownership and "car culture." While aspects of this reading of history are certainly valid, the story of suburbanization is more nuanced than this old-school view of a big bang spewing "little boxes made of ticky-tacky" outward from ground zero at the city center, pulling all the scared white folks with it. By refocusing on the greater logic of metropolitan regionalism and industrial dispersal that helped to steward extensive, dispersed residential development, we can move beyond the urban/suburban dualism and the common trope that suburbanization should be read as a rejection of the city in general (Hise 1997; Walker 1981).

16. While Oakland's industrial economy was diversified enough that it did not suffer "the urban crisis" to the same extent as the Rust Belt cities in the Northeast and Midwest (Sugrue 2005), it nevertheless followed the same trend.

17. The state of emergency led to a host of redevelopment initiatives, including the Home Maintenance and Improvement and Urban Homesteading programs.

18. The Cypress Freeway collapsed in 1989 during the Loma Prieta earthquake, killing forty-two people. In response to public outcry over the socioeconomic

impact of its original location, the new freeway was built farther west, adjacent to the Port. The old Cypress viaduct is now Mandela Parkway.

19. Calculated using data from the U.S. Census Bureau, Census of Business 1935, 1948, 1958, and 1967; and the U.S. Census Bureau, Census of Retail Trade 1977 and 1988 (see notes for tables 5.2 and 5.3).

20. A recent announcement by Kroger to open two new 72,000-square-foot Foods Co. stores in East Oakland made national news, one of them in Foothill Square where Lucky's and Albertson's stores closed their doors several years earlier.

21. A CSA is a direct-marketing arrangement that links producers and consumers. Customers purchase a share at the beginning of the season in exchange for weekly deliveries of a box of fresh produce.

References

Bagwell, Beth. 1982. *Oakland: The Story of a City*. Oakland, CA: Oakland Heritage Alliance.

Beaulac, Julie, Elizabeth Kristjansson, and Steven Cummins. 2009. A Systematic Review of Food Deserts, 1966–2007. *Preventing Chronic Disease: Public Health Research, Practice, and Policy* 6 (3): 1–10.

Boone, Christopher G., Geoffrey L. Buckley, J. Morgan Grove, and Chona Sister. 2009. Parks and People: An Environmental Justice Inquiry in Baltimore, Maryland. *Annals of the Association of American Geographers* 99 (4): 767–787.

Clark, Mike. 2001. Urban Fallow and the Surface Economy. *Futures* 33 (2): 213–218.

Cummins, Steven, and Sally Macintyre. 2002. A Systematic Study of an Urban Foodscape: The Price and Availability of Food in Greater Glasgow. *Urban Studies* 39 (11): 2115–2130.

Daniels, Roger. [1959] 1977. *The Politics of Prejudice: The Anti-Japanese Movement in California and the Struggle for Japanese Exclusion*. 2nd ed. Berkeley: University of California Press.

Davis, Mike. 1997. Sunshine and the Open Shop: Ford and Darwin in 1920s Los Angeles. *Antipode* 29 (4): 356–382.

Eisenhauer, Elizabeth. 2001. In Poor Health: Supermarket Redlining and Urban Nutrition. *GeoJournal* 53 (2): 125–133.

Foster, John Bellamy. 1999. Marx's Theory of Metabolic Rift: Classical Foundations for Environmental Sociology. *American Journal of Sociology* 105 (2): 366–405.

Fuller, Andrea. 2004. *A History of Food Insecurity in West Oakland, CA: Supermarket Location*. Oakland, CA: People's Grocery.

Gandy, Matthew. 2003. *Concrete and Clay: Reworking Nature in New York City*. Cambridge, MA: MIT Press.

Goldstein, Harold, Stefan Harvey, Rajni Banthia, Rebecca Floumoy, Victor Rubin, Sarah Treudhaft, Susan H. Babey, Allison L. Diamant, and Theresa A. Hastert. 2008. *Designed for Disease: The Link Between Local Food Environments and Obesity and Diabetes.* Los Angeles: California Center for Public Health Advocacy/PolicyLink/UCLA Center for Health Policy Research.

Groth, Paul. 2004. Workers'-Cottage and Minimal-Bungalow Districts in Oakland and Berkeley, California, 1870–1945. *Urban Morphology* 8 (1): 13–25.

Harvey, David. 1989. *The Urban Experience.* Baltimore, MD: The Johns Hopkins University Press.

Harvey, David. 2001. *Spaces of Capital: Towards a Critical Geography.* New York: Routledge.

Harvey, David. 2006. *Spaces of Global Capitalism.* London: Verso.

Henze, Laura J., Edward Kirshner, and Linda Lillow. 1979. *An Income and Capital Flow Study of East Oakland, California.* Oakland, CA: Community Economics.

Heynen, Nik. 2006. Justice of Eating in the City: The Political Ecology of Urban Hunger. In *In the Nature of Cities: Urban Political Ecology and the Politics of Urban Metabolism,* ed. N. Heynen, M. Kaika, and E. Swyngedouw, 124–136. London: Routledge

Heynen, Nik, Maria Kaika, and Erik Swyngedouw. 2006. *In the Nature of Cities: Urban Political Ecology and the Politics of Urban Metabolism.* London: Routledge.

Hillier, Amy E. 2003. Redlining and the Home Owners' Loan Corporation. *Journal of Urban History* 29 (4): 394–420.

Hise, Greg. 1997. *Magnetic Los Angeles: Planning the Twentieth-Century Metropolis.* Baltimore, MD: The Johns Hopkins University Press.

Hise, Greg. 2001. Industry and Imaginative Geographies. In *Metropolis in the Making: Los Angeles in the 1920s,* ed. T. Sitton and W. Deverell, 13–44. Berkeley: University of California Press.

HOLC. 1937. Area Description, Oakland, CA. Home Owner's Loan Corporation (HOLC) Division of Research and Statistics. Available at <http://salt.unc.edu/T-RACES> (accessed June 9, 2010).

Hondagneau-Sotelo, Pierrette. 1994. *Gendered Transitions: Mexican Experiences of Immigration.* Berkeley: University of California Press.

HOPE Collaborative. 2009. *A Place with No Sidewalks: An Assessment of Food Access, the Built Environment and Local, Sustainable Economic Development in Ecological Micro-Zones in the City of Oakland, California in 2008.* Oakland, CA: HOPE Collaborative.

Johnson, Marilynn S. 1993. *The Second Gold Rush: Oakland and the East Bay in World War II.* Berkeley: University of California Press.

Kantor, Amy C., and John D. Nyusten. 1982. De Facto Redlining: A Geographic View. *Economic Geography* 58 (4): 309–328.

Landis, John D., and Subhrajit Guhathakurta. 1989. *The Downsized Economy: Employment and Establishment Trends in Oakland: 1981–1986*. Berkeley, CA: Institute of Urban and Regional Development.

Lee, Gyongju, and Hyunwoo Lim. 2009. A Spatial Statistical Approach to Identifying Areas with Poor Access to Grocery Foods in the City of Buffalo, New York. *Urban Studies* (Edinburgh, Scotland) 46 (7): 1299–1315.

Logan, John R., and Harvey L. Molotch. 1987. *Urban Fortunes: The Political Economy of Place*. Berkeley: University of California Press.

Ma, L. Eve Armentrout. 2000. *Hometown Chinatown: The History of Oakland's Chinese Community*. New York: Garland Publishing.

Maantay, Juliana. 2002. Zoning Law, Health, and Environmental Justice: What's the Connection? *Journal of Law, Medicine & Ethics* 30 (4): 570–593.

Mamen, Katy. 2007. *Facing Goliath: Challenging the Impacts of Supermarket Consolidation on Our Local Economies, Communities, and Food Security (Policy Brief)*. Oakland, CA: The Oakland Institute.

Marech, Rona. 2002. Of Race and Place: San Antonio/Oakland, Flavors Meld in Community East of Lake. *San Francisco Chronicle*, May 31.

Marx, Karl. [1867] 1976. *Capital: A Critique of Political Economy*, trans. B. Fowkes, vol. 1. London: Penguin Classics.

Massey, Doreen. 1995. *Spatial Divisions of Labor*. 2nd ed. London: Routledge.

Massey, Douglas S., and Nancy A. Denton. 1993. *American Apartheid: Segregation and the Making of the Underclass*. Cambridge, MA: Harvard University Press.

Matsuoka, Martha. 2003. *Building Healthy Communities from the Ground Up: Environmental Justice in California*. Oakland, CA: Asian Pacific Environmental Network.

McClintock, Nathan, and Jenny Cooper. 2009. *Cultivating the Commons: An Assessment of the Potential for Urban Agriculture on Oakland's Public Lands*. Oakland, CA: HOPE Collaborative/City Slicker Farms/Food First.

McClung, William Alexander. 2000. *Landscapes of Desire: Anglo Mythologies of Los Angeles*. Berkeley: University of California Press.

McWilliams, Carey. [1949] 1999. *California: The Great Exception*. Berkeley: University of California Press.

Morello-Frosch, Rachel. 2002. Discrimination and the Political Economy of Environmental Inequality. *Environment and Planning C: Government & Policy* 20 (4): 477–496.

Nicolaides, Becky M. 2001. The Quest for Independence: Workers in the Suburbs. In *Metropolis in the Making: Los Angeles in the 1920s*, ed. T. Sitton and W. Deverell., 77–95 Berkeley: University of California Press.

Oakland Tribune. 2007. Eastmont Mall Sold to Oregon Investors, March 16.

O'Connor, James. 1998. *Natural Causes: Essays in Ecological Marxism*. London: Guilford Press.

Ong, Aihwa. 2003. *Buddha Is Hiding: Refugees, Citizenship, the New America.* Berkeley: University of California Press.

Raja, Samina, Changxing Ma, and Pavan Yadav. 2008. Beyond Food Deserts: Measuring and Mapping Racial Disparities in Neighbhorhood Food Environments. *Journal of Planning Education and Research* 27 (4): 469–482.

Rhomberg, Chris. 2004. *No There There: Race, Class, and Political Community in Oakland.* Berkeley: University of California Press.

Saxton, Alexander. 1971. *The Indispensible Enemy: Labor and the Anti-Chinese Movement in California.* Berkeley: University of California Press.

Scott, Mel. [1959] 1985. *The San Francisco Bay Area: A Metropolis in Perspective.* Berkeley: University of California Press.

Self, Robert O. 2003. *American Babylon: Race and the Struggle for Postwar Oakland.* Princeton, NJ: Princeton University Press.

Short, Anne, Julie Guthman, and Samuel Raskin. 2007. Food Deserts, Oases, or Mirages? Small Markets and Community Food Security in the San Francisco Bay Area. *Journal of Planning Education and Research* 26 (3): 352–364.

Smoyer-Tomic, Karen E., John C. Spence, and Carl Amrhein. 2006. Food Deserts in the Prairies? Supermarket Accessibility and Neighborhood Need in Edmonton, Canada. *Professional Geographer* 58 (3): 307–326.

Spiker, Steve, Ethan Sorrelgreen, and Junious Williams. 2007. *2007 Liquor Outlet Report: A Preliminary Analysis of the Relationship Between Off-Sale Liquor Outlets and Crime in Oakland for 2007.* Oakland, CA: Urban Strategies Council.

Squires, Gregory D., William Velez, and Karl E. Taueber. 1991. Insurance Redlining, Agency Location, and the Process of Urban Disinvestment. *Urban Affairs Review* 26 (4): 567–588.

Sugrue, Thomas J. 2005. *The Origins of the Urban Crisis: Race and Inequality in Postwar Detroit.* Princeton, NJ: Princeton University Press.

Swyngedouw, Erik, and Nikolas C. Heynen. 2003. Urban Political Ecology, Justice and the Politics of Scale. *Antipode* 35 (5): 898–918.

Treuhaft, Sarah, Michael J. Hamm, and Charlotte Litjens. 2009. *Healthy Food for All: Building Equitable and Sustainable Food Systems in Detroit and Oakland.* Oakland, CA: PolicyLink.

USDA (U.S. Department of Agriculture). 2009. *Report to Congress: Access to Affordable and Nutritious Food: Measuring and Understanding Food Deserts and Their Consequences.* Washington, DC: ERS/FNS/CSREES.

Walker, Richard. 1978. Two Sources of Uneven Development Under Advanced Capitalism: Spatial Differentiation and Capital Mobility. *Review of Radical Political Economics* 10 (3): 28–37.

Walker, Richard. 1981. A theory of suburbanization and the construction of urban space in the United States. In *Urbanization and Urban Planning in Capitalist Society*, ed. M. Dear and A. Scott, 383–429. New York: Methuen.

Walker, Richard. 2001. Industry Builds the City: The Suburbanization of Manufacturing in the San Francisco Bay Area, 1850–1940. *Journal of Historical Geography* 27 (1): 36–57.

Walker, Richard. 2004. *The Conquest of Bread: 150 Years of Agribusiness in California.* Berkeley: University of California Press.

Whitaker, Calvin. 1992. *The Abandonment of Housing in East Oakland.* San Jose, CA: Urban and Regional Planning, San Jose State University.

Zenk, Shannon N., Amy J. Schulz, Barbara A. Israel, Sherman A. James, Shuming Bao, and Mark Wilson. 2005. Neighborhood Racial Composition, Neighborhood Poverty, and the Spatial Accessibility of Supermarkets in Metropolitan Detroit. *American Journal of Public Health* 95 (4): 660–667.

6

Farmworker Food Insecurity and the Production of Hunger in California

Sandy Brown and Christy Getz

This chapter takes as its point of departure an apparent contradiction of contemporary U.S. agriculture, namely, that those who produce our nation's food are among the most likely to be hungry or food insecure. For those familiar with farmworker communities, this irony comes as little surprise. Yet the lived realities of farmworkers are, more often than not, rendered invisible to the vast majority of people who rely on their labor for sustenance. In an effort to address this seeming paradox, the chapter explores the concept of food security with respect to California's agricultural workforce.

Data from the Fresno Farmworker Food Security Assessment (Wirth, Strochlic, and Getz 2007) provide the basis for our analysis. Beyond simply revealing the prevalence of food insecurity and hunger however, we consider how and why this situation has come to be and, perhaps more important, why it persists despite California's highly productive and profitable agricultural landscape. In this sense, food security is more than an individual or household condition to be scientifically measured, but rather a lens through which to consider the highly unequal, uneven dynamics of global agricultural production, trade, and consumption.

Food security, as defined by the U.S. Department of Agriculture, means having access at all times to enough food for an active, healthy life for all household members. It is but one measure of the vulnerability experienced by farmworkers in their daily lives, albeit a critical one given that food is essential to human survival and, at an international level, widely recognized as an inalienable right.[1] We argue that this vulnerability has been systematically constructed within the political economy of agrarian capital accumulation, immigration politics, and neoliberal trade policy. Our goal is to expose the material relations that *produce* hunger. By choosing the term *produce* we emphasize that in a world of agricultural surpluses hunger is the result not of natural processes but rather of unequal power relations and resource access. Given this perspective, in this chapter we do not lament what people do or do not eat. Instead, we approach questions about a broader set of inequities and, indeed, injustices that characterize contemporary food provisioning within and beyond California.

Our approach to illuminating farmworker food security challenges, and situating them within a broader sociopolitical context gives us impetus to question the power relations underlying the social relations of capitalist food production. Given that a majority of farmworkers in the United States today are immigrants from rural Mexico, we explore the neoliberal domestic policies and international trade regime that privilege corporate agribusiness over small farmers in Mexico, forcing many off their land. Many of these same farmers then find themselves working as wage laborers on the U.S. side of the border, within the same agrifood regime that rendered it impossible for them to sustain their families through small-scale farming in Mexico. It is this paradox that leads us to ponder why some people remain hungry no matter how far from home they travel in search of sustenance.

While our empirical point of departure considers the food security status of farmworkers within the United States, the concept of food security, as deployed by domestic actors, has largely sidestepped a structural analysis of hunger. The result has been a focus on feeding hungry people, rather than altering the production relations and modes of governance that underpin food insecurity. In contrast, the burgeoning global food sovereignty movement posits the right of peoples to define their own food and agriculture systems, rather than being subject to the constant cycles of poverty, hunger, and migration, dictated by international market forces.

Producing Hunger, Constructing Vulnerability

At the height of United Farm Workers' (UFW) organizing, the admonitions of union founder Cesar Chavez that "the food that overflows our market shelves and fills our tables is harvested by men, women, and children who often cannot satisfy their own hunger" called attention to the marginalized position of agricultural labor in California's farm fields (National Farm Worker Ministry n.d.). The union's struggles to improve agricultural wages and working conditions mobilized white, middle-class urban consumers to support a primarily immigrant workforce, in particular through union-led boycotts (Frank 2003; Ganz 2000). Yet significant improvements in farmworkers' material conditions have failed to materialize and food insecurity and hunger remain widespread within farmworker communities.

While the reasons for the marginalization of agricultural labor are complex and contingent upon specific sociohistorical contexts of particular moments in California history, we argue that the central dynamic shaping labor relations and workers' livelihood struggles has been the development of a regime of agrarian accumulation based on capital-intensive production *and* the persistent devaluation of agricultural labor (Henderson 1998; Mitchell 2007; Walker 2004). While the often-violent marginalization of farm labor was not inevitable, the productive forces and social relations of agricultural production evolved together to make California the nation's breadbasket where farmworkers often struggle to feed themselves and their families. The chapter engages a political economy framework to understand how this devaluation of agricultural labor has been achieved through the social and political construction of a vulnerable, and therefore exploitable agricultural working class.

Of course there are many measures by which one might assess the effects of this devaluation, from poor physical and mental health (Cason et al. 2003; Villarejo et al. 2000) and lack of access to health care and affordable housing (Bradman 2005; Housing Assistance Council 2005), to unsafe and debilitating working conditions, pesticide exposure (Harrison 2008; Reeves, Katten, and Guzmán 2002) and low annual earnings, long hours, and unstable employment (Bugarin and Lopez 1998). Perhaps the most striking evidence of farmworkers' devalued position is the decline in real wages over the past several decades. Between 1975 and 1995 real wages fell at least 20 to 25 percent (Rothenberg 1998; Villarejo and Runsten 1993). While few comprehensive studies of agricultural wages have been conducted, and though accurate figures are difficult to estimate, a study by Kahn, Martin, and Hardiman (2005) utilizing California Employment Development Department data illustrated the stagnation. Between 1991 and 2001, annual agricultural worker earnings remained at $8,500 for direct-hire workers and $5,000 for workers employed by farm labor contractors, representing a 32 percent *decline* in inflation-adjusted dollars during the period (Ibid.).[2] These low annual earnings are due, in part, to the fact that unlike many occupations, farmworkers do not enjoy year-round employment or stable work schedules. According to the National Agricultural Workers Survey (NAWS), farmworkers average 190 days, or 34.5 weeks of employment per year. And one California study estimates the average number of hours at 1,000 per year, about half the hours of full-time employment in other industries (Martin and Mason 2003).[3]

The lived realities of farmworkers stand in stark contrast to a consistent expansion of California's highly productive and profitable agricultural landscape. While farmworker incomes have declined, the value of agricultural products has continued along a trajectory of expansion begun in the nineteenth century. Between 2002 and 2007 alone, California's agricultural sales increased 32 percent, from $25.7 billion to $33.9 billion (USDA 2007). In Fresno County, where almost one-half of surveyed farmworkers are food insecure at some point during the year (Wirth, Strochlic, and Getz 2007), agricultural sales increased by 32 percent, from $2.8 billion to $3.7 billion over the same period. Given these statistics, workers' loss appears to be capital's gain.

Indeed, California growers have developed many mechanisms for addressing a fundamental challenge to capital accumulation in agriculture, namely the difference in *production time,* the time that elapses between the planting and harvesting, and the actual *labor time* needed to plant, tend, and harvest a crop (Henderson 1998; Mann and Dickinson 1978). While agricultural employers invest in labor power (workers' wages) insofar as it is needed for production, workers must invest in their own reproduction (food, housing, etc.) year-round. As Mitchell notes: "The solution to the problem of the reproduction of labor power in California has been precisely to assure that it is not reproduced, but instead is continually replaced" by new groups of immigrant workers (2007, 567). Farmworkers have been recruited so as to ensure an oversupply and are systematically denied citizenship and workplace rights, due to their status as indentured servants, guest workers, contract laborers, and, increasingly, undocumented immigrants, facilitating growers' management of the "turnover time" problem. As Mitchell also suggests, "To put this in the starkest terms: the tenuous nature of making the desert bloom, to say nothing of the exceptionally high capital costs meant that the reproduction of capital—and of course the reproduction of the agricultural landscape—required driving to as low a point as possible the cost of reproducing the labor that plants and picks the desert-blooming capital" (2007, 573).

It is precisely because of workers' vulnerability that agribusiness interests have expanded investments in capital-intensive inputs, such as irrigation, machinery, seed and stock, fertilizers, and pesticides.

Throughout its hundred and fifty year history, California agribusiness has relied on a variety of mechanisms to ensure that farmworkers remain marginalized, including (1) state intervention in labor relations, primarily through immigration and labor policy, (2) the ideological construction of a racialized agricultural working class that has systematically been denied claims to better wages and working conditions, and (3) the continued availability of new groups of immigrant workers to valorize the agricultural landscape (Almaguer 1994; Daniel 1981; Galarza 1964; Makja and Makja 1982; McWilliams [1939] 1969; Mitchell 1996; Weber 1994). The production relations that characterize California agriculture today have been forged over a long history, involving local growers and transnational agribusiness interests, state and multilateral institutional actors, and modes of governance that privilege capital over labor and corporate agribusiness over small farmers in the developing world.

Today, California farmworkers hail primarily from Mexico, where the imposition of neoliberal policies has exacerbated livelihood challenges for small farmers, or *campesinos*, and led to increased northward migration (Polaski 2004; Barry 1995). Understanding the dynamics of agricultural production and the social reproduction of farm labor in California today (of which food and nutrition are clearly essential components), thus requires connecting geographies of poverty and inequality across international boundaries (cf. Mitchell 2007), from Fresno, California, to the southern Mexican states of Chiapas and Oaxaca.

The devaluation of farm labor has been, at least in part, achieved because of the invisibility of farmworkers within the material and ideological landscapes shaping California's political economy. If the basic structures of exploitation in agricultural production remain unchanged it is because they go largely unnoticed. Throughout California's history immigrants have essentially been acknowledged as a (criminal) "other" at the same time that they have been erased as value-producing workers. This invisibility is true, not only within the popular imaginary, but also within the realms of activism and academia. Unlike the UFW's consumer boycotts, contemporary alternative food movements direct surprisingly little attention to farm labor issues (Allen et al. 2003). Even where initiatives attempt to address hired labor standards, the focus has been on private voluntary programs as opposed to public regulation and collective action to promote farmworker protections and rights (Brown and Getz 2008; Getz, Brown, and Shreck 2008). The resurgence of interest in food and agriculture among academic and popular writers has likewise overlooked the role of hired labor in agrifood production, preferring to celebrate all forms of resistance to the conventional food system and to promote agrarian visions of small-scale family farms (Guthman 2004; Harrison 2008). Despite the fact that 85 percent of the labor required to produce California's field crops and livestock is performed by hired labor (Villarejo et al. 2000, 8), "the world of the worker or farm-labor activist rarely surfaces in writings about food and agriculture" (Garcia 2007, 68).

In the United States, research and activism focused on food security issues similarly overlook questions of agricultural production relations, in particular the farm labor question. The community food security movement has developed to incorporate a diverse set of both actors, from anti-hunger activists to public health practitioners, and goals, from food access and urban gardens to farmer support programs. Food security is defined as "the access for all people at all times to enough food for an active, healthy life," wording which is widely accepted worldwide by public and private actors focused on issues of poverty and hunger (FAO 1986). Proponents rightly argue that this framework allows for a more nuanced understanding of food access and deprivation, one which takes into account the nutritional content and/or quality of food, rather than just quantity (e.g., caloric intake). However, the domestic community's food security movement has been critiqued for its focus on local self-empowerment and a "do-it-yourself," voluntarist (and often antistate) approach, which eschews structural critiques (Allen 1999).

In contrast the concept of food security as it is used internationally, refers to issues such as building production capacity, promoting autonomy and self-determination with respect to food supply, reducing vulnerability to international market fluctuations and political pressures, and attaining reliability, sustainability, and equity (FAO 1986). More recently, in response to new, trade-driven notions of food security, another movement has emerged to promote food sovereignty rather than food security. The food sovereignty movement, comprising a network of NGOs, demands the removal of agriculture from the international trade system, which they deem to undermine, not support, food security for millions by allowing rich countries to maintain agricultural subsidies while forcing poor countries to dismantle farmer supports (Lee 2007).

Food sovereignty actors deploy the concept of food security to support smallholder or peasant production, local and sustainable agricultural practices, and farmworker rights through agrarian reform and other national and regional policies (<www.foodsovereignty.org>). Specifically, the food sovereignty and global peasants movement, La Via Campesina, calls for access to adequate food a matter of "simple justice," pointing out that the political-economic relations and governance structures that have increased food insecurity also lie at the root of the mass human migration taking place on a global scale, a fact that has brought many of the farmworkers in question in this study to California in the first place.

In practice the food sovereignty movement has focused much more heavily on small farmer issues while sidestepping issues of wage labor in agriculture, which we view as vital to understanding the persistence of vulnerability and hunger for California farmworkers. In the following section we explore findings from a study conducted in Fresno County, which suggest that farmworkers are indeed at heightened risk of food insecurity and hunger. We then link the question of food security to a more in-depth analysis of the dynamics underpinning vulnerability.

California Farmworkers: Hunger in the Nation's Breadbasket

Methodology

With respect to hunger and nutrition, little research has been conducted that explicitly addresses the agricultural workforce, although several recent studies suggest that farmworkers experience particularly high rates of food insecurity (Harrison et al. 2007; Quandt et al. 2004; Villarejo et al. 2000; Weigel et al. 2007). To address this lack of attention to farmworker communities, the University of California at Berkeley conducted the Fresno Farmworker Food Security Assessment (FFFSA) in collaboration with the California Institute for Rural Studies. The assessment included a survey of 454 farmworkers, as well as focus groups with farmworkers. A Farmworker Food Security Taskforce was formed to advise the assessment project and to ensure that information generated would be useful to Fresno-based organizations with a stake in farmworker food security issues.[4] Coauthor Christy Getz was the principal investigator of this two-year assessment. In the rest of this chapter, in addition to citing data and findings from the FFFSA, we draw on coauthor Sandy Brown's extensive ethnographic fieldwork with farmworkers on California's Central Coast.

The FFFSA is based on the U.S. Department of Agriculture's Household Food Security Survey Module, supplemented by a variety of questions related to food access and food assistance program utilization, employment and income, housing, family composition, age, education, remittances, period of residence in the United States, documentation, and migratory status.[5] A team of surveyors conducted the assessment in person in five areas of Fresno County that have high concentrations of agricultural workers.[6] The survey sample included 394 native Spanish-speaking agricultural workers. In addition, the survey was administered in Mixteco, to a subsample of sixty indigenous farmworkers for whom Mixteco was their native language. Mixtecos are the fourth largest indigenous group in Mexico and one of the fastest-growing and most marginalized farmworker populations in California. In an effort to capture seasonal variability in farmworkers' experiences with respect to food security and hunger, half of the surveys were conducted during the summer and half in the winter of 2005.

While efforts were made to capture a diverse sample of farmworkers and while the survey's overall demographic profile is in fact quite similar to that of farmworkers at the state level, as reported by the National Agricultural Workers Survey (U.S. Department of Labor 2003), the sample was primarily a convenience sample in one county, and is therefore not representative of all farmworkers in all regions of the state. Although geographically concentrated in one region, data from the Fresno study illuminate a persistent reality facing California's agricultural labor force, that of hunger and lack of access to quality, nutritious food. In this section we consider findings from the assessment, some of which were published by the California Institute for Rural Studies (Wirth, Strochlic, and Getz 2007). In order to present a broader overview of the socioeconomic status of agricultural workers, we supplement the food security data with additional relevant information on agricultural production, employment, wages, and workforce demographics.

Any effort to measure food security presents significant methodological challenges, particularly with respect to marginalized populations. First, the selection of a representative sample that allows one to make inferences about the larger agricultural workforce is quite difficult, due to regional variation in production relations and the instability of farmworkers' living and working situations. Even workers who do not officially migrate to follow the crops often live in temporary or informal housing and move frequently. As a result, studies that attempt to enumerate farmworkers and to measure their levels of food security likely do not capture a significant percentage of the agricultural labor force (Bugarin and Lopez 1998; Wirth 2006).[7]

Second, the process of scientifically defining and statistically measuring food insecurity as an individual or household condition hardly lends itself to an analysis of hunger as a collective, social problem related to the broader dynamics of food production, distribution, and governance.

Although the USDA's core module was developed as a tool for "community empowerment," measurement always happens at the individual or household level. Making the individual/household the *unit of measurement* poses the danger this will also become the *unit of analysis* driving policy prescriptions, thus avoiding the political economic structures underpinning highly inequitable systems of food provisioning. Indeed, much attention has been devoted to the need for education and outreach aimed at teaching people how to manage resources and to make "appropriate" food choices.[8]

In 2005, when the assessment was conducted, the USDA was still using *hunger* as part of its food security continuum. However, the USDA abandoned the term in 2006, arguing that it would achieve better measurement of the social condition, termed *food security*, rather than the physiological condition of hunger. We agree with Allen's (2007) analysis that this "innocent statistical realignment" in the name of scientific rigor "eliminates a crucial rhetorical weapon of the weak—the word hunger—in their fight against injustice" (22). Our motivation for maintaining the term *hunger* in the analysis presented here is twofold. First, it allows us to maintain continuity in communicating assessment findings. Second, and most important, it signifies an acknowledgment that hunger is not simply an individual physiological condition (as the USDA argued when it dropped the term), but rather is produced through material, social processes. Indeed, the focus on scientific rigor in measurement risks obscuring the broader contexts and material processes underlying food deprivation.

Background and Geographical Context

California's hired agricultural labor force is by far the largest in the nation, due in large part to the state's preeminence in labor-intensive fruit and vegetable crop production. Farmworkers are almost exclusively immigrants, the vast majority of whom are from Mexico. As mentioned earlier, the fluidity of the workforce and seasonality of work mean that workers often shift worksites, performing different tasks for different employers even over short time periods, making accurate enumeration of the overall workforce difficult As a result, estimates of the total number of agricultural workers living and working in California vary widely. The U.S. Census of Agriculture's official count (USDA 2007) is approximately 450,000. However, traditional census data have been found to undercount farmworkers (CRLA 2001; Sherman 1997), meaning that actual numbers are likely much higher. Martin and Mason

(2003) estimate that there are 800,000 to 900,000 individuals filling the equivalent of 350,000 full-time agricultural jobs and Kahn, Martin, and Hardiman (2005) found that, in 2001, California agricultural employers reported 1.7 million distinct Social Security numbers, or roughly 2.8 workers per year-round job in California.

Situated within California's San Joaquin Valley, Fresno County lies in the heart of the nation's breadbasket, making it an important setting for this type of research. First, it is the most productive farm county in the United States, with farm sales of over $3.7 billion in 2007 (USDA 2007). Second, it is by all accounts home to the largest farmworker population in the nation. Even conservative estimates place the population at some 52,727 workers. By way of comparison, the second-ranking California agricultural county, neighboring Kern County, counted 29,283 farm-workers (USDA 2007). The likelihood that these figures undercount the actual farmworker population notwithstanding, Fresno County remains the undeniable leader in terms of overall agricultural workforce.

Fresno is also one of the most food insecure and poorest counties in California, with 20 percent of the population living at or below the federal poverty level (Harrison et al. 2007; U.S. Census Bureau 2008). Given that agricultural wages are among the lowest of any occupation (Bugarin and Lopez 1998; Martin and Mason 2003), it is not surprising that Fresno County and, indeed, the entire San Joaquin Valley are home to some of the poorest Californians. Agriculture is San Joaquin Valley's largest private employment sector, accounting for 20 percent of employment (compared with 5.8 percent statewide) and fully half of California's farmworkers live and work in the region (Great Valley Center 2005; USDA 2007). Throughout the valley, poverty and food insecurity rates are consistently higher than the statewide average (Harrison et al. 2007). Not surprisingly, the valley also boasts one of the highest degrees of income inequality in the country (Doyle 2008).

Findings

As might be expected, FFFSA results show that respondents were more likely to experience food insecurity and hunger than the overall low-income population of Fresno County (Wirth, Strochlic, and Getz 2007; Harrison et al. 2007). Within the study sample, 34 percent of respondents were classified as food insecure and 11 percent as food insecure with hunger. As mentioned previously, in 2006 the USDA replaced these categories with low and very low food security (Allen 2007). The California Health Interview Survey (CHIS), a statewide study that focuses on health status and access that includes a section on food security, cor-roborates the finding that those who reported working in "farming, forestry, and fishing" in Fresno County reported being unable to afford enough food at higher rates than the general low-income population, 55 percent compared with 36 percent (UCLA Center for Healthy Policy Research 2009).[9] Interestingly, CHIS data reflect a higher prevalence of food insecurity than the Fresno Farmworker Food Security Assessment: 55 percent compared with 45 percent.

Inconsistencies in the limited available data notwithstanding, the finding that approximately half of the farmworker households surveyed are, in USDA parlance, *unable to access enough food for an active, healthy life*, should be viewed as nothing short of astonishing, particu-larly given its occurrence in the most productive agricultural region in the United States. In the words of one farmworker, "Something is wrong in the system. We are farmworkers, harvesting all day produce for others, and we get home and our family doesn't have enough food to eat" (Fresno Metro Ministry 2005).

The FFFSA found that income, documentation and migratory status, and food stamp use are related to food security status. Not surprisingly, income was by far the strongest predictor of food insecurity and hunger. The average monthly income for those classified as food secure was $762. For respondents categorized as food insecure without hunger, incomes declined to $542 and plummeted to an average of $319 per month for those classified as food insecure with hunger (Wirth, Strochlic, and Getz 2007, 11).

Another key finding from the FFFSA suggests that documentation of work authorization affects food security levels. The study found that, when controlling for income and other variables, farmworkers without documentation were more likely than those with legal residence or citi-zenship status to be food insecure, 55 percent compared to 34 percent (Wirth, Strochlic, and Getz 2007). Undocumented workers represent an increasing share of the agricultural labor force. The National Agricul-tural Workers Survey estimates that 53 percent of U.S. farmworkers lack authorization to legally work in the United States. However 99 percent of newcomers, a growing share of the agricultural workforce, lack such authorization (U.S. Department of Labor 2003). It is worth noting that these figures are widely considered to be conservative estimates, given that many undocumented workers are unlikely to self-identify as lacking work authorization.[10] In the following section we discuss the political construction of documentation and "illegality," highlighting the role of immigration policy and politics in maintaining farmworker vulnerability and keeping agricultural wages low. Here we simply note the correlation between food insecurity and documentation status found in the FFFSA.

Farming's seasonal ebbs and flows are an enduring reality of agricultural production. Given their shorter tenure in the United States and precarious socioeconomic and juridical status, undocumented immigrant workers are more likely to have difficulty finding work and fewer resources with which to mitigate the destabilizing cycles of seasonal employment. In addition to the more predictable seasonal ebb and flow of agricultural production, farmworker incomes can be affected by more extreme weather and production conditions, such as floods, freezes, and the loss of crops due to agricultural pests, which can delay and even end work in the fields (Bugarin and Lopez 1998).

Given high levels of variability in labor demand, some workers have adopted follow-the-crop migration strategies in order to earn enough money to support themselves and their families. Although migrant workers comprise a smaller percentage than might be expected, only 12 percent of California's total agricultural labor force (U.S. Department of Labor 2003), the FFFSA found that this group of workers was more likely to experience food insecurity and hunger (55 percent food insecure) than those who do not migrate (43 percent food insecure) (Wirth, Strochlic, and Getz 2007). As with undocumented workers, incomes for this group of workers are likely to be less stable than for nonmigrants. Furthermore, changing residence means that they may find themselves in new locations where they are unfamiliar with existing support networks and resources for accessing affordable food or emergency food assistance.

Due to a lack of legal status, undocumented farmworkers are at further risk of hunger because they are ineligible for critical public safety-net programs, including the food stamp program.[11] Within the FFFSA sample, even those who were eligible (due to legal status and income) often declined to enroll and only 48 percent of eligible respondents used the program. Some suggested that they declined to enroll due to fears about jeopardizing their immigration status, while others cited a lack of information about program requirements (Wirth, Strochlic, and Getz 2007). Such anxieties extend well beyond eligibility for public assistance programs and resonate with the broader climate of fear in which farmworkers operate. In the following section we discuss the sociohistorical role of immigration policy and the politics of backlash, which have long shaped immigrants' experiences in the United States, from a denial of labor and human rights to violent repression within communities and workplaces.

Finally, because Mixteco migrants from southern Mexico represent the most rapidly increasing share of the agricultural labor force, the UCB study included a subsample of this population, the results of which were not incorporated into the study's findings ($n = 60$). It is estimated that 20 percent of California farmworkers are indigenous, based on the proportion of farmworkers that come from states within Mexico with large indigenous populations (Wirth 2006, 42). The fact that Mixteco immigrants are more likely to be recent immigrants, and thus undocumented and migrant farmworkers, places them at increased risk of food insecurity and hunger. Respondents in this group were found to experience much greater swings in income between summer and winter months. Overall, they experienced no food insecurity during the summer but significantly higher levels during the winter months (76 percent food insecure and 48 percent hungry) (Wirth 2006, 82). While this data are considered to be preliminary and exploratory, it speaks to the changing character and implications of Mexico–United States migration. In the following section we consider how the shifts within the agricultural labor force reflect the undermining of food security in rural Mexico due to neoliberal trade policy, in particular the passage of the North American Free Trade Agreement.

Situating Farmworker Food Insecurity

If, as the data presented earlier suggests, farmworkers are at particularly high risk for food insecurity and hunger, and if, as the data also suggest, this increased risk is linked to broader political and economic insecurities, additional questions then emerge about how and why the current situation has come to be. In this section we move beyond the specific issue of measuring food (in)security, to pose a set of questions about the development of the contemporary agrarian social order, which we argue relies on the consistent devaluation of farm labor to fuel capital accumulation. To do this we consider the political economic construction of agricultural labor relations and the vulnerability of a racialized immigrant workforce, the contours of which have been constituted over California's long history of agricultural development.

Mapping the conditions underpinning farmworker vulnerability requires a consideration of dynamics that extend well beyond California's farm fields to broader agrifood commodity systems, state regulatory spaces, and the sociohistorical processes of uneven development operating at multiple scales. Indeed, on both sides of the U.S.-Mexico border, questions of food access and hunger play heavily into these dynamics. Indeed, many U.S. farmworkers find themselves facing hunger and the impetus to migrate as a result of diminishing livelihood opportunities in their primary sending country, Mexico. As domestic reforms and international trade agreements push Mexico's small-scale agricultural producers, or *campesinos*, out of production, many are increasingly pushed northward, only to become hired farmworkers in California and, more and more, throughout the United States.

California's Racialized Agricultural Working Class

From its inception, California agriculture has developed according to the logics of accumulation and competition. California growers acted, not as subsistence farmers or petty commodity producers, but rather as agrarian capitalists, pioneering the industrial farming technologies and flexible labor relations, which have made California agribusiness so profitable and which today prevail throughout the country. A historical lack of precapitalist farming (and hence, noncapitalist farmers) precipitated the development of an agrarian social order based on wage labor relations, one which has been solidified through the employment of a primarily immigrant workforce (Walker 2004).

In the previous sections, we discussed the challenges of seasonal variability in agricultural production and the fact that farmworkers have born the brunt of these burdens through low wages and unstable employment, often accompanied by food insecurity and hunger. Here we consider in further detail the strategies utilized by California growers to attenuate the gap between labor time and production time, through the maintenance of a vulnerable workforce.

Since the late nineteenth century, successive waves of immigrants have been recruited and expelled to meet growers' shifting needs for labor in the fields. Throughout the late nineteenth and twentieth centuries, different ethnic groups, including Chinese, Japanese, Filipinos, and Mexicans, have been slotted into a detailed division of labor tied discursively to concepts of racial difference (Henderson 1998). Although not present in the more common, celebratory versions of California's agrarian past, scholars have documented the often-violent process by which California's agricultural working class was formed, through the recruitment of successive waves of new immigrants and the expulsion and exclusion of groups that resisted exploitation (Almaguer

1994; Daniel 1981; McWilliams [1939] 1969; Mitchell 1996; Weber 1994).

Alongside the development of the material conditions of California's productive and profitable agrarian accumulation regime, namely techno-logical innovation, capital improvements, and constantly devalued labor, ideologies of racial difference were also being constructed. This ideological framework has served to legitimize repressive labor practices and the sys-tematic denial of rights for the agricultural workforce (Almaguer 1994; Henderson 1998). During moments of labor militancy or in cases where particular groups began moving into positions of ownership, growers were the first to mobilize racial backlash within rural communities (Almaguer 1994; Daniel 1981). More recent rounds of anti-immigrant backlash against a primarily Mexican workforce recall the historical role of racism in legitimating the exploitation and exclusion of immigrant workers.

If the experience of immigrant farmworkers has gone largely unno-ticed by the vast majority of the U.S. consumers who benefit from their labor, exploitive production relations and poverty-level wages have not always been achieved without a fight. Class struggle in the fields became increasingly protracted during the early decades of the twentieth century (Daniel 1981; McWilliams [1939] 1969; Mitchell 1996). But it was not until the end of the Bracero Program in 1964 that the United Farm Workers Union (UFW) made major inroads organizing farm labor. During the 1970s, the UFW turned its attentions to the political arena, winning passage of the Agricultural Labor Relations Act (ALRA) in 1975. This created new opportunities for farmworkers' collective action, leading to improved wages and working conditions in the fields over the following decade (Wells 1996).

However, agribusiness interests worked to ensure that the union's efforts would not deliver sustained improvements. Strategies included sourcing labor through third-party farm labor contractors, in order to avoid acting as direct employers; making modest improvements in the fields to undermine the UFW's demands; and direct retaliation against workers who attempted to organize, as well as other forms of union-busting. Additionally, a rightward political shift during the 1980s led to the failure of the ALRA to achieve its promise of facilitating worker organizing.

The Politics and Policy of "Othering"

Through high levels of class solidarity (higher than that seen among workers) across multiple scales of production and commodity categories, growers have been particularly effective at mobilizing the state to promote and protect agribusiness interests. National, state, and local governments have, from the outset, subsidized processes of land privatization and distribution, development of massive drainage and irrigation systems, and technical assistance through the land grant universities (Hundley 2001; Kloppenburg [1998] 2004). Public policy has also served as an important tool for managing labor relations and labor flow, primarily to the benefit of agricultural employers (Daniel 1981; McWilliams [1939] 1969; Schell 2002).

The significance of border and immigration politics in mobilizing anti-immigrant sentiment and undermining the bargaining power of farmworkers cannot be overstated. Immigration policy has historically served as a mechanism, not only for managing labor flow, but also for actively producing an "other," in this case a labor force that can be viewed as undeserving of the rights and benefits afforded citizen workers and that can be scapegoated during periods of economic downturn. This marginalization has relied on the mobilization of ideas, not only of class, but also of ethnic or racial difference, often couched in a framework of nationalism (Andreas 2000; Nevins 2002). Accordingly, immigration policies have excluded particular groups at various points in history, from the Chinese Exclusion Acts of 1883 to 1886, when seven of every eight California farmworkers were Chinese (NFWM n.d.) and Japanese internment during World War II, to the mass deportation of Mexican immigrants during the Great Depression and Dust Bowl era.

Through immigration policy, as well as a variety of exemptions to federal labor laws, based on notions of "agricultural exceptionalism," the state has intervened to secure a labor force for growers and to maintain its vulnerability (Makja and Makja 1982).[12] Central to the history of California agriculture is the Bracero Program (in place from 1942–1964), which brought five million Mexican workers to the fields while denying them basic rights (Calavita 1992; Galarza 1964). Throughout the Bracero period, Mexican nationals continued to enter the United States without work authorization and when the program ended the flow of undocumented immigrants increased substantially. In 1986, the Immigration Reform and Control Act (IRCA) sought to reduce the presence of unauthorized immigrants in the United States by allowing agricultural workers to apply for legal status, legalizing approximately 1.1 million farmworkers, who then moved out of agriculture and were replaced by new undocumented entrants (Martin and Mason 2003).

Since IRCA, federal immigration policy has focused almost exclusively on enforcement. Particularly striking in the current scenario is the significant public investment in enforcement technologies (border fences, detection equipment, and detention facilities) and labor (increased border patrol and other personnel to conduct workplace raids, and the stationing of National Guard troops at the border). Such techniques essentially continue to provide state-sponsored tools for employers to marginalize or eliminate recalcitrant workers. From the Clinton-era "Operation Gatekeeper," which led to a massive investment in increased border personnel and the construction of a border wall, onward, U.S. administrations have shifted to these more punitive and intrusive forms of public intervention. The events of September 11, 2001, provided new justification for the expansion and augmentation of punitive rules and programs, ensuring that the climate of fear produced *through* state activities will continue to keep immigrant farmworkers vulnerable into the indefinite future.[13]

The ethnic and class dimensions of anti-immigrant sentiment are both embodied in and obscured by the nationalist discourse over immigration and border policy (Nevins 2002). While perhaps more subtle than during periods of Chinese exclusion, Japanese internment, and the Bracero Program, the racist and classist content of anti-immigrant sentiment continues unabated. As Nevins points out, "one cannot have the national without the racialized alien" (Ibid., 157). This has generated a political context in which the immigrant workforce that valorizes California's agricultural landscape lacks rights to be present in the country, much less to take advantage of the labor rights and public supports afforded U.S.-born workers.

Yet, despite recent efforts to further criminalize immigration, immigrant laborers continue to enter the country. Current estimates place the population of undocumented Mexican nationals in the United States at 12 million, or some 9.5 percent of the current Mexican population, and 2.4 million in California, with an additional half million Mexicans entering the United States annually. This expanded wave of Latin American migration is the result of dispossession and deteriorating economic opportunities at home due to global austerity measures and trade rules, a reality that is disregarded in much of the U.S. immigration debate.

Uneven Development and Neoliberal Trade Policy

The contradictions of farmworkers' lack of access to sufficient quantities of nutritious food extend well beyond the U.S. border into the sending countries of these primarily immigrant workers. Given the makeup of California's agricultural workforce, this reality requires a consideration of the dynamics shaping livelihood struggles in rural Mexico. With diminishing livelihood opportunities at home, a growing number of people are migrating northward, to cities within Mexico and to the United States, where they work in low-wage industries such as agriculture. Ironically, many of the rural Mexicans who leave the countryside to work in the fields of U.S. agribusiness are *campesinos* who have lost their ability to farm and with it their food security (Barry 1995). Indeed, hunger is at the center of the processes of displacement and migration, which drives farmworkers to the United States in the first place.

Such displacements are a widespread sociohistorical phenomenon, compelled by centuries of uneven development and asymmetrical relationships between the United States and Mexico. At the same time that Mexico's campesinos have been "pulled" into U.S. agricultural jobs, they have been pushed off the land through international and domestic policies. However, the neoliberal reforms of the 1980s and the passage of the North American Free Trade Agreement (NAFTA) have accelerated these trends. Beginning in the 1980s the Mexican government undertook a massive restructuring of its economy, implementing a series of neoliberal reforms demanded by the World Bank, International Monetary Fund, and the United States, and shifting from import substitution strategies to an export orientation. The project included: the privatization of parastatal agencies; the rolling back of safety net programs; liberalization of trade rules and markets; privatization of communal landholdings distributed through earlier agrarian reform programs; and ending tariffs on food imports and small farmer supports, including ending price guarantees on corn, the country's most fundamental subsistence crop (Barry 1995). Based on ethnographic research within Mexico following the withdrawal of price support for corn, López (2007) found discussions about *aguantando hambre,* or enduring hunger, to be commonplace.

The Mexican state thus abandoned its regulatory role in the agricultural sector and shifted financial support from small-scale subsistence farming to production of cash crops for agro-export (McMichael 2000). The enactment of NAFTA further compromised subsistence agriculture, in particular by requiring the elimination of restrictions on the import of U.S. corn, the price of which was approximately 30 percent below the average Mexican cost of production, due to higher productivity levels and subsidies maintained by U.S. growers (Polaski 2004; Ulrich 2006). Marginal producers had accounted for 25 percent of the production of

Mexico's most important crop (Barry 1995). These are the very farmers who are now being forced out of production and into the northward migration flow. By 2002, over 1.3 million jobs had been lost in the agricultural sector (Polaski 2004). Given the central role that small farmers have played in feeding Mexico, their loss raises serious questions about the future food security of the country. Such developments, playing out on a global scale, gave rise to the international food security movement, and ultimately to the food sovereignty movement.

As immigrant farmworkers living and working in the United States struggle to meet their own needs, they send money home to their families who are struggling in the same light. Despite a precarious position in U.S. society, immigrant workers provide an indispensible source of income to 21 percent of Mexican families, amounting to almost $24.3 billion per year (Orozco 2007). Increases in remittance flows since trade and market liberalization are striking: from $1 billion in 1982 and $9.8 billion in 2002 (Polaski 2004). While remittances amount to only 3 percent of the total Mexican GDP, they constitute a much greater percentage for rural Mexican communities, up to 50 percent in some communities in the southern states of Oaxaca and Chiapas.

With respect to the broader context of food security and remittances, an FFFSA focus group we conducted in 2005, of unaccompanied men living in a California farm labor camp, yielded illustrative findings. When asked "have you ever had to send so much money home to Mexico that you haven't had enough left over to eat?" the universal response was no. The words of one participant illuminate as clear an understanding of the food security question as the "science" on which so many studies are based: "If you don't eat, you die."

Conclusion

In this chapter we have reviewed existing data on farmworker food insecurity and offered an explanation of how and why food insecurity and hunger are experienced by farmworkers. We do not suggest that this vulnerability is based on any inherent lack of ability for farmworkers to sustain themselves and their families. Instead, we have attempted to analyze this data in the context of the structural inequalities faced by farmworkers on both sides of the U.S.-Mexico border. The fundamental point is that food insecurity is produced by the global economic system in which the domestic dynamics of food production are embedded. Food (in)security is thus a lens for understanding broader processes of exploitation and inequality under capitalism. Our story extends beyond on-farm production relations and the immediate practices of agricultural employers to the broader circuits of political-economic governance, which shape the experiences of agricultural laborers on both sides of the U.S.-Mexico border and which, more often than not, heighten food insecurity. Nor is the story confined to United States-Mexico relations, but rather everywhere that people grow and eat food.

The problematic issues of food access and deprivation require an understanding that the undocumented immigrant farmworkers must be both invisible, due to their "illegal" status, and visible, in order to resist exploitation. Understanding food insecurity also necessitates a consideration of the relationship between activities of the state, including national governments' immigration laws, labor regulations, and social policies, and the international trade regimes that have privileged transnational corporate interests over smallholder agriculture. Finally, understanding food insecurity demands an analysis of the paradox of farm labor within domestic and global food production.

Since Cesar Chavez's call for "a revolution of the poor seeking bread and justice," both have continued to be denied to millions worldwide, largely as a result of the contradictions of food provisioning based on capitalist social relations. We view attempts to measure the food security status of particular groups, such as farmworkers, as critical to the process of illuminating these contradictions. However, they can only do so if they are considered within the broader contexts discussed herein. By connecting questions of food security to the underlying dynamics that produce hunger and hunger-induced migration, we hope to contribute to the opening up of more productive discussions about food security in farmworker communities.

Notes

1. Although the 1974 UN World Food Conference proclaimed people's inalienable right to freedom from hunger, the United States has never adopted this position (Allen 2007).

2. Mitchell (2007) cites the California Employment Development Department's decision to stop collecting wage information and the rise of the contract labor system as evidence that wage declines may actually be greater than official statistics suggest (also see Bugarin and Lopez 1998; Wells 1996).

3. While California's minimum wage increases during 2008 and 2009 nominally increased median hourly wages, to $8.70 per hour statewide and $8.63 for Fresno County in the first quarter of 2009 (State of California Employment

Development Department 2009), seasonal variation limits farmworkers' annual incomes and contributes to food insecurity and hunger. Furthermore, workers earning piece-rate wages, based on the quantity of produce harvested as opposed to the number of hours worked, do not realize any direct benefit from minimum wage increases.

4. The taskforce included farmworker advocates, farmers, hunger and nutrition advocates, food assistance program representatives, health care providers, nutrition educators, and members of local nonprofits and service providers.

5. The USDA defines *food security* as "access by all members at all times to enough food for an active, healthy life . . . [including] at a minimum the ready availability of nutritionally adequate and safe foods [and] assured ability to acquire acceptable foods in socially acceptable ways" and the U.S. Census Bureau uses the core module to measure this condition at the individual and household level. Households are categorized as having (1) high, (2) marginal, (3) low, or (4) very low food security (previously food insecurity with hunger) (Nord, Andrews, and Carlson 2008).

6. The five California survey sites were: two urban Fresno zip codes (93702 and 93706), two rural towns (Huron and Parlier), and one farm labor camp (Five Points).

7. Examples of such surveys include the U.S. Census, National Agricultural Workers Survey (NAWS), the U.S. Census Food Security Supplement, and the California Health Interview Survey. Because the Fresno study was conducted in person it may capture groups of farmworkers not included in surveys that focus on households with stable residences.

8. Food writers such as Michael Pollan, Anna Lappé, and Marion Nestle have devoted much attention to providing advice for readers concerned with the conventional food system. For a critique of sustainable agriculture actors' focus on the individual food choices of low-income people and people of color see Guthman 2008.

9. The California Health Interview Survey (CHIS) is a telephone survey conducted by the University of California Los Angeles Center for Health Policy Research in 2001, 2003, 2005, and 2007. The data are publicly available at <http://www.chis.ucla.edu> (accessed May 12, 2009). The project defines low income as less than 200 percent of poverty level. Unless otherwise noted all CHIS figures are at the 95 percent confidence interval.

10. The United Farm Workers suggest the figure may be as high as 85–90 percent in California.

11. The Food Stamp program is the largest of the federal food assistance programs, designed to mitigate food insecurity through subsidies for private food purchases to legal permanent residents and citizens in low-income households below 130 percent of the federal poverty level (Harrison et al. 2007). While undocumented immigrants are not eligible for food stamps, their U.S.-born children are eligible.

12. This includes exemption from the Wagner Act, National Labor Relations Act (of 1935), minimum wage and maximum hours laws, child labor restrictions,

and OSHA. Although California adopted the Agricultural Labor Relations Act to give farmworkers collective bargaining rights, it has been only mildly effective in the face of grower backlash and given the broader governance structures, which keep farmworkers vulnerable.

13. New measures include: the passage of legislation by Congress aimed at prosecuting U.S. citizens and legal residents who assist undocumented immigrants as felons, authorization of the construction of 700 additional miles of wall along the U.S.–Mexico border, increases in workplace raids and human rights violations in detention centers, and requiring some employers to utilize new and problematic instant verification systems to determine work eligibility.

References

Allen, P. 1999. Reweaving the Food Security Safety Net: Mediating Entitlement and Entrepreneurship. *Agriculture and Human Values* 16 (2): 117–129.

Allen, P. 2007. The Disappearance of Hunger in America. *Gastronomica: The Journal of Food and Culture* 7 (2): 19–23.

Allen, P., M. FitzSimmons, M. K. Goodman, and K. Warner. 2003. Shifting Plates in the Agrifood Landscape: The Tectonics of Alternative Agrifood Initiatives in California. *Journal of Rural Studies* 19 (1): 61–75.

Almaguer, T. 1994. *Racial Fault Lines: The Historical Origins of White Supremacy in California*. Berkeley: University of California Press.

Andreas, P. 2000. *Border Games: Policing the U.S.-Mexico Divide*. Ithaca: Cornell University Press.

Barry, T. 1995. *Zapata's Revenge: Free Trade and the Farm Crisis in Mexico*. Boston: South End Press.

Bradman, A. 2005. Environmental Health Perspectives: Association of Housing Disrepair Indicators. U.S. National Institute of Environmental Health Sciences, Cary, NC. <http://ehp.niehs.nih.gov> (accessed April 29, 2009).

Brown, S., and C. Getz. 2008. Privatizing Farm Worker Justice: Regulating Labor through Voluntary Certification and Labeling. *Geoforum* 39 (3):1184–1196.

Bugarin, A., and E. S. Lopez. 1998. Farmworkers in California. California Research Bureau, California State Library, CRB 98–007, Sacramento. <http://www.library.ca.gov/crb/98/07/98007a.pdf> (accessed May 10, 2009).

Calavita, K. 1992. *Inside the State: The Bracero Program, Immigration and the INS*. New York: Routledge.

California Rural Legal Assistance (CRLA). 2001. *Census 2000 Undercount of Immigrants and Farmworkers in Rural California Communities*. Marysville, CA: CRLA.

Cason, K., S. Nieto-Montenegro, A. Chavez-Martinez, N. Ly, and A. Snyder. 2003. Dietary Intake and Food Security Among Migrant Farm Workers in Pennsylvania. Harris School Working Paper, Series 04.2.

Daniel, C. 1981. *Bitter Harvest: A History of California Farm Workers, 1870–1941*. Ithaca, NY: Cornell University Press.

Doyle, M. 2008. Results Are In: California's San Joaquin Valley Is the Worst. McClatchy Newspapers, July 18. <http://www.mcclatchydc.com/251/story/44491.html> (accessed July 5, 2009).

Food and Agriculture Organization (FAO). 1986. The State of Food and Agriculture, Rome. <http://www.fao.org/docrep/003/w1358e/w1358e00.htm> (accessed June 3, 2009).

Frank, D. 2003. Where Are the Workers in Consumer-Labor Alliances? Class Dynamics and the History of Consumer-Labor Campaigns. *Politics & Society* 31 (3): 363–379.

Fresno Metro Ministry. 2005. Fresno Fresh Access: Community Food Assessment Report, 2003–2005. <http://www.fresnometmin.org> (accessed January 18, 2011).

Galarza, E. 1964. *Merchants of Labor: The Mexican Bracero Story*. Charlotte, NC, and Santa Barbara, CA: McNally and Loftin.

Ganz, M. 2000. Resources and Resourcefulness: Strategic Capacity in the Unionization of California Farm Workers, 1959–1966. *American Journal of Sociology* 105 (4): 1003–1062.

Garcia, M. 2007. Labor, Migration, and Social Justice in the Age of the Grape Boycott. *Gastronomica: The Journal of Food and Culture* 7 (2): 68–74.

Getz, C., S. Brown, and A. Shreck. 2008. Class Politics and Agricultural Exceptionalism in California's Organic Agriculture Movement. *Politics & Society* 36 (4): 478–507.

Great Valley Center. 2005. The State of the Great Central Valley of California. Modesto, CA. <http://www.greatvalley.org> (accessed June 11, 2009).

Guthman, J. 2004. *Agrarian Dreams: The Paradox of Organic Farming in California*. Berkeley: The University of California Press.

Guthman, J. 2008. If They Only Knew: Colorblindness and Universalism in California Alternative Food Institutions. *Professional Geographer* 60 (3): 387–397.

Harrison, G. G., M. Sharp, G. Manalo-LeClair, A. Ramirez, and N. McGarvey. 2007. *Food Security Among California's Low-Income Adults Improves, But Most Severely Affected Do Not Share in Improvement*. UCLA Healthy Policy Research Brief. Los Angeles: UCLA Center for Health Policy Research.

Harrison, G. G., et al. 2005. *More Than 2.9 Million Californians Now Food Insecure—One in Three Low-Income, An Increase in Just Two Years*. UCLA Healthy Policy Research Brief. Los Angeles: UCLA Center for Health Policy Research.

Harrison, J. 2008. Abandoned Bodies and Spaces of Sacrifice: Pesticide Drift Activism and the Contestation of Neoliberal Environmental Politics in California. *Geoforum* 39 (3): 1197–1214.

Henderson, G. 1998. *California and the Fictions of Capital.* Philadelphia: Temple University Press.

Housing Assistance Council. 2005. Farmworker Housing: Turning Challenges into Successes. *Rural Voices* 10 (2). <http://www.ruralhome.org/index. php?option=com_content&view=category&id=16:rural-voices&layout=blog& Itemid=33>.

Hundley, N. 2001. *The Great Thirst: Californians and Water.* Berkeley: University of California Press.

Kahn, M. A., P. Martin, and P. Hardiman. 2005 Employment and Earnings of Farm Workers: California and San Joaquin Valley, 1991, 1996, 2001. Employment Development Department, Labor Market Information Division, Sacramento. <http://migration.ucdavis.edu/cf/more.php?id=183_0_2_0> (accessed January 18, 2011).

Kloppenburg, J. R. [1988] 2004. *First the Seed: The Political Economy of Plant Biotechnology.* Madison: The University of Wisconsin Press.

Lee, R. 2007. Food Security and Food Sovereignty. Centre for Rural Studies Discussion Paper Series No. 11. Newcastle: University of Newcastle Upon Tyne, Centre for Rural Studies.

López, A. A. 2007. *The Farmworkers' Journey.* Berkeley and Los Angeles: University of California Press.

Makja, L., and M. Makja. 1982. *Farmworkers, Agribusiness, and the State.* Philadelphia, PA: Temple University Press.

Mann, S. A., and J. M. Dickinson. 1978. Obstacles to the Development of a Capitalist Agriculture. *Journal of Peasant Studies* 5 (4): 466–481.

Martin, P., and B. Mason. 2003. Hired Workers on California Farms. In *California Agriculture: Dimensions and Issues*, ed. J. Siebert, 191–214. Berkeley: Giannini Foundation of Agricultural Economics, University of California.

McMichael, P. 2000. *Development and Social Change.* Thousand Oaks, CA: Pine Forge Press.

McWilliams, C. [1939] 1969. *Factories in the Field: The Story of Migratory Farm Labor in California.* Berkeley: The University of California Press.

Mitchell, D. 1996. *The Lie of the Land: Migrant Workers and the California Landscape.* Minneapolis: The University of Minnesota Press.

Mitchell, D. 2007. Work, Struggle, Death, and Geographies of Justice: The Transformation of Landscape in and beyond California's Imperial Valley. *Landscape Research* 32 (5): 559–577.

National Farm Worker Ministry (NFWM). n.d. <http://www.nfwm.org/category/ map/learn-more/farm-worker-conditions> (accessed June 20, 2009).

Nevins, J. 2002. *Operation Gatekeeper: The Rise of the "Illegal Alien" and the Making of the U.S.-Mexico Boundary.* New York, London: Routledge.

Nord, M., M. Andrews, and S. Carlson. 2008. Economic Research Report No. (ERR-66): Household Food Insecurity in the United States, 2007. Washington, DC: United States Department of Agriculture Economic Research Service.

Orozco, M. 2007. *Sending Money Home: Worldwide Remittance Flows to Developing and Transition Countries*. Rome: International Fund for Agricultural Development.

Polaski, S. 2004. Jobs, Wages, and Household Income. In *NAFTA's Promise and Reality*, ed. J. Audley, D. Papademitriou, S. Polaski, and S. Vaughn, 11–37. Washington, DC: Carnegie Endowment for International Peace.

Quandt, S. A., T. A. Arcury, J. Early, J. Tapia, and J. Davis. 2004. Household Food Security among Migrant and Seasonal Latino Farmworkers in North Carolina. *Public Health Reports* 119 (6): 568–576.

Reeves, M., A. Katten, and M. Guzmán. 2002. *Fields of Poison: California Farmworkers and Pesticides*. San Francisco: Californians for Pesticide Reform.

Rothenberg, D. 1998. *With these Hands: The Hidden World of Migrant Farmworkers Today*. New York: Harcourt.

Schell, Greg. 2002. Farmworker Exceptionalism under the Law: How the Legal System Contributes to Farmworker Poverty and Powerlessness. In *The Human Cost of Food: Farmworkers' Lives, Labor, and Advocacy*, ed. J. Thomson, D. Charles, and M. F. Wiggins, 139–166. Austin: University of Texas Press.

Sherman, J., et al. 1997. Finding Invisible Farmworkers: The Parlier Survey. California Institute for Rural Studies, Davis, CA. <http://www.cirsinc.org/AgriculturalWorkerPublications.html>.

State of California Employment Development Department, Labor Market Information Division. 2009. OES Employment and Wages by Occupation. <http://www.labormarketinfo.edd.ca.gov/?pageid=1039> (accessed January 18, 2011).

UCLA Center for Health Policy Research. 2009. Ask CHIS 2.0. <http://www.chis.ucla.edu/main/default.asp> (accessed May 12, 2009).

Ulrich, T. 2006. Replanting People: Immigrant Workers Earn a Piece of an Organic Farm. *Orion Magazine* (Nov./Dec.). <http://www.orionmagazine.org/index.php/articles/article/183> (accessed April 12, 2009).

U.S. Department of Agriculture (USDA). 2009. Food Security in the United States: Measuring Food Security. <http://www.ers.usda.gov/Briefing/FoodSecurity/measurement.htm> (accessed June 20, 2009).

U.S. Department of Agriculture (USDA). 2007. Census of Agriculture, <http://www.agcensus.usda.gov/Publications/2007/Full_Report/index.asp> (accessed May 12, 2009).

U.S. Department of Labor. 2003. Findings from the National Agricultural Workers Survey (NAWS) 2001–2002: A Demographic and Employment Profile of United States Farm Workers. <http://www.doleta.gov/agworker/report9/toc.cfm> (accessed June 10, 2009).

Villarejo, D., et al. 2000. *Suffering in Silence: A Report on the Health of California's Agricultural Workers*. Davis: California Institute for Rural Studies.

Villarejo, D., and D. Runsten. 1993. *California's Agricultural Dilemma: Higher Production and Lower Wages*. Davis: California Institute for Rural Studies.

Walker, R. A. 2004. *The Conquest of Bread: 150 Years of Agribusiness in California*. New York: The New Press.

Weber, D. 1994. *Dark Sweat, White Gold: California Farm Workers, Cotton, and the New Deal*. Berkeley: University of California Press.

Wells, M. 1996. *Strawberry Fields: Politics, Class and Work in California Agriculture*. Ithaca: Cornell University Press.

Weigel, M. M., R. X. Armijos, Y. P. Hall, Y. Ramirez, and R. Orozco. 2007. The Household Food Insecurity and Health Outcomes of U.S.–Mexico Border Migrant and Seasonal Farmworkers. *Journal of Immigrant and Minority Health* 9 (3): 157–169.

Wirth, C. 2006. Hunger in the Fields: An Assessment of Farmworker Food Security in Fresno County, California. Unpublished master's thesis, University of California Davis.

Wirth, C., R. Strochlic, and C. Getz. 2007. Hunger in the Fields: Food Insecurity among Farmworkers in Fresno County. California Institute for Rural Studies, Davis. <http://www.cirsinc.org/Documents/Hunger_in_the_Fields.pdf> (accessed April 5, 2009).

III

Will Work for Food Justice

7

Growing Food *and* Justice

Dismantling Racism through Sustainable Food Systems

Alfonso Morales

Awareness of food and nutrition problems facing Americans has grown rapidly over the past five years, fueled by the writings of many including Michael Pollan and Barbara Kingsolver, the wide release of films like *Super Size Me* and *Fast Food Nation*, and the contributions of celebrity chefs such as Alice Waters and Odessa Piper who focus our attention on food-related inequalities and local foods. Paradoxically, many Americans, particularly low-income people and people of color, are overweight yet malnourished. They face an overwhelming variety of processed foods, but are unable to procure a well-balanced diet from the liquor stores and mini-marts that dominate their neighborhoods.

These groups are food insecure, but furthermore, they are victims of food injustice. The early-twentieth-century wave of work represented by Ebenezer Howard's *Garden Cities of To-morrow* (1902) and Upton Sinclair's *The Jungle* (1906) helped foster the Food and Drug act of 1906, but not since then has food been the subject of so much popular attention. However, for the last twenty years there has been a kind of "call and response" that has produced a web of relationships among government, scholars, nonprofit organizations, and foundations all interested in understanding food insecurity and food injustice. Definitions of important concepts like food security have been developed, organizations like the Community Food Security Coalition have grown up, foundations, universities, and government have developed programs to fund food research and practice, and nascent food justice organizations have emerged and are now populating communities around the country.

In this chapter I describe one of the newest threads in this web of activity, that of the Growing Food and Justice for All Initiative (GFJI), a loose coalition of organizations developed under the auspices of Growing Power, Inc., a food justice organization based in Milwaukee, Wisconsin, with offices in Chicago, Illinois, and a loose coalition of regional affiliates. Food justice organizations borrow from most every strand in the web of interrelated organizations and ideas, but they focus on issues of racial inequality in the food system by incorporating explicit antiracist messages and strategies into their work. This chapter chronicles in part how GFJI developed in response to the relative absence of people of color in the food system (Slocum 2006). Further, using three case studies I show how food justice organizations have responded to GFJI in different ways and how they are weaving together various threads from the larger web into their own activities and toward their own goals as they develop their own approaches to food justice.

Food Justice in Historical and Contemporary Economic Context

Over the last century or so food security (and its consequences) has ebbed and flowed in local and national consciousness. Nineteenth-century proponents of "social medicine" exposed the relationships among health, housing, sanitation, nutrition, and work conditions endured by new immigrants and the poor. Social justice was at the root of social medicine (Krieger and Birn 1998), and public health was created as an organized government response to the health-related consequences of urbanization and industrialization.[1] Part of this web was the budding urban planning movement and its work designing cities for self-sustainability, including food security. The intersection of economic, health, and food concerns prompted cities to establish local public markets, like Chicago's Maxwell Street Market, promoting food regulation and security, incorporating new immigrants, and providing employment opportunities (Morales 2000).[2] Likewise, the community gardening movement played an important role in food security until mid-century (Lawson 2005). The Great Depression produced local programs, like Detroit's Capuchin Soup Kitchen (founded in 1929) and federal policies (notably the Food Stamp program created in 1939), addressing food-related problems. In the absence of a consistent food policy, these various programs and policies developed in many directions throughout the mid-twentieth century, eventually producing some contradictory results (Amenta 2000).

Through midcentury self-provisioning decreased, population increased and the "industrialization" of food took center stage. During the 1960s and 1970s two important groups rediscovered the political economy of food. The one we know about includes the middle class and other white populations seeking to regain control over elements of the food system (Allen 2004; Slocum 2006). Indeed the work of growing, processing, and selling food began to change when they recognized "industrial" food for its effect on human health and the environment. Their critical engagement with food (re)opened niches in the food system for historical practices like community gardens and marketplaces and new practices like community supported agriculture (CSA), but it also made food more expensive just as industrial food was becoming more inexpensive and convenient. However, the very changes that made food inexpensive and convenient also put an important population, residents of the inner cities, in food jeopardy. Without the intellectual, organizational, and financial resources of the white middle class, this population was largely without the power to carve suitable niches from the food system. Instead, in the midst of plentiful and inexpensive food, this population lost access and developed food-related health problems.

How? The answer is well known to few and lies with market-driven decision making by large grocery store retailers. The transformation of the retail food sector is largely overlooked in this literature but bears significantly on the food security and health concerns of racial minority and poor populations. During the 1960s and 1970s, large retailers made decisions that helped create food access problems and have, unwittingly, helped produce the food justice movement. One recent report, *Access to Affordable and Nutritious Food: Measuring and Understanding Food Deserts and Their Consequences* (USDA 2009; see also Raja, Ma, and Yadev 2008) describes the phenomenon of food deserts and elaborates how inadequate or nonexistent transportation and limited proximity to grocery stores create the conditions for food deserts and subsequent health problems. But the report does not explicitly connect how suburbanization and decreasing profit margins prompted decisions to relocate grocery stores from inner cities to the suburbs. Some reference to history will make this clear.

Eisenhauer (2001) and Wrigley (2002) indicate how economic decisions made in corporate headquarters impact local food context. Over the last fifty years major grocery chains have sought suburban locations to accommodate larger stores, more parking spaces, and higher profits (USDA 2009). Eisenhauer (2001) refers to this trend as "supermarket redlining," or the process by which corporations avoid low-profit areas. Consider the impact on food access of such decision making. In 1914 American cities had fifty neighborhood grocery stores per square mile, an average of one for every street corner (Zelchenko 2006). Mayo (1993) documents store design and industry changes that transformed groceries from small neighborhood operations to large chains. Just as some were rediscovering healthy food in the 1960s, grocers began following the migration of the middle class from the city to the suburbs. The Business Enterprise Trust indicated the attitude, "It makes no sense to serve distressed areas when profits in the serene suburbs come so easily" (qtd. in Eisenhauer 2001). For instance, between 1968 and 1984, Hartford, Connecticut, lost eleven out of its thirteen grocery chains, and between 1978 and 1984 Safeway closed more than 600 inner city stores around the country (Eisenhauer 2001).

This mass departure reduced food access for low-income and minority people. Morland's multistate study (2002) found four times as many grocery stores in predominantly white neighborhoods as predominantly black ones, and other studies have noted that inner-city supermarkets have higher prices and a smaller selection of the fresh, wholegrain, nutritious foods (Sloane 2004). When taken with a general retreat from "hunger" by the USDA,[3] the market-driven relocation of groceries to the suburbs left behind the conditions for a public health disaster.

Food, and poor nutrition in particular, is a risk factor in four of the six leading causes of death in the United States—heart disease, stroke, diabetes, and cancer (Pollan 2008). We know that race and class inequalities produce insufficient nutrition and increase food-related disease. We know that what people eat and how they eat contributes significantly to mortality, morbidity, and increasing health care costs. By contrast, we know how food relates to good health (Institute of Medicine 2002). And when we think of access to fresh, whole foods we typically think it is dependent on income and on where one lives. Thus, decision making on locating grocery stores created "food deserts," and public health problems, but also germinated a new food coalition: the Community Food Security Coalition.

Food Security and the Community Food Security Coalition

The 1995 Community Food Security Empowerment Act document proposed community food security as the conceptual basis for evaluating and addressing food system problems. Endorsed by more than 125 organizations, the act defined *community food security* (CFS) as "all persons obtaining at all times a culturally acceptable, nutritionally adequate diet through local non-emergency sources."[4] The Community Food Security Coalition (CFSC) is the dominant private, nonprofit organization in the field of CFS. The CFSC engaged the public to create locally based alternatives and framed hunger and malnutrition as problems that should be addressed by *communities* not just by individuals, cities, or by national policy (Anderson and Cook 1999; Gottlieb and Fisher 1996). This represents a return to community context emphasized by public health, social medicine, and early public-sector planning efforts.

According to Gottlieb and Fisher (1996), environmental justice and civil rights provided important ideas for the CFS ethos. Here we stress how CFS describes food security in ways particularly germane to communities of color. These include the emphases on culturally acceptable diets and on community responsibility, not individual blame, for food security and healthy eating. Further, CFS thinking explicitly incorporated social justice as a key goal. The CFSC publishes materials in Spanish and has had notable people of color on its board of directors, including MacArthur Genius award winner Will Allen; Young Kim, executive director of Milwaukee's Fondy Farmers Market; Cynthia Torres of the Boulder County Farmers Market Association; and Demalda Newsome of Newsome Family Farms. By configuring and using the concept of Community Food Security, the CFSC has altered the discussion of food systems in the United States, particularly in federal government policy toward food security, and in how community-based organizations perceive and respond to food security problems.

While most U.S. government funding targeted food security through emergency food programs, the CFS framework was adopted in some federal agencies. Since 1996 the USDA has funded the Community Food Projects Competitive Grant Program (CFPCGP) in an effort to reduce food insecurity and boost the self-sufficiency of low-income communities. Community Food Projects (CFPs) are "designed to increase food security in communities by bringing the whole food system together to assess strengths, establish linkages, and create systems that improve the self-reliance of community members over their food needs" (USDA 2010). CFS works with individuals in the context of their community and thus is identified as a "capacity building" approach to communities, empowering communities to find healthy food solutions. This approach stands distinct to "medical" approaches toward problems of food access, hunger, and so on (Gottlieb and Fisher 1996).

Since its inception in 1996, the Competitive Grant Program has funded more than 240 projects in forty-five different states, the District of Columbia, and a U.S. territory with grants ranging from $10,400 to $300,000. Recent projects include the Tohono O'odham Community Action's "Traditional Food Project" in Sells, Arizona; the Lower East

Side Girls Club of New York's "Growing Girls, Growing Communities" project in Manhattan; Green Bay, Wisconsin's Brown County Task Force on Hunger's "Community Garden Outreach Program"; and the Friends of Bowdoinham Public Library's "Food Freaks" project in Bowdoinham, Maine (USDA 2007). This financial support has encouraged new ideas about food security, many of which were fostered by the Community Food Security Coalition.

In addition to capacity building, CFSC helped establish and promulgate the most popular definition of community food security: "Community food security is a condition in which all community residents obtain a safe, culturally acceptable, nutritionally adequate diet through a sustainable food system that maximizes community self-reliance and social justice" (Hamm and Bellows 2003). The general impulse of the CFSC community-based systems approach recovers and incorporates the "social" medicine, planning, and public health emphases by focusing on the community but understanding the individual's importance therein. Winne (n.d.) articulates the system-level approach: "in contrast to anti-hunger approaches that primarily focus on federal food assistance programs or emergency food distribution, CFS encourages progressive planning that addresses the underlying causes of hunger and food insecurity. Planning itself encourages community-based problem-solving strategies and promotes collaborative, multi-sector processes (Raja, Born, and Russell 2008). . . . While CFS embraces all approaches, even if they only provide short-term hunger relief, it places special emphasis on finding long-term, system-based solutions." This kind of systems-focused thinking means that Community Food Projects seek to improve low-income communities' access to healthy food as well as local farmers' ability to legally and competitively produce and market that food. In short, CFS has merged issues of food insecurity with local and sustainable agriculture.

The CFS movement has successfully increased the visibility of the food system through outreach and community organizing around food. This frames food as a compelling rural and urban issue, motivates policy forums for food, and impels the creation of new food research tools that, taken together, help *reconstruct* the place of food on the urban policy agenda.[5] However, despite the agenda setting and policy-making successes, food insecurity is growing in the United States. According to the USDA, in 2007 36.2 million Americans lived in food-insecure households, including 12.4 million children. Approximately 22 percent of black households and 20 percent of Hispanic households are considered food insecure, compared with 8 percent of white households (USDA n.d.). In this context it would be no surprise for organizations to emerge with an explicit interest in ameliorating food security problems of people of color in the United States.

Food Justice and the Growing Food and Justice for All Initiative

The effort to reconstruct the foodscape for people of color has augmented the discussion of food security with organizing around the concept of food justice. This idea grew from racial inequality in food access and its accompanying public health problems. In the same way that the civil rights movement grew from racial inequalities in housing, voting, transportation, and the like, new voices are naming the racism in food, but they are not alone. Tom Vilsack expressed these racial inequalities, in his first major speech as the U.S. secretary of agriculture under President Obama. Vilsack wondered aloud what the founder of the USDA, Abraham Lincoln, would find if he walked into the USDA building today and asked, "How are we doing?" "And he'd be told," said Vilsack, "'Mr. President, some folks refer to the USDA as the last plantation.' And he'd say, 'What do you mean by that?' 'Well it's got a pretty poor history when it comes to taking care of folks of color. It's discriminated against them in programming and it's made it somewhat more difficult for some people of color to be hired and promoted. It's not a very good history, Mr. President" (Federation of Southern Cooperatives 2009).

Although the CFS framework was explicitly dedicated to social justice, the early CFSC leadership was predominantly white while a number of coalition members were people of color. The justice orientation of the CFSC combined with demands from its membership led to incorporating people of color into the board and committees. But some people of color working within the CFSC expressed frustration that many white members were unwilling to examine issues of racial privilege within the organization (Slocum 2006).[6]

Erika Allen, a person of biracial decent, recalled the initial disjunction and how the CFSC responded: "Initially, people of color were not well represented on CFSC committees, many people of color attended the conferences, but we were not represented on board. Slowly the membership pressed the organization, but some of us were frustrated by the slow movement of the CFSC on those problems associated with communities of color." Aware of the food access problems in their communities and discerning the linkages between race and food, Allen and others sharing her views began to couch food security problems as problems of *food justice*. For these leaders food justice is on par with food security. They recognize CFSC success with food security and with identifying problems in communities of color, but, according to Allen, "the results of CFSC policies, helpful though they are, are not reflected in those communities' particular context."

Impelled by slow movement they perceived on their issues in the CFSC, and desiring a more complete translation of their issues to their communities, a small group of like-minded people, mostly of color, decided to sidestep the CFSC and move on to do the hard work of growing an initiative (the GFJI). They did not seek a new organization per se; rather, according to Allen, "we decided to do it ourselves, create a system that works for us, something we do together, we learn from each other. We have not created a new organization but we are teaching each other to see and dismantle food injustice and use our organizations to reconstruct parts of the food system without destroying the CFSC and the good work they do." In this view, the food justice movement is more about complementing food security, not supplanting it, creating working relationships between organizations, and embracing what Allen calls "the dynamic tension" between CFSC and the emergent food justice movement. The GFJI works to promote individual and organizational empowerment through training, networking, and creating a supportive community. Erika Allen points to the fact that many people keep their feet in both worlds, and contribute to both communities as a hopeful sign of how they can work together.

While racially motivated food justice has scant and scattered organizational infrastructure, its current manifestation does have a name and a place of origin.[7] The GFJI was established under the auspices of Growing Power, Inc., the Milwaukee-based organization founded by MacArthur Award winner Will Allen, Erika Allen's father. Since 1993, Will, Erika, and Growing Power staff have worked with diverse local communities to develop community food systems responsive to the circumstances of people of color. At about the same time as the USDA and CFSC initiatives began, Will Allen started a two-acre urban farm in a food-insecure Milwaukee community. The produce is sold in the community at affordable prices. But Growing Power is much more than an urban farm: it sponsors national and international workshops on food security, maintains flourishing aquaponics and vermicomposting programs, helps teach leadership skills, and provides on-site training in sustainable food production. The organization embodies the community-based, systemic approach of CFS by reconnecting vulnerable populations to healthy food and by developing empowered individuals in economically and socially marginalized communities.

Erika Allen has played a key role in establishing the Growing Food and Justice Initiative, which promises to expand Will Allen's vision (Allen n.d.). The GFJI has its roots in the critique of the CFSC, but came to fruition as part of Growing Power's 2006 strategic planning process. Erika Allen pushed hard to add "dismantling racism" to other, more mundane and typical organizational goals found in a strategic planning process. Jerry Kaufman, emeritus professor of planning at the University of Wisconsin and president of the board supported her goal, but also described it as "close to the foul line." He argued, "Not many nonprofit organizations would seek such a broad goal." Will Allen was also initially reluctant, as he was burdened with day-to-day organizational operations, but Erika Allen insisted on amending the plan to include "dismantling racism." Kaufman refers to her as "a real driver" in dismantling racism throughout organizational processes, but also in seeking that goal in the larger community and society. She "set that issue on a pedestal" and increased awareness in an organization already dedicated to inclusivity, he said. However, the Growing Power board recognized that the additional activity accompanying this expanded mission might overwhelm the organization, so instead of internalizing the initiative they decided to support the creation of the GFJI, which would focus most explicitly on using the food system as a means to dismantle racism.

Growing Power diverted some resources to the GFJI, and over many months Erika Allen and her colleagues developed the vision for the first GFJI conference in September 2008. Supported by the Noyes Foundation of Milwaukee and the USDA Risk Management Agency, the first GFJI conference attracted more than four hundred people. The conference was designed to bring people together to network, learn from each other, and forge new partnerships around food system self-determination for low-income communities and communities of color. Following the first conference Allen couched her reminiscences in civil rights language: "Food is the next frontier of the civil rights movement. As a child of that movement, I think about it a lot." Indeed we can see how her less prominent but vital efforts behind the scenes recall the role of women of color in earlier civil rights movements.[8] In fact, were it not for her efforts GFJI and the first gathering would not have come to be.

By choosing to focus explicitly on racism and sustainable food systems, GFJI created space for a diverse community to join together and support one another in the eradication of racism and the growth of sustainable food systems. As Pothukuchi points out, "in the 1990s, the community food security concept was devised as a framework for integrating solutions to the problems faced by poor households (such as hunger, limited access to healthy food, and obesity), and those faced by farmers (such as low farm-gate prices, pressures toward consolidation, and competition from overseas)" (2007, 7). By adding the additional concern of racism, GFJI has effectively tightened the connections already implicit in the concept of community food security: racially diverse households and farmers are some of the most at-risk groups in the communities targeted by USDA Community Food Projects. GFJI demonstrates the widespread appeal of an organization that combines the challenging topics of racism, sustainable agriculture, and community food security. But as Will Allen points out, these problems cannot be untangled, and must be tackled simultaneously. He notes, "We are all responsible for dismantling racism and ensuring more sustainable communities, which is impossible without food security."

For its member organizations, GFJI acts as a coordinating body, a source of emotional and spiritual sustenance, and a site for germinating and sharing fresh ideas. Each month one or more "germinators" convene a conference call, on a subject of interest to member organizations. Topics range widely, from the impact of the Obama administration on food-related problems to power sharing; and from developing effective multicultural leadership to sharing strategies for getting food to low-income communities. These monthly calls, which sustain the initiative without having to support an organizational infrastructure, are an important and ongoing source of ideas and support for these organizations.[9]

Often led by people of color, food justice organizations see dismantling racism as part of food security. By taking an explicitly racialized approach, the food justice movement moves away from the colorblind perspective described by Julie Guthman in chapter 12 of this book. The food justice approach aligns movement organizations explicitly with the interests of communities and organizations whose leaders have felt marginalized by white-dominated organizations and communities. By creating a space explicitly intended for the exploration of the particular challenges facing communities of color, the food justice movement has encouraged these communities to get the help and the support that they require to continue their work. The GFJI provides some logistical support, an annual conference, and networking opportunities, but perhaps most important, a sense of community, continuity, and connection among colorful communities working to improve their food security.

By bringing the individual organizations together in a cohesive body, GFJI relates racial social critique to antiracist tools of sustainable agriculture in the service of creating a nationwide network dedicated to implementing food justice strategies. In addition, GFJI provides its members with the important sense of belonging to an organization whose goals are directly aligned with their own: although many sustainable agriculture and CFS organizations have fostered the ambitions of people and communities of color over the years, they have not always shared the antiracist agenda. GFJI places racism front and center in the context of food and agriculture, allowing its members to feel confident that they are engaged in a community that shares their ideals and aspirations, and creating an infrastructure through which these organizations can support each other. This is a particularly important point in light of the overwhelming challenges involved in challenging the deeply entrenched and richly funded agribusiness industry, in addition to government agencies with deeply racist histories, such as the USDA. In the first coalition conference—GFJI I—organizations from around the country presented their work across race and the food system, and not surprisingly responded to the agenda in different ways, making clear two things: first, the variety of approaches and practices there are among antiracist food system organizations; and second, how the GFJI fulfilled its purpose by becoming an opportunity for each organization to articulate its unique approach, learn from the nuances others described, and even find that the GFJI might not be what they need.

Growing Food and Justice for All—The First Conference

The first GFJI conference, GFJI I, held September 18–21, 2008, was significant work, yet demonstrated that a light infrastructural touch could create a national forum for hearing from and supporting food justice organizations from around the country. At GFJI I people would hear their stories of particular struggles and successes, have their identity reinforced, and their practices celebrated and even reconstructed with support from like-minded organizations. Broadly speaking, attendees presented three types of work at the conference: identifying and combating racism; describing political efforts on behalf of immigrant workers or other disenfranchised communities of color; and reimagining the participation of immigrants, indigenous peoples and other communities of color within the food system. In the remainder of this chapter I offer three case studies of projects that comprise this final category of work.[10]

Case Study 1: Creating Opportunities for Hmong Farmers

Young Kim is the executive director of the Fondy Food Center in Milwaukee Wisconsin—a nonprofit "farm to fork" food system agency working to bring locally grown produce into Milwaukee's inner city. A second-generation Korean American, he was born in Tuscaloosa, Alabama, and raised in Louisiana and North Carolina. Bridging traditions was not easy, but became useful: "Growing up in a traditional Korean household in the Deep South was weird," he says, "I learned to have two identities—at home I was a good Korean boy and at school I was a good ol' boy. The whole time I hated myself for doing it, but now I realize that learning to navigate between the two cultures is what I had to do to fit in and survive." Young's experience is similar to that of many professionals of color traversing relationships and experiences as they create rich identities capable of interfacing with various stakeholders. This interpersonal capacity has served him well as a board member of the CFSC.

When Young took the reins of the Fondy Food Center in 2003, he found an organization incompletely filling its mission to create family-sustaining jobs and increase the number of healthy and affordable food options in the inner city. The Center relied on a Milwaukee food assessment to legitimize efforts to increase employment and access to healthy food, but was not systematically incorporating the elements available to fulfill its mission. In assessing the situation, Young was drawn to the agency's thirty Hmong-American farmers and their struggle to find an economic niche and make a profit. He visited their operations and found disturbing trends. Almost all Hmong immigrants rented land using only oral agreements, often paying up to $450 per acre per season—four times the going market rate for white farmers. The Hmong farmers also made do without amenities such as tractors, irrigation, cold storage, and greenhouse access that other more traditional American farmers in the region take for granted.

Young immediately implemented a systems-level approach to addressing these problems facing this immigrant population. He began orchestrating three elements: his staff, his time and role, and his vision for a way to link farm to inner-city consumer. First, he directed his staff to coordinate training for these farmers in the off season on topics such as marketing, crop insurance, and record keeping. Second, he focused the Fondy Farmers Market to simultaneously improve access to consumers and provide consumers with healthy produce choices. Third, he began to establish relationships with the local philanthropic community to secure resources and further legitimize his efforts. He then prepared a talk for GFJI I.

While Young is a veteran of the Community Food Security Coalition, a member of their board of directors, and a longtime supporter of CFS, he was cautiously hopeful about presenting at GFJI I. But he was disappointed by the first conference: "To be quite frank, the presentation that we did at GFJI left a lot to be desired. I think we had one person from the conference attend our workshop on Hmong immigrant farmers, and there were no relationships or networks that emerged from the GFJI conference." No conference is perfect for every participant and though disappointed by the turnout at his session Young was happy he attended GFJI I because on further reflection, he noted one potentially useful relationship. "We did co-present with a Minnesota farming organization, and while that interaction was useful, follow up networking has been minimal." For Young Kim, GFJI is still a work in progress, something he would admit is a two-way street and his current CFSC and Fondy commitments limit the amount of time he can devote to GFJI. However, his work with Fondy and the CFSC represent the spirit of GFJI, in the same way that GFJI is in some ways an echo of the CFSC.

Young is clearly concerned with antiracism and justice in the food system. His most prominent contribution to antiracism is helping the Hmong establish their reputation as hard-working new immigrants contributing to the community. But he has worked hard to secure equitable access to resources so that food security is better realized. His most prominent success in that area was during the summer of 2009, when a leading Milwaukee philanthropist allocated forty acres from her family farm for Fondy farmers to rent at reasonable rates starting in 2011. Thus, in orchestrating his and his organization's efforts Young promotes food justice and food security in various ways. He is hopeful, but he also recognizes the explicitly racialized process facing new immigrants when he contemplates the future for Hmong farmers in Wisconsin. "The Hmong are some of the most talented farmers because I've seen them grow things that aren't supposed to grow in Wisconsin," he goes on, "and if we can guide them through this uniquely American immigrant process where they go through the wringer and emerge as Hmong

Americans with their farming traditions intact, our country will be stronger for it. But in the meantime, it is a real struggle" (emphasis in original). Clearly Young's commitment to food justice is related to his commitment to food security and the broader relationship between agricultural practices and community identity.

Like the examples that follow, Young's analysis of the Hmong farmers' circumstances and his actions on their behalf demonstrates the interconnectedness of food, agriculture, community, health, and employment—in other words, an expansive constellation of ideas and behaviors in practice. This recalls the early historical work of planning, social medicine, and public health that understood communities as independent in some ways, but interrelated as well. Young's work relies on relationships that he creates and fosters, but it is also anchored in, and legitimized by, the ongoing work of Milwaukee's government—in providing assessments and other tools for him to do his work. At one and the same time, Young and others like him are replicating and updating the historical planning efforts that linked food, cities, and employment.

Case Study 2: Indigenous Farming Traditions
While the Hmong are among the newest Americans facing challenges to their food security, Native Americans have been struggling with this issue for centuries. Ben Yahola is co-director of the Mvskoke Food Sovereignty Initiative & Earthkeepers Voices for Native America. For him, co-director Vicky Karhu, and their organization in Oklahoma, the challenge is to reintroduce farming and food preparation methods that are more culturally and ecologically appropriate and contribute to building a sense of community and a healthier lifestyle. Yahola and Karhu feel that the community they build is explicitly political, economic, social, and spiritual.

Yahola represented the organization at GFJI I. He directed his first thoughts to the political history of food and the environment. "The indigenous in America embraced the early immigrants. As Europeans exported back to their homelands metals, herbs and spices, they also took with them democracy, the power of the people," he said. "This people power we had was based on nature and the spiritual relationship to the ecology and the food system, and both of these continue in the indigenous communities today." Yahola, like other indigenous or "first peoples" displaced from their original homes, embodies a curious interweaving of person and place, having both a specific and a more generalized respect for the land, but moreover, an intense connection to the systemic interrelationships binding together people, place, and process. At the same time his language indicates how this interwoven reality is readily adaptable to new concepts such as the "food system."

His political view extends to a critique of the early European immigrants and ties institutional racism to democracy of food and environment: "The newcomers have for centuries been unable to exercise the true meaning of democracy in food and environment so the authorities opted to establish in federal law the Christian Doctrine of Discovery, a philosophy legalizing the removal of indigenous people from their homelands. It is currently the foundation of the American Indian Policy and Law." Again, this direct linkage of racial critique to democratic processes to food and environment, and to institutional racism, is at the heart of food justice thinking. Yahola focused on the political component of the extraordinary disconnection between food and politics, economics, and ecology, a disconnection with real and severe consequences for indigenous communities, as is described in chapter 2 in this book.

For Yahola, participating in GFJI was another step toward reclaiming native food and cultural traditions through the decolonizing process of truth telling. As he put it, "With allies we are now reconnecting with the land both experientially and ideologically. The indigenous of the Americas are improving the health and food/cultural sovereignty of their communities." Food and culture are strictly woven together in many indigenous traditions. Absent one means erosion of the other, and Yahola understands well how closely articulated food and cultural practices can help empower individual identity and increase community political power. Further, recovering farming practices implies a close connection to context. The truth of contextualized experience is important to indigenous people who for centuries have honed plant species to fit particular growing conditions, and have developed culinary methods suited to what is grown and when.

As described in chapter 2, Native Americans experience extremely high rates of diabetes. Yahola recognizes the food-related roots of this disease in his own work: "The metabolic disorders have caused our cultural wisdom keepers to die in the early stages of their life. Now the youth are affected by the dominant society's corporate influences promoting unhealthy foods and lifestyles. At times we feel powerless, but I found new ways of changing our internalized oppression into physical action, ways that awaken the spirit to do something positive to improve our health." The challenge for Yahola and his organization was to find a tool to help critique this oppression and develop the knowledge and practices to replace it with more healthy expression.

At GFJI I, he found his tool. He attended the Theater of the Oppressed workshop, which is dedicated to the mission of using theater to help people understand, critique, and transform their context (Boal 1985). From participating in the workshop he says, "I discovered an alternative method of physical expression related to our current cultural conditions." Recently the Mvskoke Food Sovereignty Initiative has incorporated the Theater of the Oppressed methods in its annual Mvskoke Creek Nation Diabetes Summit. Doing so helps members form a critique of oppression by establishing a critical distance through analysis and representation of their circumstances and their goals.

Yahola's presentation and his work reflect both antiracism and a community food-security perspective. His work against racism also emphasizes self-determination and cultural sovereignty. Though the ideas are allied, he deploys them with different emphases contingent on his purpose. He uses the ideas together to make a political critique, one that is both antiracist and community empowering. His work on food security embraces the CFS perspective and is reflected in his ongoing work on food sovereignty and the linkage of food security to health. This food/health connection is important to the tribe because the relationship between food and health is central to maintaining the strength of the tribe, the culture-keepers, and others who help preserve the tribe and whose wisdom helps it adapt to changing circumstances. Like Young, Yahola mobilizes elements of both the CFS and the GFJI perspective, but unlike Young, his experience at GFJI bore immediate fruit, useful to the goals he has for his tribe.

Case Study 3: The Rural Enterprise Center

Between the experience of the most recent group (the Hmong) and the oldest group (the Mvskoke) lies the Latino experience, one constantly renewed by new immigrants even as some Latinos can trace their history in North America for hundreds of years. GFJI I helped organizations that are reconfiguring the Latino immigrant experience. Many farm- and food-related organizations around the country recognize that immigrant farmers face the same challenges that all small farmers do, but have additional burdens of cultural, racial, and social barriers, together with limited experience and knowledge of the agricultural system and farming methods in America. Likewise, these groups have group-relevant resources they can draw upon in reconfiguring their experience. Thus, while carving a niche for themselves, these groups are also exploiting new production practices and models of organization that specifically draw on ethnic resources, constituting in effect a new hybrid organization.

In Northfield, Minnesota, a Guatemalan immigrant named Regi Haslett-Marroquin is organizing a largely Mexican group of immigrants in adopting their agriculture experience to the growing conditions of the upper Midwest. To better understand his work, we can review how most people start a farm. Typically, they follow one or a combination of the following paths: marry into a farming operation, inherit a farm, or purchase a farm. None of these options is particularly viable for new immigrant families in Minnesota. Consider first the case of Latino/Hispanic families marrying into a farming operation. This uncommon occurrence does not constitute a systematic option for creating new agricultural enterprises. Second, inheriting a farm requires preexisting farmers and children who desire to inherit the farm. Even if there would be willing children, there are virtually no farmers of color in Minnesota, and fewer new immigrant families or Latino/Hispanic families who have established farms that they might eventually leave behind. Finally, in terms of purchasing a farm, one must have access to credit, which is not the case for the largest percentage of the Latino population that also has a farming vocation. Most immigrants flock to the perceived opportunities of cities, seeking escape from the perceived disadvantages and inconveniences and the diminished status associated with agriculture.

On top of all of these issues, Latinos (and Young could echo this for Hmong farmers) often face the same crude racism and discrimination in rural areas where land is still affordable. A well-meaning network of farmers and a supporting community can surround a Latino family, but it takes only one or two individuals who are racist or discriminatory to turn the situation into a nightmare. Regi's own story, narrated on American Public Media's *The Story* radio program[11] reveals the bigotry he experienced while founding a sustainable farm in rural Minnesota. On the basis of his work with immigrant farmers, he knows about how immigrants new to a community are pestered by prejudiced neighbors whose complaints might include calls to county officials reporting anything from a dog barking too much, to roosters making noise too early, to where cars are parked, to unfounded claims that there are "illegal" people living on the farm, and so on. Reflecting on his experience and learning the stories of other immigrants prompted Regi to initiate the Latino Enterprise Center, now the Rural Enterprise Center (REC), a program of the Mainstreet Initiative, which is a rural downtown development program in Minnesota.

Regi's impulse was both social and economic. He recognized that the immigrants laboring on farms and in processing facilities would produce more economic value if they owned their own operations. Once these operations were established, the economic benefits produced by the new farmers would reverberate though the community. Furthermore, Regi reasoned that a new and better social place in society could follow the farmers' economic success. Instead of being castigated for not contributing to society, immigrants establishing their own ventures could gain praise in learning to play useful roles in the community. The REC model assumes the immigrant is willing to take on risky business ventures, and that the immigrant's experience is valuable to the success of that venture. Many immigrants reject agriculture because it provided no opportunities for mobility in their country of origin. But Regi and the REC want to help immigrants to reframe their role as not only a source of labor, but as potential contributors to many important parts of economic and social life.

At GFJI I Regi shared the REC model he developed for establishing new farmers and for improving the prospects for new immigrants in rural Minnesota. The key to the REC strategy is building relationships: between people, between immigrants and their experience, their new circumstances, and between organizations. An infrastructure of supportive relationships reduces the risks and barriers to farmers entering the food and agriculture sector. The following review of the overall strategy details the four phases from REC documents,[12] and shows how each phase helps transform the immigrant's economic identity and how they view their experiences.

In phase I immigrants from a community are gathered together, and their history and knowledge are assessed as well as their interest in farming or processing agricultural products. This reconfigures the immigrant experience in an important way. Previously immigrants were sought solely for labor, not for their unique entrepreneurial contributions, but in phase I they are encouraged, even expected, to trade their identity as labor for an entrepreneurial identity. Phase I creates relationships among immigrant families and identifies or establishes networks among them. In this way the REC discovers who is serious about farming. Potential leaders are identified among the beginning farmers in each community, and the REC maps other assets and the power structure in the community. Using this information leaders reach out to potential farmers stuck, for instance, in factory jobs that prevent them from flourishing and engaging their skills and assets associated with their potential farming vocation.

Phase II is an extensive screening process, which incorporates the potential farmer into large community garden plots, market gardening plots (up to one acre in size), or existing poultry production in association with other immigrant families. After a season the potential farmer is familiar with seasonal growing patterns , the rigorous planning needed in the northern climate to avoid natural risks and crop failures, and so on. In phase II the transformation process picks up speed as promising participants are filtered from those who lose interest or choose other paths. Over the course of a growing season these potential farmers are provided with concrete experience to reestablish their skills in this new context. But no farmer goes it alone: the immigrants work together and support each other and at the end of the season evaluate their experience (see figure 7.1).

Formal training in phase III then begins. This training is year-round, hands-on work in an ongoing enterprise. Here the immigrants learn the tools of the trade, as well as the variety of roles associated with different enterprises. In this way their transformation from laborer to farmer is almost complete in that they are enabled to think and speak like existing farm/food processing operators about the same problems of credit and management that all farmers face. But the big difference is that they have

Figure 7.1
Through the Minnesota Project, immigrant farmers work together in small operations to raise chickens

their own experiences to reflect on and use to reconfigure operations in light of their unique resources and knowledge. Phase III is the formal training and technical assistance stage. The REC is engaging this stage by launching an Agripreneur Training Center, where up to twelve specialized prototype enterprises are developed for training purposes. Each training enterprise unit is designed with real farming conditions in mind, an incubator site compatible with the economic and social conditions of Latino families and with finance, farm management, and long-term farm planning components necessary to graduate farmers compatible with the existing farmland ownership, financing, and management systems.

Finally, in phase IV new farms or processing operations are launched. By this time a large supportive infrastructure of relationships are in place. Farmers also have significant knowledge of what they will grow or process and how to market it. In short, by phase IV recent immigrants will be well along the process of reconstructing their economic identity, from laborer, to entrepreneur, whose pre-migration experience is no longer a stigma, but is instead an important resource, relevant in their new business activities. Here, new farms/farmers are launched. In this phase the large support infrastructure described will engage financing, technical assistance, continued training, family support, transition counseling, product marketing, cooperative organization and networking services, product distribution and value added, product branding, financing, growth planning and management, and other ongoing support needed to keep farmers engaged and innovating their way into the future.

This process is still incomplete. The REC will be working with phase III participants starting in winter 2009–2010.[13] By 2011 new businesses will be launched. However, the incremental approach of this process makes the realistic assumption that identity (re)formation and business incubation are not overnight processes, but demand time for reflection and interaction with others in order to assimilate new roles, to test new ideas and practices, and to launch new businesses.

In one sense, the antiracism of the REC approach assumes a good offense is the best defense. In other words, bigotry and prejudice are harder to practice as immigrants accommodate themselves across a variety of roles in economic, social, and political life. Thus, the REC strategy is to tie food security and its economic elements in a reciprocal relationship to GFJI and its antiracism, successes in each feeding back to reinforce the other.

In terms of the GFJI conference, Regi believes that "engaging the Growing Food and Justice Initiative was a critical step in building a regional support infrastructure for our organization and farmers." Furthermore, he has used the tools GFJI is developing, such as a listserv and monthly phone conferences on minority farm/food related topics. Regi observed,

Since we participated in the 2008 conference in Milwaukee, we have engaged the listserv, individuals and organizations. We have learned about Chuck Weibel's operation in Milan, MN, and a lot about Will Allen's operation in Milwaukee. What our friends are doing in urban areas is absolutely inspiring and more than anything else, I personally feel free to speak out about racism and discrimination because of the broad support and accompaniment that I have received from some many hundreds of individuals and organizations.

It is this power to speak and the power to farm that REC seeks in fostering new farmers.

Concluding Thoughts and New Research Opportunities

The emergence of GFJI is the culmination of a long process in which like-minded people disagreed about process and priority in the CFSC, disagreements that led Growing Power to create the GFJI and to support it, not solely as an organization per se but as a coalition of like-minded organizations seeking food security and food justice in diverse communities. Food system problems are simultaneously local and global: solutions for production agriculture that work one place can be adopted to work in other places; organizational designs that work in some conditions can be adopted to other circumstances. Yes, each social, environmental, ecological, economic, and political matrix of race and food is unique. But food security and food justice are woven together by individuals and organizations who recognize a problem, reconstruct it as an opportunity, and organize around it while at the same time empowering communities in agricultural production, healthier consumption, local politics, and economic self-determination. A vision of self-sustaining, independent, yet interdependent community and local economic activity etches itself in different ways in distinct communities, not so much as resistance to industrial agriculture, but more toward establishing resilient and sustainable communities.

Though this vision may be clear to some, to make it actionable and understand how it can be successfully operationalized requires further research and experimentation. Federal level policy *may* support such a vision, yet it may not. What is clear to me is that a robust system is like a robust ecology it should have many niches. It may appear disorganized, or perhaps economically inefficient, but investigation can reveal dimensions of the "social" infrastructure that nourishes organizations and can also comprehend how economic goals fit with other noneconomic goals and benefits. I would suggest that research proceed in three avenues: food production and distribution; the GFJI and its prospects; and the articulation and use of these distinct frameworks—CFS and GFJI—by different organizations for various purposes. I offer the following research questions for other investigators engaging this process.

First, further historical and contemporary research should be conducted on production and distribution systems; here I have in mind three kinds of work, on the immigrant farmer/processor, the food distribution system, and the small grocery and corner store. We need more actionable knowledge about all three in different contexts. Immigrant farmers range broadly, from berry growers in Michigan (Santos 2009), to produce production in the northwest (O'Hagan 2009), and more. Some people of color (my family, for instance) have farmed and ranched in the United States for more than a hundred years. I can say that the practices of "colorful" farmers, especially when part of the industrial food system, are not always distinct from those of majority farmers. Yet there are still interesting questions about internal household relationships and organization, as well as questions about the relationships between farmers of color and their larger socioeconomic and political settings.

Without access to markets, farmers are out of business. Small farmers often face particular difficulty accessing markets, and alternative distribution arrangements are developing to create access to markets for farmers of color.[14] Corner stores are rightfully castigated for the product mix they provide, but they too are caught in the logic of profit optimization with limited choices of suppliers. There are initiatives underway to reconstruct the corner store as a source of healthy food, reasonably priced. Small groceries initiatives are also ongoing around the country. We need more historical and contemporary research to understand the immigrant farmer, the limited-resource farmer (who is typically a person of color), the role of the small grocery or corner store, and how these are all tied into the food system in distinct and often marginal ways.[15] Furthermore, we need applied research to discover and advance policy objectives related to the antiracist and economic objectives espoused by GFJI and its participant organizations.

Second, like any organization, we should be interested in and understand the organizational and institutional elements of the GFJI. How knowledge diffuses in the organization, how growth and change occur in the initiative and influence it, and how the GFJI members and the initiative itself pursue antiracism, are all very important questions. Though the GFJI is more a loose coalition than a full-fledged organization, it does produce a conference, monthly calls, and support for organizations; therefore how it is organized, and how that organization changes over time is important as an alternative approach to weaving together and multiplying the influence of a variety of existing organizations.

Third, how does the antiracist agenda of the GFJI articulate with other kinds of food-systems thinking, and with what implications for various demographic groups and the places where they live? Here I have in mind a set of general questions about the relationship between the CFSC and the GFJI, but I am also thinking about those organizations and their relationship to how we think about food in society. "Industrial" food, urban agriculture, sustainable and local food, community gardens—each has its history, ideas, and particular practices. Each is also associated with values we often think of as incommensurate, but I would argue for research that uncovers comparability in these practices and fosters dialogue among the practitioners.

However, today we recognize that these various food-system activities and how they are organized have significant implications for places.[16] Thus I would argue for more research on the integument of place and person. Beyond the significant work on food deserts are many research questions about community formation, political activism, and approaches to leveraging food-based economic development in marginalized communities. Many GFJI organizations are pointing in this direction; we need to tell their stories.

One path to a resilient and sustainable food system lies in the work promoted by the GJFI and in the GFJI itself. The GFJI did not invent the ideas of cultural and political-economic autonomy that it espouses, but it recognizes these are central components of viable communities of color. It also acknowledges the importance of sharing across distinct local practices, celebrating success, commiserating in struggle, and envisioning a more just future. The multiple forms this struggle takes should not confuse us; rather, we should take heart in how organizations like those discussed here are pursuing self-determination, food security, and food justice. Furthermore, these multiple models are an invitation to the academic and policy-making communities to understand the variety of activities taking place and to shape policy to enhance the chances for economic, political, and social success. The emerging strategy to build local, cultural, economic, and political assets in support of indigenous or ethnic farms is historic. It is a path inspired by local culture to confront and overcome the political economic force of capital that has torn the commons asunder. As such, these organizations and indeed the GFJI represent important examples of dialogue between the localized and situated, and the national and overarching, while indicating the importance of both.

Acknowledgments

The author appreciates the many comments on prior drafts of this chapter provided by Jessica Saturley, Elise Gold, Alison Alkon, Laura Tolkoff, and CeCe Sieffert.

Notes

1. Rudolf Virchow ([1848] 1985), a German physician and the "Father of Modern Pathology" wrote: "Public health and medicine is a social science, and politics is nothing else but medicine on a large scale. . . . If medicine is to fulfill her great task, then she must enter the political and social life. Do we not always find the diseases of the populace traceable to defects in society?"

2. Tangires (2003) writes of the longer history of marketplaces in the United States.

3. Allen (2007) notes this significant change in the rhetoric of food security "was eliminating the word *hunger* from its official assessment of food security in America and replacing it with the term *very low food security*" (19).

4. CFSC 1994.

5. However, Kami Pothukuchi (2007) argues it was never there. I would suggest that food access was important to early city government, but its importance has changed over time.

6. Though this history is not the direct concern of this chapter, I suggest that researchers interested in the subject focus on both individual and organizational elements of the CFSC, as well as contextual elements of its time and place. Organizational factors include the difference between named and funded CFSC priorities, committee and board memberships and how board and committee mandates or pronouncements were received and prioritized, and what resources were allocated to various priorities. Individual factors include interpersonal relationships, but with a focus on how the CFSC prioritized individual training and professional development needs of leaders and community members.

7. Religiously motivated food justice intersects with racially motivations and has a deep history in the United States. See, for instance, the work of Catholic Relief Services, <http://crs.org/about/history/> (accessed May 20, 2011).

8. See, for instance, Pardo 1990 and Barnett 1993.

9. These monthly calls attract from twelve to thirty people at a time, and more than fifty organizations have been represented. A list of call topics is available from the author, Alfonso Morales (morales1@wisc.edu).

10. More men than women made presentations at GFJI I. And many more men than women made presentations on the subjects of this chapter. Still, women were prominent in a number of other sessions; see the notes from the conference (GFJI 2008).

11. See <http://thestory.org/archive/the_story_824_The Fight_To_Farm.mp3/view> (accessed October 4, 2009); see also the news article on immigrants' rights at <http://www.tcdailyplanet.net/news/2009/09/14/immigrants-rights-activists-vie-attention> (accessed October 4, 2009).

12. For details, see the pages at <http://www.ruralec.com/ (accessed June 14, 2009).

13. Ibid. Some sixty people representing more than twenty households are creating new businesses. Enterprises involved meat poultry, poultry processing, grain processing, grain production (as in small grains), vegetable production (garlic), and black beans (as in edible dry beans).

14. Particularly useful in this regard is the report "Satiating the Demand" found at <http://urpl.wisc.edu/extension/reports.php> (accessed June 1, 2009) and the report "Supply Chain Basics" found at <http://www.agmrc.org/markets_industries/supply_chain/> (accessed June 1, 2009).

15. For examples, see <http://www.foodsecurity.org/LRPreport.pdf> (accessed June 1, 2009); <http://www.thefoodtrust.org/php/programs/store.network.php> (accessed June 1, 2009); and <http://www.policylink.org/site/c.lkIXLbMNJrE/b.5136687/k.61DA/Healthy_Food_Retailing.htm> (accessed June 1, 2009).

16. See, for instance, the report "Why Place Matters: Building the Movement for Healthy Communities" at <http://www.policylink.org/site/c.lkIXLbMNJrE/b.5137443/apps/s/content.asp?ct=6997411> (accessed June 14, 2009).

References

Allen, Patricia. 2004. *Together at the Table: Sustainability and Sustenance in the American Agri-Foods Movement.* University Park: Pennsylvania State University Press.

Allen, Patricia. 2007. The Disappearance of Hunger in America. *Gastronomica* (Summer): 19–23.

Allen, Will. n.d. A Good Food Manifesto for America. <https://www.growingfoodandjustice.org/uploads/Will_20Allen_27s_20Good_20Food_20Manifesto-1.pdf> (accessed June 24, 2009).

Amenta, Edwin. 2000. *Bold Relief.* Princeton, NJ: Princeton University Press.

Anderson, M. D., and J. T. Cook. 1999. Community Food Security: Practice in Need of Theory? *Agriculture and Human Values* 16:141–150.

Barnett, Bernice McNair. 1993. Invisible Southern Black Women Leaders in the Civil Rights Movement: The Triple Constraints of Gender, Race, and Class. *Gender & Society* 7 (2): 162–182.

Boal, Augusto. 1985. *Theatre of the Oppressed*. Trans A. Charles and Odilia-Maria McBride-Leal. New York: Theatre Communications Group.

Community Food Security Coalition (CFSC). 1994. *A Community Food Security Act: A Proposal for New Food System Legislation as Part of the 1995 Farm Bill*. Hartford, CT: Hartford Food System.

Eisenhauer, Elizabeth. 2001. In Poor Health: Supermarket Redlining and Urban Nutrition. *GeoJournal* 53:125–133.

Federation of Southern Cooperatives. 2009. Transcript of U.S. Agriculture Secretary Tom Vilsack's speech to Federation of Southern Cooperatives/Land Assistance Fund's Georgia Annual Farmer's Conference, January 21. <http://www.federationsoutherncoop.com/albany/Vilsackspeech.pdf> (accessed June 26, 2009).

Gottlieb, Robert, and Andrew Fisher. 1996. First Feed the Face: Environmental Justice and Community Food Security. *Antipode* 28: 193–203.

GFJI. 2008. First Annual GFJI Gathering, Friday Sessions Notes. <https://www.growingfoodandjustice.org/1st_GFJI_Gathering.html> (accessed June 27, 2009).

Hamm, M. W., and A. C. Bellows. 2003. Community Food Security and Nutrition Educators. *Journal of Nutrition Education and Behavior* 35: 37–43.

Healthy Food Healthy Communities: A Decade of Community Food Projects in Action. 2006 (developed in cooperation between USDA, CSREES, Community Food Security Coalition, and World Hunger Year), <http://www.foodsecurity.org/pubs.html#cfpdecade>.

Howard, Ebenezer. 1902. *Garden Cities of To-morrow*. London: Swan Sonnenschein.

Institute of Medicine. 2002. *The Future of the Public's Health*. Washington, DC: National Academy Press.

Kloppenberg, Jack, Jr., J. Hendrickson, and G. W. Stevenson. 1996. Coming in to the Foodshed. *Agriculture and Human Values* 13 (3): 33–41.

Krieger, Nancy, and Anne-Emmanuelle Birn. 1998. A Vision of Social Justice as the Foundation of Public Health: Commemorating 150 Years of the Spirit of 1848. *American Journal of Public Health* 88: 1603–1606.

Lawson, Laura J. 2005. *City Bountiful: A Century of Community Gardening in America*. Berkeley: University of California Press.

Lyson, Thomas A. 2004. *Civic Agriculture: Reconnecting Farm, Food, and Community*. Medford, MA: Tufts University Press.

Mayo, James M. 1993. *The American Grocery Store*. Santa Barbara, CA: Greenwood Publishing Group.

Morales, Alfonso. 2000. Peddling Policy: Street Vending in Historical and Contemporary Context. *International Journal of Sociology and Social Policy* 20:76–99.

Morland, Kimberly, Steve Wing, Ana Diez Roux, and Charles Poole. 2002. Neighborhood Characteristics Associated with the Location of Food Stores and Food Service Places. *American Journal of Preventive Medicine* 22 (1): 23–29.

O'Hagan, Maureen. 2009. Refugees Find a Chance in a Field in Kent. *Seattle Times*, September 27.

Pardo, Mary. 1990. Mexican-American Women Grass-roots Community Activists: The Mothers of East Los Angeles. *Frontiers* 11 (1): 1–8.

Pothukuchi, Kami. 2007. *Building Community Food Security: Lessons from Community Food Projects*, 1999–2003. Community Food Security Coalition, <http://www.foodsecurity.org/BuildingCommunityFoodSecurity.pdf>.

Raja, Samina, Branden Born, and Jessica Kozlowski Russell. 2008. *A Planners Guide to Community and Regional Food Planning*. American Planning Association, Planning Advisory Service Report No. 554.

Raja, Samina, Changxing Ma, and Pavan Yadav. 2008. Beyond Food Deserts: Measuring and Mapping Racial Disparities in Neighborhood Food Environments. *Journal of Planning Education and Research* 27 (June): 469–482.

Santos, Maria Josepha. 2009. Knowledge and Networks: Mexican American Entrepreneurship in Southwestern Michigan. In *An American Story: Mexican American Entrepreneurship and Wealth Creation*, ed. John S. Butler, Alfonso Morales, and David Torres, 151–174. West Lafayette, IN: Purdue University Press.

Sinclair, Upton. 1906. *The Jungle*. New York: Doubleday.

Sloane, David C. 2004. Bad Meat and Brown Bananas: Building a Legacy of Health by Confronting Health Disparities around Food. *Planners Network* (Winter): 49–50.

Slocum, Rachel. 2006. Anti-racist Practice and the Work of Community Food Organizations. *Antipode* 38: 327–349.

Tangires, Helen. 2003. *Public Markets and Civic Culture in Nineteenth-Century America*. Baltimore, MD: Johns Hopkins University Press.

United States Department of Agriculture (USDA). n.d. Food Security in the United States: Key Statistics and Graphics. Economic Research Service. <http://www.ers.usda.gov/Briefing/FoodSecurity/stats_graphs.htm> (accessed June 27, 2009).

United States Department of Agriculture (USDA). 2007. Healthy Food, Healthy Communities: A Decade of Community Food Projects in Action. <http://csrees.usda.gov/newsroom/news/2007news/cfp_report.pdf> (accessed January 11, 2011).

United States Department of Agriculture (USDA). 2009. *Access to Affordable and Nutritious Food: Measuring and Understanding Food Deserts and Their Consequences*. Washington, DC: Economic Research Service.

United States Department of Agriculture (USDA). 2010. Cooperative State Research, Education, and Extension Service. Community Food Projects Competitive Grants. <http://www.csrees.usda.gov/nea/food/in_focus/hunger_if_competitive.html> (accessed January 11, 2011).

Virchow, Rudolf. [1848] 1985. *Collected Essays on Public Health and Epidemiology*. Cambridge, UK: Science History Publications.

Winne, Mark. n.d. Community Food Security: Promoting Food Security and Building Healthy Food Systems. <http://www.foodsecurity.org/PerspectivesOnCFS .pdf> (accessed August 12, 2009).

Wrigley, Neil. 2002. "Food Deserts" in British Cities: Policy Context and Research Priorities. *Urban Studies* 39 (11): 2029–2040.

Zelchenko, Peter. 2006. Supermarket Gentrification: Part of the "Dietary Divide." *Progress in Planning* 168 (1): 11–15.

8

Community Food Security "For Us, By Us"

The Nation of Islam and the Pan African Orthodox Christian Church

Priscilla McCutcheon

If Black people could secure sufficient power to maintain a balance it might be possible for Black people and white people to live together as two separate peoples in one country. Black people must remain separate using the separateness that already exists as a basis for political power, for economic power, and for the transmission of cultural values. . . . The only hope for peace in America depends upon the possibility of building this kind of Black power. . . . A Black Nation within a nation must come into being if we are to survive. For the Black man everything must be judged in terms of Black liberation. There is but one authority and that is the Black experience. (Cleage 1972, xxxvii)

This excerpt is from *Black Christian Nationalism* written by Albert Cleage. In it he describes his beliefs and methods on how blacks should empower themselves in what he deems to be a white supremacist society. This book also provides an ideal starting point for exploring how two Black Nationalist religious organizations are using food not only as a means to address hunger, but also as a tool of empowerment among blacks. In 1999, the Pan African Orthodox Christian Church (PAOCC) completed the purchase of over 1,500 acres of farmland on the border of Georgia and South Carolina to build a self-sustaining community called Beulah Land Farms. Their land acquisition has now grown to over 4,000 acres. Albert Cleage (1972), the founder of the PAOCC, envisioned a place where members of the church would come together, but also a place where inner-city youth would be exposed to nature and farmland.

The Nation of Islam (NOI) purchased over 1,556 acres of rural South Georgia farmland in 1994, naming it Muhammad Farms. Its expressed purpose is to feed the forty million black people in America (Muhammad 2005). The NOI was founded by Master Fard Muhammad and led for decades by Elijah Muhammad. Its most notable leader, Malcolm X, strongly believed in black landownership and its importance in achieving self-sufficiency. While both organizations' work around food and health is heavily influenced by their self-proclamation as Black Nationalist religious organizations, the NOI and the PAOCC define Black Nationalism in two *distinct* ways. Food is not simply used to address hunger, but also to build community among blacks. Food is a part of a larger ideology of Black Nationalism, in which self-reliance and the individual achievements of blacks are linked to the "black community" at large.

Black Nationalism is one of the six political ideologies that Michael Dawson (2001) describes in his seminal work *Black Visions*. Dawson argues that "popular support for black nationalism continues to be based on the time-tested skepticism in black communities that, when it comes to race, America will live up to its liberal values" (Ibid., 86). Black Nationalism is the belief that race and racial discrimination are at the center of the black experience in America and must be addressed by blacks if this group is to ever achieve any tangible progress. Elijah Muhammad, one of the key founders and leaders of the NOI, fervently preached the need for blacks to look beyond the dream of integrating into a white society that devalued every aspect of black life. Instead, he asserted that blacks should develop their own racial identity, and effectively form a psychological and geographically separate nation from whites to promote the true liberation of blacks (Muhammad 1997a). Historically, the PAOCC does not define black liberation as a geographic separation from whites. Instead psychological separation is key so that blacks can develop their own institutions that serve their identities and interests (Cleage 1972). When blacks secure a sufficient amount of equality and power, Cleage argues that only then will it be possible for blacks and whites to live together in the United States but still utilize separate institutions. The recent philosophy of the PAOCC builds on the organization's historical belief of blackness as a unifying factor. The PAOCC asserts that blackness alone should not be the sole organizing factor, but issues including class must also be considered. Blackness then, is complicated and cannot be portrayed as simply white vs. black or us vs. them. Both the NOI and the PAOCC are using their work around food and agriculture to promote some aspects of Black Nationalism achievable through self-sufficiency.

The purpose of this chapter is to delve into the relationship between food and racial identity and religion through the lens of these two Black Nationalist religions. I intend to explore this concept of just sustainability by investigating the ways that race and racial identity are at the heart of the ideological justification and actions of both the NOI and the PAOCC's work around food and health. More important are the ways

psychological not geographical separation

in which self-sufficiency and community contribute to just sustainability not only in both religious organizations but also among blacks as a collective. This chapter reflects archival and textual research for the NOI. I also engaged in archival and textual research for the PAOCC, but had the added opportunity to do participant observation at Beulah Land Farms. While my access to both organizations differs, the amount of textual information on each is extensive.

To begin, in the first section, I give a brief history of these organizations and their food programs ending with exploring why these two cases might offer unique insights into questions of race in alternative food movements. Second, I bring to light the insistence on self-reliance, borrowing the name FUBU to illuminate the For Us, By Us principle that guides these Black Nationalist religions' work around food and health. Third, I unpack the word *community*, keying in on how both groups define this term racially, socially, and spatially and noting both similarities and differences. I conclude by offering some recommendations on how the broader community food movement might utilize these two examples in the quest to understand the linkage between race and food provisioning practices, which can only aid in efforts to diversify all levels of the movement.

Organizational Histories

Nation of Islam

The story about the founding of the NOI centers on "the prophet" who appeared on the streets of Detroit in 1930. Fard Muhammad traveled from house to house initially using the Bible to teach blacks because this, to him, was the only religion that they knew. Food and health were central to the NOI's teachings from its inception. NOI members preached that blacks should not eat unhealthy food, equating such food to poison.

The NOI was founded during the Great Depression when racial barriers prevented progress, and some blacks felt that the black Christian church was not forceful enough in its stand against racial inequality. Elijah Muhammad was one of the first followers of Master Muhammad. He preached fervently about what blacks could do to gain freedom on this Earth. Malcolm X is perhaps the best-known member of the NOI. He was a vocal minister who furthered the message of the NOI preaching to masses of blacks both inside and outside of the NOI about self-sufficiency, the evilness of whites, and the importance of a geographically separate nation. Though Malcolm X eventually broke away from the NOI and changed his views on race drastically in his later years, the NOI and its teachings are often still associated with views he espoused earlier in its history (Lincoln 1994).

The NOI, commonly characterized as a Black Nationalist organization (Harris-Lacewell 2004; Squires 2002; West 1999; Dawson 1994), contends that the only path to black liberation is in forming an autonomous nation separate from whites (Marable 1998). The NOI rejects Christianity as a direct manifestation of white racism, which is appealing to many members. Even so, some of the NOI's teachings resemble those of some black churches and other groups that share its strict ethical and moral guidelines. The NOI has had a profound influence on the religious experience of blacks both inside and outside of this religious sect (Curtis 2006; Dawson 2001; Muhammad 1997a).

The NOI's teachings have received criticisms from blacks and non-blacks alike. During the 1950s and 1960s, some prominent black leaders denounced the NOI as hateful and divisive, and some black intellectuals ignored this group altogether in their discussion of black religious life. Despite this, the NOI has retained relationships with some black Christian ministers, politicians, and businessmen. Among blacks, the relationship with the NOI is a complicated one. Many acknowledge and often share the NOI's frustrations, but differ on their methods for achieving equality. The NOI's insistence on black unity to achieve progress appeals to some blacks both inside of and outside of the religious organization (Lincoln 1994).

Lincoln (1994) argues that most average whites know little about the NOI. "Those who learn of the movement tend to consider it an extreme and dangerous social organization" (172). Some even compare the NOI to the Ku Klux Klan. For the most part, the NOI is more concerned with its perception among blacks and expanding its membership to one day include all blacks. Any objection by whites to its teachings is seen as a natural extension of a white supremacist society.

Food plays a prominent role in the NOI's teachings. In the excerpt that follows from *How to Eat, How to Live*, Elijah Muhammad emphasizes the importance of food in maintaining physical, mental, and spiritual health among blacks. He preaches: "Many years [can] be added to our lives if we only knew how to protect our lives from their enemies . . . food keeps us here; it is essential that we eat food which gives and maintains life . . . we must protect our lives as well as possible from the destruction of food. If we eat the proper food, and eat at the proper time, the food will keep us living a long, long time" (Muhammad 1997b, 1).

The NOI's rhetoric articulates a clear connection between black liberation and the production and consumption of food. The NOI's religious and racial goals influenced its decision to produce and distribute food as a means to increase black autonomy. Elijah Muhammad speaks at length about the appropriate food intake needed to completely emancipate blacks from racial oppression. He admonishes black Christians for continuing to consume pork, alcohol, and tobacco, connecting their consumption to mental bondage, and argues that they are tools used to oppress the black community (Muhammad 1997b; Rouse and Hoskins 2004). According to scholars studying the NOI, proper consumption signifies purity and a commitment to the teachings of Allah (Curtis 2006; Rouse and Hoskins 2004; Witt 1999).

The current work and rhetoric of the NOI on food and health suggests a commitment to the principles of the organic food movement. The NOI believes that food should ideally be chemically free. Though the NOI is not currently farming organically, its Web site includes information about the dangers of pesticides, insecticides, genetically modified seeds and food, and the general danger of large agribusiness. The NOI's minister of agriculture, Dr. Ridgely Abdul Mu'min (Muhammad), states that monetary limitations and the lack of labor prohibit the NOI from growing solely organic produce ("Background of Farm" n.d.).

The NOI's goal of promoting black liberation is seen broadly in its priority of landownership to increase food production and autonomy among blacks (Malcolm X 1965). Perhaps the most widespread effort of the NOI to promote self-reliance among blacks is the purchase of Muhammad Farms in Bronwood, Georgia to grow a variet of fruits and vegetables. Further investigation into the publicly available archives of Muhammad Farms reveals that its leaders' goals are to "develop a sustainable agriculture system that would provide at least one meal per day . . . [for] 40 million black people" (Muhammad 2005, 1). This land would not only be used to feed people, but also to eventually develop a separate nation for blacks. The NOI contends that much of the land in southern states was cultivated by black slaves and rightfully belongs to them. Though the majority of the NOI's membership is located in major cities including Chicago, Detroit, and Atlanta, the NOI purchased land that is generally outside of its reach to further members' efforts around food, health, and black autonomy. Food grown on this land is distributed mainly to NOI mosques in major cities and also to some black community members outside of the NOI located near mosques in these cities.

Pan African Orthodox Christian Church

The Pan African Orthodox Christian Church is a lesser-known organization founded in Detroit. Reverend Albert Cleage founded Central Congregational Church in 1956 while he was diligently working for racial equality in the city. Throughout the 1960s, Cleage's religious ideology became increasingly Afrocentric, and he renamed the church the Shrine of the Black Madonna. He went on to form congregations in Atlanta and Houston (Shrine of the Black Madonna 1). Cleage is most known for developing black liberation theology as a Black Nationalistic tradition throughout Christianity. He argued that black people, particularly black ministers, were ordained to teach liberation theology and that the black experience should be central to the teachings of all black churches (Cleage 1972).

Central to Cleage's (1972) teachings are ideas of black nationhood as a way for blacks to cope with their marginalized position and exclusion from a white society. Cleage and the church support the notion that God believes in the freedom of black people, and Jesus is also called the black Messiah. The churches are named The Shrine of the Black Madonna to further emphasize that Jesus is black and to empower black women. Albert Cleage, the founder of the PAOCC, created Black Christian Nationalism. Among the many positions of Black Christian Nationalism is a rejection of the individualism that, in Cleage's opinion, permeates white society. According to an article written by the Holy Patriarch, Jaramogi Menelik Kimathi, communalism, when used in accordance to God's will, is key to helping the greatest number of people possible (Kimathi n.d.). Individualism is not useful to blacks and contradicts the communal nature of African religion, the black church in the United States, and the black nation more broadly. Authentic black liberation requires that black people recognize that race is the overarching way that they are defined in this country and take ownership of this definition. Their oppression is group oppression that cannot be overcome by the individual success of blacks or individual acts of kindness by whites. Cleage says:

White people are in one group and Black people are in another, and the interests of the two groups are in basic conflict. You may be Black and confused and not realize that you are a part of a group, but if you are Black you are in the Black group. Everyone knows that but you. The whole white group . . . all work together to keep us powerless and they have no choice . . . White people have little value for black people other than the wealth that they can extract from us. (Cleage 1972, 84)

An important aspect of Cleage's explanation of Black Nationalism as a necessity is his discussion of power and group recognition. Despite the fact that some individual blacks may view themselves as better off economically or socially than other blacks, Cleage argues that all are in the struggle together. Equally important is Cleage's description of power and Black Christian Nationalism. Cleage does not argue that whites are inherently evil. He instead argues that the actions of the white power structure derive from power itself. Black people and white people are not different in nature, but rather the differences come from experience and environment. He argues that "white people have been ruthless in their use of power. Under the same circumstances the Black man *will* act the same way" (Cleage 1972, 102). He goes on to say that unless black people separately define a value system based on community, perceived equality will only bring forth similar individualized actions.

The PAOCC has recently expanded their philosophy. Kimathi notes that "blackness alone is no longer a sufficient basis for unity" (Kimathi n.d., 9) because we live in a world where class conflict and racism—to offer one example—cannot be divorced from one another. Blackness is still key, but it is complex, and a host of other complexities must be recognized.

It is useful here to quote Kimathi's article in more detail. He says, "Our struggle is not simply for Black power but also for righteousness, justice, communalism, and goodness-power that is used in compatibility with the will of GOD. We seek always to do the greatest good for the greatest number of people. . . . This is the evolution of Black theology." It is important that this evolution of theology does not signal a departure from the needs of black people, but rather, that these needs are understood within a global context that is interconnected with other important issues.

Beulah Land Farms has now grown to an approximately 4,000-acre lakefront site in Calhoun Falls, South Carolina, on the Georgia/South Carolina border. Its purpose is to promote communal living and self-sufficiency among blacks (New Georgia Encyclopedia 2006). It is a physical setting in which residents are removed from the stimuli of their normal environment to allow a spiritual and physical transformation to occur (Kimathi n.d.). The original tract of land was purchased by Albert Cleage and is now a productive farm. Members of the PAOCC describe it as a "multi-faceted agricultural complex" (PAOCC 2009, 1). On this land, members grow vegetables and fruit but also raise poultry and cattle, and use aquaculture. This land is also a retreat for members of the

PAOCC (Ibid., 1). The name *Beulah* itself is taken from the Book of Isaiah in the Holy Bible; it means to be married to the land. Biblically, Beulah is seen as an environment that stimulates a spiritual transformation and awakening (PAOCC 2009).

Why the NOI and PAOCC

Both the NOI's Muhammad Farms and the PAOCC's Beulah Land Farms are attempts, in distinct ideological and action-based ways, by Black Nationalist religious organizations to develop an idea of black nationhood based on community.

Though the NOI has received a considerable amount of attention in the media, both it and the PAOCC have received limited scholarly attention. Even less recognition has been paid to their work around food and health. Dorris Witt (1999) discusses the NOI's focus on the dietary and spiritual harms that come from eating foods deemed impure. Witt focuses specifically on the connections that the NOI makes between food and agriculture and black male masculinity. Rouse and Hoskins (2004) describe the dietary guidelines that NOI members follow, including how improper dietary intake is attributed to poor health but also to "negatively valued modes of behavior" (Ibid., 237). Scholars have also investigated black settlements in the United States, including freedom farms after slavery and during the Civil Rights movement (Lee 1999). However, there are no works to date that specifically focus on both the NOI's and the PAOCC's actions around food and how they contribute to the formation of a community and race-based identity designed to influence not only their members but also blacks outside of these distinct religions and denominations.

Because of each group's existence as a Black Nationalist organization, it is unlikely that either the NOI or the PAOCC's work will ever gain widespread attention or appeal among the dominant food movements. This is not simply because some portions of Black Nationalism may be offensive to some, but also because, while in different ways, both the NOI and the PAOCC seem to desire to be somewhat on the outside of these movements. In very distinct ways, they aim to build self-sufficiency and a reliable food source for blacks in the United States. Furthermore, both groups aim to expand this self-sufficiency and ownership to even more institutions. The PAOCC is using its farm to grow cattle, but members also have the vision of one day producing an environmentally sustainable electricity source and already have their own water source.

While both the NOI and the PAOCC are on the outskirts of the community food movement, many of their beliefs about the dangers of pesticide and genetically modified food are strikingly similar to the beliefs of the community food movement. I would argue that their reasoning is slightly different, often echoing race-based and black liberation-themed theology. The goal of both religions is to achieve a vision of community improvement and autonomy among blacks. Race and community are at the heart of both organizations' food work. Increasingly, antiracist practices and white privilege are becoming the focus of scholarly work on community food.

Both the NOI and the PAOCC are Black Nationalist religious organizations, which adds an interesting perspective to addressing whiteness and white privilege in the community food movement. Most groups in the community food movement assert that they are "colorblind" and do not view race as relevant to their work (Guthman 2008). Slocum (2006) finds that some of these groups are uncomfortable addressing white privilege, feeling that racism will always be a constant in society. Both the NOI and the PAOCC, to varying degrees, have been formed around race as one of their key group identifiers. These organizations obviously differ from many of the organizations that Slocum describes in her work, and not only address race directly, but also use it as a key organizing principle. In the PAOCC's case, blackness is not enough. Included in blackness is a concern for blacks across the world and for problems that affect a variety of different groups. I would argue that neither of these organizations see themselves as a part of a largely white community food movement or white society as a whole. And to be certain, the goal of neither organization is to be a part of a movement or society that they largely see as unjust and without true opportunities for blacks to succeed. Their intent is for food production to be a part of a broader goal of building community. Regardless, I argue that because the NOI and PAOCC do not want to assimilate into a broader and whiter movement, they can be useful for understanding the importance of racial identity in the community food movement.

When thinking about race and community food, the important question seems less how these two organizations should strive to become a part of the community food movement, but rather how the community food movement can transform itself *and* a broader society that thus far has not been inclusive. It would be easy to come to the conclusion that since these groups "peacefully" lie on the outskirts, there is no point in even investigating them alongside other groups within the movement. I argue that scholars should instead delve into both groups and the communities that they create to get a sense of what inclusion means to these two Black Nationalist organizations. This may in turn provide suggestions on how the broader community food movement can become more inclusive. These groups are not simply looking to provide sustainable food, but also a communal religious experience for blacks that is just and empowering. Both the NOI and the PAOCC in distinct ways are getting at deeper questions of race and racial identity in their community food work. Those looking to diversify the movement and bring marginalized groups in might use these two cases as examples that growing and distributing food is not enough. Social justice must be at the heart of community food work because the relations occurring in it are a microcosm of the relations occurring in society. It is in this tenor that I proceed to the next section on self-reliance. Specifically, I describe how this principle is evident in both the NOI's and the PAOCC's work around food and health.

The FUBU Principle

In 1992, five black men started FUBU (For Us, By Us), a shoe company geared toward inner-city youth but more important, one that is operated and was conceived of in the inner city (Nash 2010). The founders felt that inner-city American communities and cultures were being exploited and commodified by popular shoe companies without having a sense of ownership. They were concerned with the psychological benefits that a sense of ownership has within the concept of "the black community." While community food security could not be further away from the initial conceptualization of a hip hop sneaker company, the emphasis on self-reliance is central to conceptualizing both the NOI and the PAOCC's work around food and health.

Self-reliance is a key tenet of Black Nationalism. Put simply, its definition is ownership and control for blacks over economic, political, and social institutions within the black community. It is the need for blacks to look to themselves for their means of survival. Crucial to the concept of self-reliance is that it is not individualistic, but rather "self" is seen as a community made up of blacks (Dawson 2001). Melissa Harris-Lacewell brilliantly sums up the concept of self-reliance when she defines it as the belief that while many sectors of society may by law be integrated, no one is going to look out for the best interest of blacks but blacks themselves (Harris-Lacewell 2004).

Following slavery and as a result of segregation, many blacks developed their own institutions that included farms started by groups such as the NOI, black-led religious denominations, and black social organizations. (Dawson 2001). At the core of this emphasis on self-reliance is land, which Black Nationalist discourses relate directly to power. Dawson states that "many black nationalists of the modern era continue to promote Garvey's view that black liberation could only come with the reclaiming of the land and the creation of the revolutionary Afrikan state" (Ibid., 94). This concept of a black nation adds yet another layer to both organizations' work around food and health. The NOI believes that this black nation should be a geographically separate nation within the United States; the PAOCC argues that this separation is a mental separation that does not necessitate physical boundaries.

Preaching Self-reliance through the NOI's Muhammad Farms

Self-reliance is at the core of the NOI's teachings around food and health. The definition of self-reliance comes from the core belief that "any ideological perspectives are considered to be the tools of white oppression" (Dawson 2001, 88). Many Black Nationalist organizations believe that because of this, actions must be for blacks and by blacks. The NOI argues that blacks must take care to consume an appropriate diet, but more important, one that is grown by blacks that have the best interest of their own people in mind. In *How to Eat, How to Live,* Elijah Muhammad admonishes black Christians for continuing to consume poisonous foods (Muhammad 1997b). NOI doctrine and leaders argue that self-reliance first begins by purifying "one's own body and controlling the food that enters it." They contrast this self-reliance with the "slave diet" (Curtis 2006, 102) in which blacks could not control what they put into their body. Self-reliance then begins with limiting food intake and eliminating the consumption of poisonous food.

The NOI's understanding of self-reliance based on food also includes the establishment of grocery stores and restaurants in inner-city neighborhoods. One example is Salaam Restaurant in Chicago, which did not use white flour and served mainly fish. Jesse Jackson even used this establishment for weekly meetings of his "Operation Breadbasket." Some would argue that by doing this Jackson was making a regular statement that his organization's goal of feeding people was consistent with the NOI's focus on developing self-reliance and community building through food.

Perhaps the most widespread effort of the NOI to promote self-reliance among blacks is the purchase of farmland in Bronwood, Georgia. As a part of its 3-Year Economic Savings Program, the NOI purchased 1,556 acres of land to grow a variety of organic fruits and vegetables and to eventually provide adequate housing for blacks. Again, the goal of the NOI is to ensure that all 40 million black Americans are able to have one healthy meal a day ("3-Year Economic Savings Program" 2005). Dr. Ridgley Abdul Mu'min (Muhammad), the minister of agriculture for the NOI, argues that this will increase community autonomy among blacks. Knowledge plus land plus capital equals power (Muhammad 2005). He argues that at one point blacks were beginning to create a nation within a nation, citing black landownership in the forty-five years following slavery.

The NOI's insistence on landownership is consistent with many Black Nationalists. While Black Nationalists often debate over the location, many agree that it should be in the Black Belt of the South (Dawson 2001), known as the homeland for blacks in the United States. This is partially why the NOI acquired land in southern Georgia; it is a step toward eventually building this black nation. The NOI argues that since slaves labored on this land, it is rightfully theirs. They also cite black labor through sharecropping and free blacks' farm ownership in the south in the years following slavery. Due to the USDA's unfair lending practices, much of this land has been lost (Grant 2001).

Self-reliance and Beulah Land Farms

Albert Cleage, the founder of the Pan African Orthodox Christian Church, argues for black separation as one of the only avenues for blacks to achieve self-reliance. He acknowledges that this separatism has been forced upon blacks living in a segregated and unjust society. Blacks must however use the separation that already exists and take ownership of it. The promises of integration have not been fulfilled and only require blacks to reject their blackness and adopt all aspects of a white society (Cleage 1972). Self-reliance is one of the major themes of Black Christian Nationalism. Cleage preaches that in "desperation, we turn away from the white world and to that Black Nation for strength" (62). The ultimate goal of the PAOCC's Black Christian Nationalism is Pan Africanism, but more specifically liberation. To do this, black "counter institutions must be built as a basis for black unity and power" (Ibid.,

206). Such institutions enable the achievement of tangible liberation through mental and spiritual transformation.

Beulah Land Farms is a step in this spiritual transformation. The purpose of Beulah Land Farms is to build self-sufficiency and provide this spiritual transformation here on earth. Members of the PAOCC take a holistic approach to agricultural production, not only growing fruits and vegetables but also raising poultry and cattle. An interesting production method of the PAOCC is the use of aquaculture (PAOCC 2009, 1). Keeping in line with the spiritual aspects, Beulah Land Farms is seen as an environment that stimulates a spiritual transformation and awakening. Similar to the NOI, the PAOCC is using land to increase self-reliance among its members and eventually spread this message to all blacks. Self-reliance through a community of blacks is what members of Beulah Land Christian Center are attempting to achieve.

While Beulah Land Farms is a separate physical retreat for members of the PAOCC, the ideological beliefs of the PAOCC is Black Christian Nationalism, which does not insist on the need for a true physical separation of whites from blacks. Cleage argues that if one simply identifies with the promised land geographically, one is missing its true purpose. Instead, Beulah Land Farms is less about a geographic space, and more about a community of people that share common values and a common way of life (Cleage 1972). Therefore the first step to achieving this promised land is for blacks to ascribe to a communal way of life.

"Community" in the Nation of Islam and the Pan African Orthodox Christian Church

When describing how he believed blacks should develop a new consciousness, Stokely Carmichael says that "this consciousness might be called as sense of peoplehood, pride, rather than shame, in blackness, and an attitude of brotherly, communal responsibility among all black people for one another" (Carmichael and Hamilton 1967, vii). Though Carmichael had no official affiliation with either organization, his emphasis on community responsibility and collectivity is prevalent in the religious doctrine of both the NOI and the PAOCC. More important, both groups' take on communal values and responsibility are mirrored in their work around food and health. Who is involved in the community and what goals this community must have influence their actions. From the outset, it is clear that both groups have an expressed purpose of developing communal values among black people.

Minister Louis Farrakhan wrote in the most recent *Final Call News* that "we must accept our responsibility to build our community" (Farrakhan 2009, 1). This message of communal responsibility is a part of the NOI's insistence on acquiring land for food production to eventually build a separate nation for blacks. Part of the NOI's mission on Muhammad Farms is to expose all black people to healthy food. The NOI regularly attends conferences including the Federation of Southern Cooperatives, comprised of black farmers from throughout the southern region. According to one representative of Muhammad Farms, their message is not about conversion, but rather convincing blacks that healthy eating is essential to self-improvement (Muhammad 2005).

While much of the NOI's early rhetoric focused on condemning blacks who practiced Christianity and consumed alcohol and pork, later messages attempted to reach blacks where they are. This is not to suggest that the NOI is abandoning its belief that Islam is the true and only religion of black people, but rather members argue that they must reach blacks in black Christian churches along with the nonreligious. Food is an interesting avenue to build a more inclusive community of blacks who might one day become members of the NOI. Like many Black Nationalist teachings, there is a focus on *we* and *us* in describing the problems that black people still face. When the NOI speaks about using Muhammad Farms for the purpose of nation building, they are speaking of a black nation that will support the "masses of our people" (Muhammad 2005, 1). They even go on to describe blacks in urban centers including Atlanta and Cincinnati who are using their collective efforts to provide better communities for future generation of blacks. The NOI minister of agriculture ends this message by preaching that "now we must take the lessons from our past and use the courage from the faith in God and buy land, develop our skills, save and invest our money to build a glorious future for ourselves and our children" (Ibid.).

Community is also at the forefront of the PAOCC's race-based and religious messages around food. The mission of Beulah Land Farms is to "build a community for all people that is a microcosm of the Kingdom of God" (PAOCC 2009, 1). Beulah Land Farms was only recently completed, though the land has under the PAOCC's ownership for years. The farm includes a Christian Worship Center and camp for inner-city youth. On the Web site, they are insistent on Beulah Land Farms being a place where all are welcome. Furthermore, they work with federal agencies including the United States Department of Agriculture to secure training and funds to continue their work around agriculture and aquaculture (Maxwell n.d.).

The PAOCC is a lesser-known Black Nationalist religion. From reading the sermons and writings of their founder and first Holy Patriarch Albert Cleage, it is obvious that black pride and empowerment are at the center of this Christian community experience. Cleage, who passed away in 2000, argued that all black churches and ministers must recognize that black liberation *is* the true teaching of Jesus and should be taught in all black churches. He says that "the white man's declaration of the black inferiority has served to provide for Black people a unique experience" (Cleage 1972, xxxi). Cleage goes on to preach that this unique experience calls for blacks to develop their own set of values that are not based on individuals, but are rather community values that black people develop collectively. This is what is being done at Beulah Land Farms through a holistic approach geared toward having a transformative and communal spiritual experience for blacks on earth (PAOCC 2009).

I witnessed Beulah Land Farms as described earlier by participating in farm activities. All of the PAOCC workers live on the land. Communal activities include eating together at sunset and worshiping together. Individual growth is also crucial at Beulah. Members often use their time on the farm to reflect on their spirituality. In a publication written by the PAOCC, members described their experiences at Beulah Land Farms (Djenaba n.d.). One member (Anonymous) says that "Beulah Land is a red clay soil, fresh vegetables all the time, dusty pink sunrises, and having everyone show up for all assignments and meetings. It is also driving more than ten miles to the 'neighborhood grocery' store, not to mention the new-born calves almost daily and the sweet well water to drink. Above all it is the deep refreshing sleep at night." To this member, the land and the work are peaceful and somewhat removed from society.

Exploring aspects of community is a key focus of much scholarly work about blacks. Many scholars explore community through the concept of the black counterpublic. The Black Public Sphere Collective is a group of scholars who met and wrote a book about the experience of blacks within what they call the black public sphere of the black counterpublic. They describe the black counterpublic as "a sphere of critical practice and visionary politics, in which intellectuals can join with the energies of the street, the school, the church, and the city to constitute a challenge to the exclusionary violence of much public space in the United States" (Harris-Lacewell 2004, 5). There is this recognition that blackness is a political category that was formed based on exclusion from mainstream white society. Institutions were developed in response to this exclusion, and these institutions remain relevant today. Both the NOI and the PAOCC were formed partially in response to this exclusion, but also out of a desire to build institutions that would serve the needs of blacks and help develop a separate and community-based belief system. Their work around food and health reflects the desire to build this community on earth. This community is not simply reactionary, but is a community that must be built based on the specific needs and desires of blacks. There is agency among blacks in this formation of community through food. The NOI has been actively working to create a geographically separate nation from whites, while the PAOCC believes that nation building is key, but does not entail geographic separation. It is instead psychological in an attempt to develop and maintain black communal values. Reed (1999) takes on what he calls a "communitarian mythology," saying that it causes leaders to make generic appeals to race and is not important to blacks achieving economic, political, and social progress. Reed states that blacks must go beyond an empty racial identity and look at the processes that reify racial inequalities. This idea of community or nation building by the NOI and the PAOCC seems to be an effort to do just that. Their beliefs are not based on empty rhetoric but rather a strong belief that community building is necessary for blacks to survive in a white society. The NOI and the PAOCC are actually building the values and common interests that Reed suggests blacks do through their work around food.

Community in the Community Food Movement

In "Reweaving the Food Security Safety Net," Patricia Allen delves into the concept of community. She discusses this idea of "mythical community interests" that tend to dominate community food movements, which often ignore cultural differences and assume cooperation between different groups (Allen 1999, 121). The community food movement's sense of community elevates shared geographic boundaries. An insistence on community is complicated. Similar to Allen, I argue that a community is not simply geographic. Furthermore, the NOI and the PAOCC are examples of two groups that insist (in similar and different ways) that the cultural values of blacks should be central to these organizations' agricultural work. They are using a specific geographic space but appealing to a nationwide constituency of blacks far outside of the geographic boundaries of their respective farms.

Both the NOI and the PAOCC are attempting to build community based on the common identifier of race and racial identity. Both groups recognize that blacks are not a homogenous group but still see race as a common and important connector. Albert Cleage (1972) preaches that being an individual is a choice that does not exist for black people. This is what both the NOI and the PAOCC are attempting to do through these food programs. Their methods are slightly different, but they revolve around this notion of building a community based around food that not only acknowledges blackness but also uses it as a source of empowerment.

According to both organizations, justice and sustainability can only be achieved through building a self-sustaining community. Key to this community is its members' work around food and agriculture. The NOI argues on the one hand that this can only be done if blacks form a separate geographic nation through Muhammad Farms, because no black institution can truly sustain itself when geographically integrated into a white supremacist society. The PAOCC, on the other hand, wants to use the farm to build a just and sustainable community in which members provide for themselves their own sources of water, food, energy, and peace. The food from this farm must be able to sustain black people if there are food shortages. Both groups are reconceptualizing just sustainability through their focus on racial identity in their food programs.

Conclusion

The Nation of Islam and the Pan African Orthodox Christian Church are two groups using food not only as a source of nourishment, but also as a means to define and uplift the black community. They have established farms in rural areas of South Carolina and Georgia with the intention of growing healthy food for their communities. Volunteers on the NOI's Muhammad Farms are largely NOI members who work to provide food for blacks in major cities throughout the nation. Those who work on the PAOCC's Beulah Land Farms are also largely members of this religious body who are working to address hunger worldwide and build community.

While both organizations have different religious ideologies, what ties them together is the thread of Black Nationalism and empowerment. The NOI argues that for blacks to truly achieve this power, they must form a separate geographic nation comprised only of black people in what is known as the Black Belt of the South (Muhammad 2005). The PAOCC, on the other hand, aim for peace on earth for blacks at Beulah Land Farms, and do not specifically advocate a geographic separation (New Georgia Encyclopedia 2006). Both groups are working to achieve self-reliance and feel that blacks must work as a community for this to truly be accomplished. Self-reliance is the belief that individual blacks must rely on each other to achieve true progress. The NOI and the PAOCC aim to do this through their food programs.

The food programs of the NOI and the PAOCC offer insight into the intersection of race, religion, and food. This is an important intersection that helps to illuminate the social justice components of these food programs. Future investigation is needed to determine how these food programs fit into the community food security movement as a whole. I am currently conducting interviews with individual members and engaging in participant observation at Beulah Land Farms. I hope to do the same at Muhammad Farms. The PAOCC and the NOI tell us a great deal, not only about the activities of two black religious food programs, but also how black people in these programs are addressing race and building on racial identity through food. There are undoubtedly other, similar food programs that exist that might contain differences, but also the threads of similarities. For now, these two programs offer a starting point to investigate this nuanced intersection of race, religion, and food.

Acknowledgments

This material is based in part on work supported by the National Science Foundation under award number 0902925 and also the Association of American Geographers through a Dissertation Research Grant. I would like to acknowledge participants at the 2008 Race and Food Conference at the University of California Santa Cruz, the 2009 Association of American Geographers Annual Meeting, and numerous professors at the University of Georgia for their willingness to talk these ideas through with me. Last, I would like to thank my family for their continuous support.

References

Allen, P. 1999. Reweaving the Food Security Safety Net: Mediating Entitlement and Entrepreneurship. *Agriculture and Human Values* 16 (2): 117–129.

"Background of Farm." n.d. *Visit Muhammad Farms.* <http://www.muhammad farms.com/background_of_farm.htm#Background> (accessed January 2011).

Carmichael, S., and C. V. Hamilton. 1967. *Black Power: The Politics of Liberation in America.* New York: Random House.

Cleage, A. 1972. *Black Christian Nationalism.* New York: William Morrow & Company.

Curtis, E. E. 2006. *Black Muslim Religion in the Nation of Islam, 1960–1975.* Chapel Hill: The University of North Carolina Press.

Dawson, M. 1994. A Black Counterpublic? Economic Earthquakes, Racial Agenda(s), and Black Politics. *Public Culture* 7 (1): 195–223.

Dawson, M. 2001. *Black Visions: The Roots of Contemporary African-American Political Ideologies.* Chicago: The University of Chicago Press.

Djenaba. n.d. "I've Seen the Promised Land." *Shrine of the Black Madonna 4th Pan African Synod Souvenir Booklet.* Beulah Land: Fulfilling Our Founder's Vision.

Farrakhan, L. 2009. The Final Call. FinalCall.Com News. <http://www.finalcall. com/> (accessed March 2009).

Grant, G. 2001. Letter to Bush. *The Farmer.* March 29. http://www .muhammadfarms.com/Letter%20to%20Bush-Mar29.htm (accessed December 2007).

Guthman, J. 2008. If They Only Knew. *Colorblindness and Universalism in California Alternative Food* 60 (3): 387–397.

Harris-Lacewell, M. 2004. *Barbershops, Bibles, and BET: Everyday Talk and Black Political Thought.* Princeton, NJ: Princeton University Press.

Kimathi, M. n.d. Synod 2000: Great Transitions. Shrine of the Black Madonna 4th Pan African Synod Souvenir Booklet. Beulah Land: Fulfilling our Founder's Vision. Detroit: Pan African Orthodox Christian Church.

Lee, C. K. 1999. *For Freedom's Sake: The Life of Fannie Lou Hamer.* Urbana: University of Illinois Press.

Lincoln, C. E. 1994. *Black Muslims in America.* Boston: Beacon Press.

Malcolm X. 1965. *Malcolm X Speaks.* New York: Pathfinder Books.

Marable, M. 1998. *Black Leadership.* New York: Columbia University Press.

Maxwell, A. n.d. Beulah Land Farms Come to Life in Abbeville County. <http:// www.sc.nrcs.usda.gov/news/beulahland.html> (accessed March 2009).

Muhammad, E. 1997a. Message to the Black Man in America. <http://www .seventhfam.com/temple/books/black_man/blkindex.htm> (accessed November 2007).

Muhammad, E. 1997b. How to Eat to Live. <http://www.seventhfam.com/ temple/books/eattolive_one/eat1index.htm> (accessed November 2007).

Muhammad, R. 2005. The Farm Is the Engine of Our National Life. *The Farmer.* <http://www.muhammadfarms.com/Farmer-Feb28-2005.htm> (accessed November 2007).

Nash, S. N. 2010. FUBU Founder Daymond John Stages His Next Act. Daily Finance. <http://www.dailyfinance.com/story/fubu-founder-daymond-john -stages-his-next-act/19557498> (accessed January 2011).

New Georgia Encyclopedia. 2006. Shrine of the Black Madonna. The University of Georgia Press, Athens. <http://www.georgiaencyclopedia.org/nge/Article .jsp?id=h-1630> (accessed March 2009).

Pan African Orthodox Christian Church (PAOCC). 2009. Beulah Land Christian Center. <http://blcc20.com/> (accessed March 2009).

Reed, A. 1999. *Stirrings in the Jug: Black Politics in the Post-Segregation Era.* Minneapolis: University of Minnesota Press.

Rouse, C., and J. Hoskins. 2004. Purity, Soul Food, and Sunni Islam: Explorations at the Intersection of Consumption and Resistance. *Cultural Anthropology* 19 (2): 226–249.

Slocum, R. 2006. Anti-racist Practice and the Work of Community Food Organizations. *Antipode.* <http://www.rslocum.com/Slocum_Antipode_2006.pdf> (accessed February 2009).

Squires, C. 2002. Rethinking the Black Public Sphere: An Alternative Vocabulary for Multiple. *Public Spheres* 12 (4): 446–468.

"3-Year Economic Savings Program." 2005. The official site of the Nation of Islam. <http://www.noi.org/3year-econ.html> (accessed November 2007).

West, M. 1999. Like a River: The Million Man March and the Black Nationalist Tradition in the United States. *The Journal of Historical Sociology* 12 (1): 81–100.

Witt, D. 1999. *Black Hunger: Food and the Politics of U.S. Identity.* New York: Oxford University Press.

Environmental and Food Justice

Toward Local, Slow, and Deep Food Systems

Teresa M. Mares and Devon G. Peña

Recently, the second author had a fascinating conversation with an acquaintance who identifies as a vegan activist. Living in the Pacific Northwest, she is highly committed to the Slow Food Movement and explained her philosophy of the connections between slow and local food:

If you go slow that means you also go local. Slow leads to local. I only eat local grains, veggies, fruits, and nuts. Every meal is slow-cooked from organic ingredients grown slowly by farmers that I know personally. Many are close friends and I often work on their farms for the food I need. I have become self-reliant and I have helped the local farmers become self-reliant. This unites slow and local food ethics. Together with my vegan diet, I am reducing my own carbon footprint. . . . The vegan philosophy means I am not guilty of inflicting pain on others including animals or the people who go hungry because so many of us still eat dead animal protein.

The second author then asked this vegan friend to explain more about the communities where her farmer friends live and work. All are white farmers who live in the Skagit watershed north of Seattle or the Chehalis watershed south. When asked if the vegan activist knew the names of the Native American first nations inhabiting these watersheds, her response was a disappointing surprise:

Well, in the Skagit, you know, there are a lot of multigenerational farmers who are not Native American. They have been here a long time and have as much stake in this watershed as any one else. But I don't remember the names of, you know, any tribes. I haven't met any Indians myself, so I really can't tell you much about the cultural history of the area. . . . It is also a problem with, or because of the conflicts over salmon recovery. The Indians and the farmers are fighting it out but I am not that well read on the matter.

This response came as a surprise because we naively expected that anyone with the values and ethics to become an advocate for local and slow food would also be concerned with the foodways of Native communities in a given locality. Surely, one *must* be aware of the deep history of places to practice a politics of consuming local and slow food. Is it not essential in supporting local food systems to consider the severely crippled state of local Native food systems and the forced disappearance of heritage cuisines, resulting from the impact that even the most organic, vegan-friendly settler-farmers might be exerting on indigenous resource rights?

Our vegan friend lacked knowledge of Native ethnobotany, the rich traditions related to the collection and use of wild plants recognized and valued for their nutritional, medicinal, and spiritual properties. She did not know any of the wild mushrooms in the Skagit or Chehalis watersheds that are still harvested by Native people. Camus bulbs and huckleberries? Not aware. Further, she did not seem to fully realize the impacts of modern forestry, agribusiness (including organics), and urban sprawl on the habitat of native species in the area. By only considering the direct impacts of her food consumption practices, she drastically overestimated and simplified the degree of reducing her personal ecological footprint. Lacking depth about the environmental history of the lands of the Skagit and Chehalis, she assumed that organic farmers were necessarily sustainable and equitable. Lacking deep local knowledge, she could not estimate a more accurate rendition of the "ecological footprint" she partakes in by being a beneficiary of generations of structural violence and intergenerational historical trauma experienced by Native peoples and their floral and faunal kin in the Puget Sound bioregion.

While we both respect the commitment and self-reliant ethics that often accompany attempts to eat locally and slowly, and embrace the critique of corporate globalization that originally spurred the slow food movement abroad;[1] this exchange leaves us with many questions. First, is it *deep* enough merely to consider our carbon footprints, or must we consider the broader societal and cultural footprints that we leave behind? Second, should we not also consider how a call to eat locally invokes spaces that have been settled, colonized, ruptured, and remade through complex processes of human movement and environmental history making? And finally, is it not necessary to stand in solidarity with those communities that are disallowed from celebrating *their* local food because of forced displacement at the hands of multinational trade agreements like the North American Free Trade Agreement (NAFTA) or settler-led or corporate-engineered takeover of rural lands, seeds, and livelihoods?

This introductory vignette, while indeed provocative, is only one indication of the incompleteness and imperfection of alternative food movements and the need for transformative work and critical thought in developing more just food systems. In this chapter, we seek to advance both the concept of and the movement for food justice by exploring how diasporic and immigrant gardeners mobilize deep senses of personal and collective identity while employing place-based agroecological knowledge in urban spaces. We begin with a brief critique of mainstream movements for alternative and local foods and their adherence to a food security discourse in order to problematize discussions of the local and the global. We then turn to the principles and history of the environmental justice (EJ) movement to consider how the food justice movement would be well served to integrate frameworks of sovereignty and autonomy developed by EJ activists and scholars. Through an analysis of our ethnographic data from field sites in Los Angeles and Seattle, we demonstrate how Mesoamerican diasporic and migratory peoples engage in a phenomenon we describe as *autotopography* or the grounding of self and communal identities through place making. In these cases, we consider the cultivation and celebration of meaningful food to be central to place making. We also consider how those with whom we work see food as more than a mere commodity, instead envisioning it as a relationship that forces us to stretch our understanding of what it means to grow and eat food justly.

Local Food in a Global World?

Vegans and other enthusiasts of what has often been termed *alternative food systems* do not necessarily embrace concepts of social justice or food sovereignty in their discourse and practice. Instead, the dominant constructs that direct and constrain the practices of many alternative farmers that produce for local food enthusiasts are tightly bound with the governmentalized USDA concept of "food security" and "organic certification."[2] Most relevant to our immediate concerns in this paper is the dominant concept of food security and its inability to account for an understanding of food as more than just a nutritional commodity but rather, a set of social relations and cultural practices, including foodways and heritage cuisines that constitute a larger whole (Allen 2004; DeLind 2006; Esteva and Prakash 1998; Mares and Peña 2010).

The Community Food Security (CFS) movement in the United States has answered some of the critiques raised about the concept of food security developed at the international scale, elevating concerns over the cultural appropriateness and relevance of foods and the need for systems thinking. Prominent food researcher Patricia Allen finds promise in the movement, but also raises concerns about the effects of alternative economic strategies that are found in community supported agriculture (CSA) and farmers market models and the possibility that these types of "designer" food production schemes may create a two-tiered food system built upon class differences.[3] She also critiques the movement's view that using food assistance programs is "dependence," pointing out that in antihunger perspectives food is viewed as a *right* to be fulfilled by the state if the market, or for us the self-reliant community, fails. We commend Allen's contributions in pushing the CFS movement to take into better account class inequalities and the material realities of those who are unlikely to benefit from the alternative economic arrangements most common in today's urban food landscapes.

Other researchers have contributed thoughtful critiques and reflections on the CFS movement. In her article "Whiteness, Space, and Alternative Food Practice," Rachel Slocum discusses how whiteness is produced and embodied in U.S. alternative food practices, focusing specifically on those practices that are framed within a discourse of CFS. She rejects the notion that whiteness is inherently negative, but rather questions how the ethics and politics embedded within alternative food practices might move "the US, collective, toward joy through food" (Slocum 2007, 521). While we echo Slocum's desire to imagine new possibilities and go "beyond oppositional politics" (522), we question the ways that her argument serves to reify mainstream alternative food systems as the center by which all other practices might be judged. We propose that it is essential to open an inquiry into sustainable food practices that do not operate in opposition to, but rather *autonomously* from the mainstream alternative foods movement.[4]

Slocum outlines four broad types of alternative food projects, among them efforts that seek to "protect heirloom seed stock, native plants and soil fertility in addition to advocating in-season-eating and the promotion of groups' food heritage" and those that "advocate for social justice for oppressed groups, bifurcated into producer/worker rights on the one hand and hunger and food insecurity on the other" (2007, 522). Slocum, in her much-needed interrogation into whiteness, argues, "The desire for good and sufficient food and jobs and thriving economies is not white. It becomes white through what white bodies do in this effort" (521). In reducing power dynamics to what "white bodies do," Slocum fails to provide an analysis of structural violence and its relationship with state power and its practices and technologies of governmentality. Slocum overlooks the fact that many of the practices she sees as "alternative"—including the preservation of heirloom seeds stock and native plants, regeneration of soil fertility, seasonally-oriented diet, and promotion of heritage cuisines—are precisely the traditional place-based practices of Native, Chicana/o, and other marginalized communities that we have borne witness to through our fieldwork or direct lived experiences. These are the same practices that are consistently celebrated in the discourse on local food, and claimed as "alternative," but this involves the articulation of a broader social movement based primarily on "what white bodies do." In the process, Slocum and other advocates apparently forget that these practices are already fully "alterNative"—in the sense of the deeply rooted practices of Native peoples that *alter* and challenge the dominant food system. Failing to acknowledge this alterNative source, white advocates of local food assert their privileged positionality and marginalize those who are most vulnerable to the enduring and cumulative effects of the structural violence and intergenerational historical trauma that have undermined local food systems.

Engaging communities that have been historically excluded from the mainstream alternative foods movement is critical in the movement for food justice. Within food justice, it is simply not enough to examine the ethics of going slow to go local. One *has to go deep*, and this means respecting local knowledge, wherever and whenever it is found. As discussed in our opening vignette, there is a wealth of multigenerational place-based agroecological, ethnobotanical, and gastronomical knowledge within Native communities in the United States. However, there is also a wealth of this knowledge in diasporic and immigrant communities that have faced parallel histories of colonization, displacement, and environmental racism.

We live in a time of neoliberal globalization and the mass displacement of rural place-based peoples who have been shoved away into what has been aptly described as a "planet of slums" (Davis 2006). This is a world that invokes the "end of the local and place-based" (cf. Appadurai 1996). The "end of the local" is said to be a result of the perpetual process of structural violence experienced by peoples and communities displaced from the land and into the migratory streams that bring, in our case, Mesoamerican native farmers into every major metropolitan center from Los Angeles to Seattle. A central component of this violent process derives from the effects of international trade agreements including NAFTA that have prompted massive increases in the prices of food staples, devalued local currency, opened up avenues for the dumping of

genetically modified and heavily subsidized food commodities, and devastated local communities (Patel 2007). Indeed, as a result of NAFTA,
it is estimated that anywhere from 1.3 million to more than 2 million
Mexican farmers were forced off their lands since the agreement went
into effect on January 1, 1994 (Campbell and Hendricks 2006; Patel
2007).

While trade agreements are not the only contributing factor to transnational migration and movement, they exacerbate an appalling set of
inequalities that transect the U.S.-Mexico border. These structural
inequalities continue to threaten the wellbeing of working-class, poor,
and laboring communities on both sides of the border through repeated
attacks on their natural resources, access to food, and cultural traditions.
The resurgence of far-right groups and ideologies seeking to "criminalize" transnational workers further cements the precarious and vulnerable
position of people in the new Mesoamerican diaspora (see Chacon and
Davis 2006); but it is also now leading to a mass mobilization of Mex-
ican-origin and other Latina/o people in direct-action protests and legal
challenges recently spurred by the draconian "Show Me Your Papers
Law" (SB1070) in Arizona.[5] Despite this, through our fieldwork we have
been fortunate to witness amazing examples of resiliency, autonomy, and
strength in the food practices of diasporic Mesoamerican peoples. Learning from these experiences, we believe that the food justice movement
should adopt an organizing frame of food sovereignty—including the
notion that food is not just about nutrition, it is also about culture. Food
sovereignty implies a radical ethics that derives from a commitment to
the defense and resurgence of already existing local, slow, and deep food
practices in marginalized communities. This would allow activists, scholars, and cultivators to depart from focusing on issues of access (as dictated by a food security approach), to a more comprehensive focus on
entitlements to land, decision making, and control over natural assets,
structural conditions that would allow for the process of developing
autotopographies that tie individual and collective identities to deep
senses of place and healthy, culturally appropriate food practices.

Environmental Justice Principles and Food Justice: A Necessary Connection

The Principles of Environmental Justice (PEJ) were drafted and adopted
in 1991 by delegates to the First National People of Color Environmental
Leadership Summit that took place in Washington, D.C. These principles

do not explicitly address issues related to environmental racism in our food systems. Nonetheless, movement activists have been involved in struggles for access to adequate and safe food since the earliest days of this movement through struggles against hunger and for food security. However, a more holistic understanding of urban agriculture has been a major concern of some EJ activists who have worked to connect issues of racism, food, and urban spaces since at least the 1980s (Pinderhughes 2003). Despite its exclusion from the PEJ, the theme of sustainable agriculture did appear as an area of concentration at the Second National People of Color Environmental Leadership Summit in 2002. This conference featured three separate agriculture and food-related "expert" panels and the publication of at least one preconference resource paper (Peña 2002). Prior to EJ Summit II there had been little systematic reflection and analysis of these issues within the movement's itinerary of conferences, organizational meetings, and workshops.

The main points raised by the EJ Summit II discussions on sustainable agriculture were mainly related to the loss of local food security in low-income communities of color. However, at these discussions, Peña argued for a broader framework that also addressed how policies favoring globalization and concentration of agriculture create uncertainty for local food sovereignty:

[The EJ movement] can support local struggles to establish frameworks for local participation and control of the management of these [agricultural] lands. It is important to promote a movement that focuses not just on the restoration of land rights but, equally important, the recovery of traditional systems of local natural resource management . . . To support sustainable agriculture in indigenous [and diasporic] communities, the EJM must continue to support campaigns that link policies for the restoration of indigenous land and traditional resource rights with the theory and *practice of sovereignty* (i.e., self-governance according to customary law). From the vantage of indigenous [and diasporic communities] . . . there can be no sustainable agriculture without cultural survival and political autonomy. (Peña 2002, 24; italics and brackets added)

Peña's vision of indigenous autonomous food systems moves the EJ movement beyond a narrow definition of *food security*, which treats food as a nutritional commodity, and toward a broad ideal of *food sovereignty* that encompasses the deeper social and cultural meanings indigenous and diasporic communities assign to food.

On the global stage, food sovereignty is a vision developed through and inspired by the work of La Via Campesina, an international peasant [sic] movement with members around the world. Rather than regarding food as a mere nutritional commodity, as the food security approach does, food sovereignty posits food as a fundamental human right. In doing so, the movement places food systems in the contexts of a critique of and direct resistance against neoliberalism, the processes of displacement, and the inequitable distribution of land and other resources (see Brown and Getz, chapter 6, this volume; Holt-Giménez, chapter 14, this volume). In discussing the contributions of La Via Campesina, *Stuffed and Starved* author Raj Patel argues that food sovereignty is "important not only because it has been authored by those most directly hurt by the way contemporary agriculture is set up, but also because it offers a profound agenda for change for everyone [and that it] aims to redress the abuse of the powerless by the powerful, wherever in the food system that abuse may happen" (Patel 2007, 302; brackets added). The concept of food sovereignty is closely aligned with the broad vision put forth at EJ Summit II. We believe food sovereignty's attention to power inequalities makes it the best starting place for re-envisioning food justice rooted in the practices of diasporic and immigrant communities in the United States, and a motivating challenge for food justice activists. The central rallying point of food justice should be to identify power dynamics in the food system with the goal of restoring self-determination, control, and autonomy to eaters and growers alike.

Our work with immigrant and diasporic indigenous communities in Los Angeles and Seattle revealed that practices of food sovereignty provide opportunities for the creation of autotopographies, the creation of deep senses of place. Those with whom we have worked use the cultivation of food to recreate their place-based cultural identities in the context of new landscapes. In doing so, they regard food not as a nutritional commodity but as that which encompasses a set of deep social and cultural relationships that foster community, cultural, and place-based identities.

Decommodifying Food in Autonomous Spaces: Lessons from the (Corn)field

The spaces of autonomy dedicated to building local food sovereignty are opening in thousands of local places across the world. The alterNative institutions that grassroots social movements are creating can bridge the divide separating producer from consumer while relying on collective intellectual, material, and cultural assets in order to decommodify food. As anthropologists and just food advocates, we have located our recent research on food systems within diasporic and migratory communities along the West Coast of the United States, with people who share common stories of displacement and struggles to build community and maintain place-based identities amid structural violence and oppression. Employing community-based ethnographic techniques has proven useful to understand, learn from, and participate in these communities. Recognizing and engaging local knowledge has been instrumental in struggles by EJ activists who often *contest* the superiority of Western scientific expertise to engage local and indigenous knowledge as science in its own right. As anthropologists who choose to support and make connections between these movements, it is necessary that we do precisely what we would encourage the vegan activist whose words begin this chapter to do: de-center and question our own expertise, and critically engage the expertise and knowledge systems of immigrant, Native, and diasporic communities.

In this section we present ethnographies we have generated in collaboration with Mesoamerican and Latina/o farmers in two urban locales on the West Coast of the United States. These cases illustrate the role of autotopographical practices in sustaining vigorous and culturally appropriate local food systems that address underlying issues of violence, resilience, and autonomy. In a sense, the process of crafting place-based identities determines the "depth" of local food systems. The "thicker" a sense of place, the deeper the food-related practices that a community sustains. The cases—both in urban settings—involve diasporic Mesoamerican and mestizo/a peoples who have had to adapt to massive displacements from their origin communities. In each case, communities have established autonomous spaces in which the cultivation of food becomes a way for displaced farmers to weave their place-based identities into new landscapes and to negotiate their "social citizenship" in a "safe" and "self-made" space that can offer a buffer against elements of a nation that are increasingly hostile or ambivalent about their presence. Here, we focus on the possibility that autonomous food cultivation practices enable the families and communities working in these landscapes to create and sustain decommodified relationships to food. Thus, these examples have much to teach those whose food activism is limited to their own individual consumption.

The South Central Farmers
The South Central Farm (SCF) in Los Angeles was established following the Rodney King trial and subsequent uprising in 1992 and demolished in 2006 after a three-year campaign by the farmers and a global coalition to prevent eviction and enclosure by a private land developer. At fourteen

Figure 9.1
Alameda transit corridor connecting Port of Los Angeles/Long Beach with Island Empire intermodal hub

acres, ten of them intensively cultivated, it was one of the largest urban farms in a core inner city in the United States. The former farm site is located in an area of South Central L.A. zoned for industrial and residential uses; however, the site is presently surrounded by warehouses and wrecking yards. The site is framed on its west and east sides by one of the nation's principal railroad lines that links the Port of Los Angeles in Long Beach to regional freight transit hubs and a six-lane arterial viaduct (see figure 9.1).

Originally comprised of 360 families, the SCF included U.S.-born Chicana/os and people from indigenous diaspora communities originating in communities across Mesoamerica. For thirteen years, the farmers—including families of Mixtec, Nahua, Tojolobal, Triqui, Tzeltal, Seri, Yaqui, and Zapotec descent—relied on a unique piece of urban space to grow food as they worked toward self-reliance and conviviality. In the process of creating a Mesoamerican agroecological landscape in a U.S. inner city, they developed a collective system for local food sovereignty that fostered a strong sense of place for community mobilization.

Los Angeles is a dynamic city where ancient heirloom seeds of land race *maíz*, *calabacita*, and *frijol* (corn, squash, and beans) have found their way north from Mexico along with farmers from the southern Mexican states of Oaxaca or Chiapas. These seeds trace back five thousand years to the heart of Mesoamerica and have come to grow amid the hot pavement of the urban United States, thriving in vibrant inner-city cultural landscapes across North America.[6] Family plots at the South Central Farm are perhaps best understood as the efforts of diasporic people to replicate the *huerto familiar* or hometown kitchen gardens in Mexico, Central America, Puerto Rico, Cuba, and the Dominican Republic. A comparison of the Maya kitchen garden and the typical modern family plot at the SCF reveals that Mexican gardeners continued to grow the sacred trinity of maize (*Zea mayz* L.), beans (*Phaseolus vulgaris* L.), and squash (*Cucurbita pepo* L.). They also grew avocados (*Persea americana*), bananas (*Musa sapientum* L.), and traditional aromatic and medicinal herbs that are central to the classic Mexican *hortaliza* or herb patch (see figure 9.2).

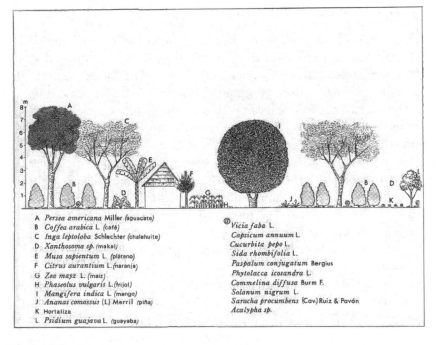

A *Persea americana* Miller (aguacate)
B *Coffea arabica* L. (café)
C *Inga leptoloba* Schlechter (chalahuite)
D *Xanthosoma sp.* (makal)
E *Musa sapientum* L. (plátano)
F *Citrus aurantium* L. (naranja)
G *Zea mayz* L. (maíz)
H *Phaseolus vulgaris* L. (frijol)
I *Mangifera indica* L. (mango)
J *Ananas comossus* (L) Merril (piña)
K Hortaliza
L *Psidium guajava* L. (guayaba)

Vicia faba L.
Capsicum annuum L.
Cucurbita pepo L.
Sida rhombifolia L.
Paspalum conjugatum Bergius
Phytolacca icosandra L.
Commelina diffusa Burm F.
Solanum nigrum L.
Saracha procumbens (Cav.) Ruiz & Pavón
Acalypha sp.

Figure 9.2
Classic Maya home kitchen garden (*huerto familiar*)

Repeatedly, farmers explained to the second author that many of the plants they have used for generations in Mexico are now only grown in urban farms and kitchen gardens in the United States. Indigenous diaspora families that were involved at the South Central Farm described how they no longer have relatives farming in Mexico and that these heirloom land race cultigens are therefore being preserved only through seed saving, planting, and cultivation practices in the United States and Canada (see Mares and Peña 2010). By asserting autonomy over their own food systems through cultivating foods they ate in Mexico, the South Central Farmers preserved both heirloom crop varieties and their own cultural identities.

The struggle of the South Central Farmers was an important example of an emerging, grassroots restoration ecology that produced formidable resistance to neoliberal enclosure and privatization of potential urban common spaces. As such, it represents an important turning point in the history, organizational forms, and terrains of struggle of the U.S. environmental justice movement. The South Central Farmers represent an example of a grassroots ecological democracy based on the integration and use of both material practices and biotic baggage (heirloom seeds and other plant germplasm) from point-of-origin communities and the reproduction of village-based forms of community self-organization. The SCF represents transnational diasporic people who were not only reshaping urban landscapes but also challenging the politics of urban planning and policy through the autonomous cultivation of food.

The vernacular foodscapes created at the South Central Farm are results of communities appropriating spaces to support urban agriculture, a pattern that is particularly important for low-income immigrant communities.[7] *El jardín* (garden) is a space for the charting of individual autotopographies—self-telling through place shaping. This is certainly true of the classic home-based kitchen gardens that were grown at the SCF, and those that continue to spread across the urban United States. These *jardincitos* are spiritual and political symbols of a process involving nothing less than the *re*-territorialization of place as a home by transnational communities.

In the second author's interviews with farmers at the South Central Farm about why they garden, many replied with the same set of reasons: to supplement the family food budget; grow ingredients for traditional recipes; grow organic (meaning to grow one's own food in order to know where it comes from and that it is fresh); visit with friends and family members, learn about traditional foods from elders; feel more at home;

and to grow herbs and vegetables to supply family businesses. These kinds of gardens that are appearing across the West Coast are examples of emergent forms of urban spatial resistance. These struggles emerge through the process of autotopography—and unfold in communal spaces that nurture conviviality. One gardener at the SCF, a thirty-year-old Zapotec woman, described her involvement at the farm in the following way: "I planted this garden because it is a little space like home. I grow the same plants that I had back in my garden in Oaxaca. We can eat like we ate at home and this makes us feel like ourselves. It allows us to keep a part of who we are after coming to the United States."

However, the production of heritage foods and familiar landscapes is only one part of the practice of conviviality in this urban agroecosystem. The SCF illustrates the importance of the "production of meaning." The second author was introduced to Mixtec (and Zapotec) traditions of storytelling underneath one of the urban farm's *pochote* trees (*Chorisia speciosa* or silk floss tree). The pochote is considered a sacred tree among the indigenous peoples of the Mixteca bioregion. Every weekend as families gathered, the children sat under the pochote to hear stories narrated by elders. Many of the stories were related to deer and deer hunting (deer are also considered sacred beings and spirit guides to other dimensions of reality). When the bulldozers arrived in June 2006, the pochote trees were among the first of the profuse vegetation to be protected by protestors. The second author asked a young Nahua woman involved in the SCF protests why they were protecting the trees. Her response illustrates our point about the need to connect local and slow food with a deep sense of place, community, and agroecological practices: "This tree is sacred. . . . The sacred tree is where we can gather to pray and share stories. Without our gathering under these trees, the garden cannot be happy. The corn needs the pochote tree to be happy so the corn silk will not wither. Our children learn the ways of our people by making this tree part of their place in the world." Again, cultivation is tied to a deep sense of place and community identity.

Local autotopographical spaces like the *huertos familiares* at the South Central Farm are constructed in conscious opposition to the global commodity chains that constitute the dominant food system. But this process is both internally heterogeneous and highly contested. In the case of the SCF, which until its destruction was still officially administered by the regional food bank, one example of the contested nature of a communal space was seen in the challenges of managing a few acres of urban land to support the food production activities of some 360 families. The result at the SCF was the division of the limited acreage by family plots of fairly uniform size (each approximately 200 to 260 square meters). These were divided from each other by a maze of lower-grade chainlink fencing that has been improvised over time.

One remarkable feature of the South Central Farm was the profusion of cactus corridors or cacti fencing growing alongside the chainlink fencing, representing a transition to a more culturally appropriate division of the space through a permaculture feature similar to the *nopal* (a type of edible cactus) fences that are common traditional fixtures across the rural landscapes of northern Mexico, especially the states of Chihuahua, Sonora, Coahuila, and Tamaulipas. One farmer reported that many involved at the SCF wanted to bring down the chainlink fences altogether and were beginning to replace them not just with cactus but also sugarcane, banana, avocado, and other fruit-bearing trees and shrubs. Other growers were using edible vines to cover the fencing. These efforts allowed for a more natural (vegetation-based) set of boundaries that all gardeners could enjoy and use, in effect, transforming the fence into an element of an edible landscape rather than a set of boundaries. "Edible fences make good neighbors," one farmer told the second author. These features were not only a means to enhance the edible landscape and challenge an exclusionary understanding of space and boundaries, but also to integrate meaningful and familiar foods into U.S. soil in recognition of their uses beyond mere nutrition (see figure 9.3).

The struggle of the South Central Farmers was at the center of a widening conflict over an urban commons that arose from the political economic context of contested urban land-use politics. The misguided tendency of municipal planning authorities to overvalue urban spaces for commercial/industrial uses over all other uses is the deeper cause of this conflict (see Diaz 2005). The farmers at the SCF faced a crisis embedded in the contested legal status of the land as property, which defined it as a space that should be developed for commercial and industrial uses, discounting the economic, ecological, and cultural value of this place to the community.

Some 360 families from the urban farm were finally evicted in June 2006. This was the second time that many of the South Central Farmers experienced this sort of violent displacement from the land. However, and this is the point of the "resilience" of Mesoamerican diaspora communities, instead of "disappearing" after eviction, the families maintained a weekly vigil and monthly tianguis (farmers market) on a site across from the old garden. More still, the SCF has developed into a new

Figure 9.3
Garden plot at South Central showing plants and features similar to the classic Maya garden in figure 9.2

501(c)(3) organization that includes an agricultural cooperative producing "Food for the Hood" on eighty acres of irrigated land outside Bakersville, California. The resurgence of the farm, while now physically relocated from its original central urban locale, means that the SCF organization is still effectively pursuing food sovereignty for the member families and their urban communities. This farm-to-table connection remains a vital force today in sustaining the heritage cuisine practices of the displaced families.

Puget Sound Urban Farmers
The Seattle area is quickly becoming a vibrant center of alternative food movements. With nearly ninety community gardens coordinated by the city's Department of Neighborhoods, fourteen farmers markets operating throughout the city, and a city council that approved a Local Foods Action Initiative in 2008, it is clear that there is strong institutional support for transforming the food system into one that is more sustainable and profitable for local food producers. In the Seattle foodshed,

there are also fascinating examples of the decommodification of food operating against and within mainstream alternative market frameworks. Diasporic, migrant, and immigrant communities from Latin America, East Africa, Russia, and all regions of Asia are transforming the urban landscape into spaces that look more like *home* through practices of autotopography and place making. Sometimes these spaces are state- or city-sanctioned spaces for growing food (like Department of Neighborhoods P-Patch community gardens, sometimes they take place in home kitchen gardens, and sometimes they operate as guerrilla gardens—what we have previously described as insurgent uses of public space (Mares and Peña 2010).

This transformative use of these spaces includes cultivating food crops that are "culturally appropriate," but equally important, cultivating them through practices like intercropping, biointensive gardening, and terracing. These are all agroecological techniques that the permaculture-embracing, mainstream-alternative community yearns to learn and employ. The key challenge that the authors see is the need for clearing a space for celebrating and honoring the rich sets of knowledge that immigrant communities possess without allowing for the possibility of cooptation or appropriation. These growers are scientists and experts in their own right, and if maintaining autonomous spaces is necessary for the continuation of these practices, then standing in solidarity with these growers should be the priority of just food activists and scholars.

Three key ethnographic vignettes will help to demonstrate the ways that autonomous spaces can be sites of decommodification and autotopography. Since 2005, the first author has been involved in ongoing field research that explores both the strategies and networks that Latino families and individuals use to define and meet their food needs, and how nonprofit and governmental institutions and agencies are conceptualizing and responding to or neglecting the needs, or both, of this community. This research involved participant observation and informal interviews with Latino growers who are using community gardens as spaces of food production, conviviality, and community building. In one attempt to ground the field study as an applied project relevant to the local community, the first author volunteered with a local nonprofit organization to coordinate organic gardening classes in Spanish, though with a less than adequate degree of success.

During the first year of research, the first author was chatting with Octavio,[8] a gardener from Mexico, at a community event. The first author asked him if he had attended any of these gardening classes in the past, hoping that he might be interested in the upcoming classes. His response was significant, as it reveals a direct example of the need for legitimizing and honoring agricultural knowledge that runs counter to the mainstream alternative foods movement. Coming from a small agricultural town in central Mexico, Octavio's identity has long been shaped by his relation to the land. When recalling his participation in the gardening classes the previous year, he explained how he was a bit offended that they tried to teach him, *un Mexicano*, how to grow corn. He pointed to a small child around four years old while saying that he had been growing corn all his life, and that his family had done the same before he was born. In fact, the seeds of the corn that he was growing in his garden plot had been sent to him through postal mail by his mother in Mexico. In the class, he was instructed to plant his seeds much deeper than he *knew* was necessary, and with great pride, he took the first author over to see his corn plants that were taller than any other plants in the garden. It was not just corn that Octavio was growing, but corn that was familiar, necessary, and meaningful. The fact that these plants were thriving in an environment drastically different from the one where they originated not only illustrated his deep knowledge, but also that these seeds were more than food commodities in waiting, but rather, kin to be nurtured. As importantly, his refusal to follow directions, deciding instead to employ his own place-based knowledge demonstrates that autonomous food practices are at work in the Seattle food system. The knowledge systems and autotopographical practices of these autonomous growers must be recognized if the movement for local food is to successfully integrate demands for food justice.

A second vignette closely follows and brings the story full circle. At this same community event, there was a massive table of snacks and refreshments, largely donations from local businesses, but also some foods that were grown in the community garden where we were celebrating. Piles of sugar cookies, hot dogs, salmon burgers, and prepackaged chips and salsa dominated the table, but it was the homemade offerings that were the most celebrated and valuable. Nearby, a cider press was cranking apples to mush and a grill was being fired with the sporadic help of neighbors. After roasting a few ears of his much-prized corn on the grill and sharing them with the first author and his young daughter, Teresa's newfound friend Octavio revealed how proud he was that his daughter preferred the corn, *his corn*, over the highly sweetened and processed "American" foods available. He looked at her with joy as she skipped away to join the other children, cob clutched tightly in hand.

This exchange reveals the importance of shaping and passing down heritage cuisines that are tightly bound to cultural identity. It is this deep connection with food that we must cultivate if food justice, as a movement, is destined to succeed.

The third vignette connects our field sites in a deep and inspiring way. In the winter of 2006, deep in the midst of the struggle at the South Central Farm, Tezozomoc, a central organizer with the South Central Farmers, was invited as an honored guest to speak at several engagements at the University of Washington. After these events were over, all parties involved were aching to get out of the academic confines of the university and visit gardens in the Seattle area. The two authors of this piece coordinated a visit to an urban farm in Seattle so that Tezozomoc could see the work of his colleagues in Seattle. During a particularly rainy day, we were guided through the farm by Mauricio, a young father who had moved to Seattle from Mexico several years prior, along with his two children. While the children raced through the damp beds and made plans for what they would plant the next spring, Tezozomoc told Mauricio about the events that were coming to pass at the South Central Farm, urging him to do whatever he needed to do to protect this parallel space in Seattle in the event that the farm ever faced similar pressures to those of the South Central Farmers. Mauricio responded that he would do so, and with great seriousness, asked Tezozomoc if he had been growing any *papalo* in his garden in Los Angeles. Mauricio, while struck by the events taking place in Los Angeles, was equally concerned that he was having a hard time getting hold of any seed for this plant in Seattle. Tezozomoc laughed, saying that he was growing acres and acres of *papalo*,[9] and that he would be sure to send some seeds up north for Mauricio and his compadres to grow.

A few months later, the second author of this piece presented the first author a large bag of *papalo* seeds that Tezozomoc had passed along to him during a recent trip to Los Angeles. It was an impressively heavy bag of seeds, definitely enough for the farmers in Seattle, and then some. The first author couldn't wait to take these seeds to Mauricio, to complete the next step in a long line of seed sharing that had begun in the fields in the South Central Farm, and possibly, to a shared homeland even further back. As she handed this bag of seeds imbued with deep meaning and significance to Mauricio, she realized that her research was about so much more than just food—*alimento*. It was about *comida*. Gustavo Esteva and Madhu Suri Prakash have this to say about *comida*: "There is no English word for *comida*. It is not easy to explain why.

Thinking of that makes us feel sad. While "feast" comes closest in its implication of eating together, it refers only to a special occasion, while *comida* is eating by the "social majorities" in the "normal" course of every day. Perhaps we need to recall that the Anglo-Saxon world was the cultural space in which the industrial mode of production was established first and foremost. There, vernacular activities related to *comida* have been suffocated or suppressed" (Esteva and Prakash 1998, 59).

In some small way, despite the hundreds of miles that separated them, Tezozomoc and Mauricio were sharing *comida*, bound together in a global diaspora that is seeking new homes in faraway places. With their communities, they are struggling for food justice through building networks of food sovereignty.

As with the South Central Farmers, these urban growers in the Seattle area reveal an alterNative and decommodified relationship with food. Interestingly, all three vignettes took place in an urban site that has been formally protected by the city of Seattle for the purpose of growing food in community garden spaces. This space is one of two remaining pieces of farmland remaining in the city, and it has a long history of being cultivated by immigrant gardeners from all ends of the earth. These growers mentioned here are just a few of the many immigrant gardeners across the Puget Sound region that are using urban spaces in deeply autonomous ways to both create and maintain close cultural ties through food.

Conclusion: Rebalancing Power in the Global Food System

Scholars and cultivators need to depart from focusing on issues of equal access as dictated by the food security paradigm. Instead, we must employ a more comprehensive food sovereignty framework that allows us to support autonomous struggles for the exercise of entitlements to land and community-based decision making, along with democratic and participatory control over our local natural assets. Doing so will enable a deep connection with what we eat, learning from the autotopographical practices and deep senses of place like those we have discussed here.

The central claim of a food sovereignty framework is that food is a right, not a commodity. What would it look like if we issued this statement as the first demand of our food justice movement? What if we thought of this not as an individual right, but rather, took an alterNative approach to embrace the self-provisioning of food through locally grounded cooperative union and mutual aid? What if we followed the path of movements like La Via Campesina, the South Central Farmers, and the actions of Latino growers all over Seattle to reclaim space, identity, and food sovereignty? Perhaps then we wouldn't ask permission from the state to be "free" and instead we would create our own sovereign freedoms through direct organizing and community-based action.

The demands of the food justice movement should necessarily resist further industrialization and globalization of our food system since the emphasis should remain on place-based self-provisioning and demands to restore more autonomous forms of food sovereignty. Reestablishing and reinventing heritage agroecosystems would entail a reduction in the production of exotic crops for cash-export markets and a prioritization of local food self-sufficiency. It would also entail elevating the knowledge and expertise of those growers who have autonomously maintained these heritage agroecosystems to the same level of the "experts" who insist on technological and scientific remedies to food problems in the United States and abroad. The decommodification and relocalization of food systems are two critical elements of any truly just and sustainable agriculture and food policy. Of course, this will also require that we punch deep holes in the arguments of the naysayers who claim we cannot feed the world without a reliance on industrial mass production of food (for an excellent example, see Lappé, Collins, and Rosset 1998).

Perhaps the most enduring way to rebalance power in the global food system is by supporting struggles that move us toward the decommodification of food through support for marginalized peoples' autonomous cultivation practices. Both the farmers at the South Central Farm and the Latino growers in Seattle have demonstrated to us the viability of nurturing sustained and deep connections with all that nourishes us, tying their place-based community identities to new landscapes through processes of autotopography. In following their example, we should shift toward the "local"—in the sense of a spatial reorientation of the food system from global commodity chains toward local, bioregional food systems that both follow and facilitate deeper senses of place. We must also consider how to make resistance against the global systems that commodify food. This will require combining the proactive forces that are already rebuilding local food systems with political pressure for the United States to rebalance global power.

Our proposals for the food justice movement require a more radical set of practices that lead not so much to a restructuring as to an autonomous and reiterative geography of relocalization that supplants the dominant global food system. They require that we collectively strive for deep connections—with our food, with the places we live, and with each other. Finally, they require that we simultaneously challenge the avarice-driven hunger for profit of transnational agribusiness corporations while consciously rebuilding our place-based local food systems. This must be done in solidarity with others around the world who share our hunger and thirst for justice.

Acknowledgments

We would like to everyone who has contributed to this chapter, including our research participants, our students who continue to push our thinking into new directions, and the editors of this volume. Special thanks go to Yecelica Valdivia, whose undergraduate honors thesis offered us a new way of thinking about the possibilities of food justice and food sovereignty. Teresa Mares's dissertation research for 2009–2010 is generously funded by the Stroum Dissertation Fellowship from the University of Washington and an American Fellowship from the American Association of University Women. The views expressed by the authors do not necessarily reflect the official policies or views of the University of Washington or the Acequia Institute.

Notes

1. Here, we refer to the reaction against the opening of a McDonalds restaurant in the Pizza di Spagna in Rome, widely cited as the motivating factor behind the Slow Food movement abroad. For an excellent discussion of Slow Food's more radical beginnings, see Katz 2006.

2. Many writers have offered thoughtful critiques of the limitations of and contradictions within organic certification, including Guthman (2004), Katz (2006), and Patel (2007).

3. In Seattle, one only has to visit the neighborhood farmers markets to see Allen's concern manifested in five-dollar loaves of organic artisan bread.

4. Here, we are following Peña (2003) who first proposed that environmental justice theory and practice need to consider differences between equity-based models of environmental justice and an alternative approach that emphasizes autonomy. This argument suggests that equity-based theories of environmental justice are reformist and integrationist and that they fail to rupture or challenge the dominant system that emphasizes individual rights to equal opportunity. This requires the cooperation and legitimation of the state, and the evidence suggests that this is insufficient as a basis for the attainment of environmental justice. The autonomy model emphasizes cooperative and communal orientations, rather than individualistic ones, and encourages perspectives derived from place-based

knowledge that can promote environmental self-determination as a matter of local practices that emerge from place and are free of the need to gain the acceptance or endorsement of the state. These "alterNative" food systems developed in and continue to exist within the "margins" among those displaced, native and ethnic or working-class, communities that have always valued food self-sufficiency as a matter of heritage or survival or both, and have usually thus of necessity maintained the knowledge, methods, and materials—if not always the land base—to sustain their local, slow, and deep foodways.

5. SB 1070 is an Arizona law that criminalizes the undocumented as felons and imposes requirements for the proof of citizenship or legal status that have been challenged in the courts. For more on the emerging mass-based social movement against SB1070 and for immigrant rights, see the various news and blog entries at <http://mexmigration.blogspot.com>.

6. In Vancouver, British Columbia, Mayan diasporic people have created an impressive home kitchen garden project that utilizes unused open space by collaborating with University of British Columbia faculty and students and the city's elected officials.

7. See Klindienst 2006, Peña 2002, Pinderhughes 2003, and Saldivar-Tanaka and Krasny 2004.

8. All names of the growers in Seattle are pseudonyms.

9. Also called *papaloquelite* or summer cilantro, *porophyllum ruderale* is an herb commonly used in Mexican cooking.

References

Allen, Patricia. 2004. *Together at the Table: Sustainability and Sustenance in the American Agrifood System*. University Park: Penn State University Press.

Appadurai, Arjun. 1996. *Modernity at Large: Cultural Dimensions of Globalization*. Minneapolis: University of Minnesota Press.

Campbell, Monica, and Tyche Hendricks. 2006. NAFTA and Dumping Subsidized Corn on Mexico Has Driven 1.5 Million Farmers off the Land & Forced Millions to Migrate. *San Francisco Chronicle*, July 31. <http://www.organic consumers.org/articles/article_1371.cfm> (accessed March 2, 2010).

Chacon, Justin Akers, and Mike Davis. 2006. *No One is Illegal: Fighting Violence and State Repression on the U.S.-Mexico Border*. Chicago, IL: Haymarket Books.

Davis, Mike. 2006. *Planet of Slums*. New York: Verso Press.

DeLind, Laura B. 2006. Of Bodies, Place, and Culture: Resituating Local Food. *Journal of Agricultural & Environmental Ethics* 19 (2): 121–146.

Diaz, David R. 2005. *Barrio Urbanism*. New York: Routledge.

Esteva, Gustavo, and Madhu Suri Prakash. 1998. *Grassroots Postmodernism: Remaking the Soil of Cultures*. London: Zed.

Guthman, Julie. 2004. *Agrarian Dreams: The Paradox of Organic Farming in California*. Berkeley: University of California Press.

Katz, Sandor. 2006. *The Revolution Will Not Be Microwaved: Inside America's Underground Food Movements*. White River Junction, VT: Chelsea Green Publishing.

Klindienst, Patricia. 2006. *The Earth Knows My Name: Food, Culture and Sustainability in the Gardens of Ethnic Americans*. Boston: Beacon Press.

Lappé, Frances Moore, Joseph Collins, and Peter Rosset. 1998. *World Hunger: 12 Myths*. New York: Grove Press.

Mares, Teresa, and Devon Peña. 2010. Urban Agriculture in the Making of Insurgent Spaces in Los Angeles and Seattle. In *Insurgent Public Space*, ed. Jeffrey Hou, 241–254. New York: Routledge Press.

Patel, Raj. 2007. *Stuffed and Starved: The Hidden Battle for the World Food System*. Brooklyn, NY: Melville House Publishing.

Peña, Devon G. 2002. Environmental Justice and Sustainable Agriculture: Linking Social and Ecological Sides of Sustainability. *Occasional Paper Series*, Second National People of Color Environmental Leadership Summit, Washington, DC, October 23–27. <http://www.ejrc.cau.edu/summit2/SustainableAg.pdf> (accessed May 5, 2009).

Peña, Devon G. 2003. Autonomy, Equity and Environmental Justice. In *Power, Justice, and the Environment: A Critical Appraisal of the Environmental Justice Movement*, ed. David Pellow and Robert Brule, 131–152. Cambridge, MA: MIT Press; London: Earthscan.

Pinderhughes, Raquel. 2003. Poverty and the Environment: The Urban Agriculture Connection. In *Natural Assets: Democratizing Environmental Ownership*, ed. James K. Boyce and Barry G. Shelley, 299–312. Washington, DC: Island Press.

Saldivar-Tanaka, Laura, and Marianne E. Krasny. 2004. Culturing Community Development, Neighborhood Open Space, and Civic Agriculture: The Case of Latino Community Gardens in New York City. *Agriculture and Human Values* 21 (4): 399–412.

Slocum, Rachel. 2007. Whiteness, Space, and Alternative Food Practice. *Geoforum* 38 (3): 520–533.

10

Vegans of Color, Racialized Embodiment, and Problematics of the "Exotic"

A. Breeze Harper

In 2007, Johanna, a vegan and woman of color, established an online forum called *Vegans of Color*. Resisting mainstream notions of veganism as separate from race/racism/racialization, the founder of the group stated: "This blog was started to give a voice to vegans of color. Many vegan spaces seem to be assumed (consciously or not) to be white by default, with the dialogue within often coming from a place of white privilege. We're not single-issue here. All oppressions are connected." From dealing with the problematic of "exotifying" nonwhite vegan foods as objects for the "white Eurocentric gaze," to sharing narratives about the difficulties that vegans of color (VOCs) have in largely white vegan and animal rights events, the *Vegans of Color* blog and forum represent the racialized experiences of nonwhite as well as white vegans. Since its inception, eleven more vegans of color, including the author, have joined Johanna as contributors. What is it about racialized and class experiences in the United States that lead some proponents of veganism to take a "colorblind" (Bonilla-Silva 2006) approach to alternative food philosophies such as veganism, while for others issues of race and class are not separate from food politics and practices?

In the global West, white-identified people who engage in activism such as alternative foodways are stereotyped as radically leftist and progressive, incapable of participating in the overt racism one can normally find within radical right extremist white-bodied organizations (Poldervaart 2001; Clark 2004). However, deeper scholarly research has found that, collectively, "good whites" tend to shy away from antiracism and reflections on white and class privilege within alternative food movements (Slocum 2006) and animal rights activism (Nagra 2003; Poldervaart 2001). This is noteworthy, as white middle-class individuals dominate the alternative food movement in the United States. Previous research, however, has ignored veganism and animal rights as important components of alternative food movements. In this chapter, I will analyze the ways that racial privilege manifests in the vegan and animal rights movement through discursive analysis of online dialogues found on the *Vegans of Color* blog. I am particularly interested in the ways that racial privilege can affect the experiences of vegans of color living in white-dominated societies.

Practitioners of veganism abstain from animal consumption and a majority support animal rights. However, the culture of veganism itself is comprised of many different subcultures and philosophies ranging from strict vegans for animal rights, to people who are dietary vegans for personal health reasons, to people who practice veganism for religious and spiritual reasons (Cherry 2006; Iacobbo and Iacobbo 2006). The Vegan Society, the organization that coined the term *vegan*, states that the heart of veganism is the practice of *Ahimsa*, or "compassion, kindness, and justice" for all living beings:

Ahimsa is a Sanskrit word for non-killing and non-harming. It is not mere passiveness, but a positive method of meeting the dilemmas and decisions of daily life. In the western world, we call it *Dynamic Harmlessness*. The six pillars of this dynamic philosophy for modern life (one for each letter: A-H-I-M-S-A) are: ABSTINENCE FROM ANIMAL PRODUCTS; HARMLESSNESS WITH REVERENCE FOR LIFE; INTEGRITY OF THOUGHT, WORD, AND DEED; MASTERY OVER ONESELF; SERVICE TO HUMANITY, NATURE, AND CREATION; ADVANCEMENT OF UNDERSTANDING AND TRUTH. (American Vegan Society n.d.; emphasis in original)

Despite this inclusive definition, veganism is associated with white people of privilege. According to Andrew Rowan, a vice president at the Humane Society of the United States, surveys indicate that the Animal Rights movement is "less than three percent" people of color (Hamanaka and Basile 2005). A 2005 Grassroots Animal Rights Conference in New York drew 316 attendees, but the people of color caucus numbered a mere eight (Ibid.). And vegan/vegetarianism ranks no. 32 on Christian Lander's (2009) satirical list of *Stuff White People Like*. Observers have noted a similar preponderance of whites within other radical forms of activism including the antiglobalization movement and resistance to the prison industrial complex (Appel 2003; Clark 2004; Nagra 2003; Poldervaart 2001; Yancy 2004).

Most vegan texts ignore issues of race and class completely. For example, the *New York Times* best-selling vegan book series *Skinny Bitch* promotes veganism as a way for women to lose weight, be healthy, and alleviate suffering of nonhuman animals. These books' assumed audience is white middle-class heterosexual females living in locations where a whole-foods vegan diet is easily accessible (geographically and financially). The book *Skinny Bitch: Bun in the Oven,* which focuses on pregnancy, begins each chapter with a picture of a white skinny pregnant woman. Throughout the text, the authors implicate personal laziness as the reason people are overweight. There is never any reflection on how (1) class and the racialized experience in the United States affect a pregnant woman's access to healthful food and nutritional information, and (2) how the authors' white, racialized, and class-privileged consciousness influence their perception of veganism as the answer to obesity problems. Though the authors' intent for the book was not to focus on race- and class-based experiences of veganism and pregnancy, the absence of this personal reflection and assumptions made about their audience are intriguing and quite telling. The most widely read vegan magazine in the United States, *VegNews,* declared Rory Freedman, one of the coauthors of *Skinny Bitch,* as vegan "person of the year."

Texts such as the *Skinny Bitch* series engage in a "colorblind" approach to food politics that ignores the affects of race and class on an individual's circumstances. In a colorblind or "post-racist" society, it is believed that racism no longer exists because skin color no longer has social significance. Throughout this chapter, "colorblind" will be written in quotations to reflect Jerry Kang's theory of colorblindness equaling "expected whiteness" (Kang 2000). For example, if a white person were to tell his or her Chinese friend, "I don't think of you as Chinese, I am colorblind," Kang would argue that this Chinese friend would not be seen as race-neutral, but in fact would seen by the colorblind friend as "any other [white] person." This concept is part of a larger body of scholarly work around the issues of whiteness and white privilege. Whiteness is

The ability of Whites to control the cultural discourse of racial equality—colorblindness rhetoric and "individual-group sleight of hand"—as well as Whites' socialization to, and insistence upon, social preeminence. Whites operate within a "comfort zone" that renders Whiteness "normal." And when displaced, Whites often employ strategies that reinstate Whiteness at the center. Here the metaprivilege of Whiteness resides in the "absence of awareness of White privilege". . . . Whiteness does not acknowledge either its own privilege or the material and sociocultural mechanisms by which that privilege is protected. White privilege itself becomes invisible. (Flagg 2005, 5–6)

Whiteness is to know and move through a white-dominated nation such as the United States in a manner in which culturally familiar objects, spaces, and places are deemed racially "neutral." However, within nations that have a history of racialized colonialism, such objects, spaces, and places are often culturally specific to whites. Dwyer and Jones explain whiteness to be a type of white sociospatial epistemology. Such an epistemology is the ability for the white collectivity of the United States to speak from and to "survey and navigate social space from a position of authority" (2000, 210), with the assumption that their epistemologies are applicable to all peoples. Geographer Julie Guthman refers to the ability of whites to ignore the cultural specificity of their histories and experiences as universalism (2008; Guthman, chapter 12, this volume).

In contrast to such "colorblindness" is an approach to food politics grounded in an understanding of "racialized consciousness" (Farr 2004). Racialized consciousness recognizes that decisions about ethics are influenced by an individual's consciousness, and that consciousness is affected by lived experiences of sexual orientation, gender, socioeconomic class, and race (Mills 1997; Salamon 2006). In societies in which racialization is part of a nation's foundations, including its social contracts, economic systems, and citizenship, a person's consciousness and how one creates philosophies will be significantly shaped by one's lived experience of race within that society (Mills 1997; Yancy 2004; Sullivan and Tuana 2007). Hence, even engagement in a consumption practice intended to promote justice, such as veganism, will be touched by one's racialized experience. One of the key benefits of white privilege is that its bearer need not recognize the affects of race and racism on their own and others' life circumstances. It is this very white racialized consciousness of the overwhelmingly white U.S. vegan movement that guides the assumption that serious dialogues around race, racism, whiteness, and racialized colonialism are unrelated to its goals. Although there are many books about veganism and animal rights that speak of speciesism[1] as comparable to racism (Patterson 2002; Spiegel 1996; Torres and Torres 2005), the vegan and animal rights movements have been reluctant to question how racism and racialization influence an individual's perception and praxis of veganism and animal rights.

In opposition, the nonwhite racialized consciousnesses of nonwhite vegans has led them engage in vegan food politics that intimately intertwines factors of race/racism/racialization. In the book *Sistah Vegan: Black Females Speak on Food, Identity, Health, and Society*, many of the volume's contributors aim to disrupt the myth that veganism is not solely the domain of white people (Harper 2009). Similarly, Bryant Terry's (2009) *Vegan Soul Kitchen: Fresh, Healthy and Creative African-*

American Cuisine approaches vegan cooking through a racialized lens. *Vegetarians and Vegans in America Today* contains a brief discussion of racism and the racialized experiences surrounding veganism and animal rights among African Americans (Iacobbo and Iacobbo 2006, 60–70). These texts provide a starting point from which vegans of color can explore the intersections of vegan philosophy, whiteness, and race/racism/racialized consciousness.

This chapter traces various ways that whiteness manifests in the vegan/animal rights movement, as well as responses to whiteness from vegans of color. The next two sections analyze dialogue from the popular forum *Vegans of Color*. In the first, I explore the various racialized and "colorblind" meanings attributed to the word *exotic*. Following that, I examine a conversation about how the embodied experience of being nonwhite in white-dominated spaces is connected to emotional distress and discomfort. The blog offers ample examples of vegans of colors' collective discomfort with, and resistance to, the manifestations of whiteness and white privilege. In contrast with important previous work on whiteness that aims to understand "what whites think about being white" (McKinney 2005, 4), this chapter examines the effects of colorblind discourses on activists of color who have embraced veganism.

Eating the "Exotic"

The *Vegans of Color* blog has featured a number of discussion threads in which writers who experience the objectification of their bodies, cultures, or homelands, or all of these, have heavily scrutinized and critiqued notions of the "exotic." On August 15, 2008, Johanna began one such discussion thread, called "The Exotifying Gaze," with the following statement: "I am really uncomfortable with how a lot of vegan cooking is described as 'exotic' (to whom?). It assumes so much about the audience racially & culturally, & as well is loaded with really creepy connotations—the exotic is there to be conquered, mastered; it's there purely to titillate your (white/Western/etc.) self."

Another blogger, who goes by the initial S, shares a similar critique: "My BIG ANNOYANCE is that I get to see my food, the food I grew up with, labeled as 'exotic' and I get that exotic means different but it's usually a white vegan blogger writing for a white vegan audience and so MY FOOD is being called different." These bloggers argue that the word *exotic*, when assigned to foods from communities of color or the global south or both presumes a white audience, marginalizing the subjectivities of vegans of color. Describing such foods as exotic signifies a "color-blind" approach in which racial consciousness is ignored, and is read by these bloggers as presumed whiteness (Nakamura 2002).

White vegans participating on the VOC listserv, however, had very different reactions. Some, like the post that follows from Gary, a white-identified male blogger, professed their naïveté to the preceding concerns and asked for further clarification.

I was looking back over my recent writings, and I think sometimes I was a bit careless in my use of "exotic." I will watch out for that in the future. What would be examples of acceptable uses of "exotic"? In a couple of instances I used "exotic" in a tongue-in-cheek manner, e.g., calling sweet potato chips "exotic" because so many people don't vary from plain old potato chips. (Not exactly a knee-slapper, I know.) But does even using it like that subtly reinforce the negative connotations of the word?

Such naive comments align with the complaint, often raised by antiracist activists, that whites expect people of color to teach them about the effects of race and racism.

Other comments by white bloggers, however, are read by vegans of color as attempts to undermine the validity of their experiences. Kram, for example, a white female blogger who had stated in a previous post that she was unfamiliar with critical race or postcolonial theory, offered an example of this perspective:

Take the case where someone prepares something for a potluck not from their culture [, i.e.] Mexican food and they label it "Mexican." What if there are Mexicans there and realize . . . oh that's Incan Food, or Indian food, or Mayan Food. Then the label actually offends someone because it is attributing something to their culture/race that is not theirs. In that case the person bringing their dish offends people and if they are called out on it . . . might not bring a foreign dish again due to potential conflict.

My curiosity is this. Is the labeling and importance of every word or term hurting the causes of veganism?

Again my example of AR events in the previous post . . . if I were to ever be called out on terms of "white guilt" or "colonialist" or other terms for trying to go to events that are more inclusive of POC [people of color] or run/by or sponsored by POC, then I will not be inclined to participate in those events.

From an antiracist or decolonial perspective, Kram's post is problematic for several reasons. She begins by equating the labeling of a food as exotic with merely stating where the particular cuisine is from. But labeling food as *Mexican* carries none of the problematic connotations associated with the term *exotic*. Referring to a dish as *foreign*, however, is similar to exotic, as both construct a normal, or native notion of food, presumably those foods eaten by white Americans and Europeans, in contrast with foods eaten by people of color. It is, of course, quite possible to argue that Mexican cuisine is far more native to the Americas than food from Europe, though it seems unlikely that Kram would label sauerkraut or bratwurst as foreign. Last, Kram clearly denies any possibility that her behavior could be incorrect or offensive and, by stating that if she were ever "called out" she would discontinue her participation, places the responsibility for her inclusion on people of color. Indeed, she implies that her inclusion is essential to the vegan movement without attributing the same regard to the inclusion of the vegans of color to whom her post is addressed.

Gary's and Kram's relationships to the word *exotic* parallel Lisa Heldke's (2003) book, *Exotic Appetites: Ruminations of a Food Adventurer*. Heldke, a white-identified class-privileged woman and university professor, examines a type of unacknowledged "cultural food colonialism" among her particular demographic. Heldke describes the process through which she goes "culture hopping" in kitchens and restaurants, searching for "'newer,' ever more 'remote'" cuisines from which she can borrow. She calls this phenomenon cultural food colonialism because her attitude toward food bears an "uncomfortable resemblance" to the conquering attitudes of European explorers. The word *exotic*, for people of color, can be associated with being objectified as a thing to be discovered and manifested. This is at the root of why so many white Westerners, such as Gary, Kram, and the white middle-class demographic under scrutiny in Heldke's book, are confused, surprised, or even annoyed that people of color do not feel the same pleasure for the "exotic" as they do. They are confused that people of color in white-dominated spaces generally do not find it to be a compliment when referred to as exotic.

Vegans of color on the listserv had a variety of responses to Kram's post. Carolina, for example, offers the following experience:

I am Latina and Native. I had been vegetarian for several years and recently turned vegan. I have been to a few vegan and vegetarian events. I did not go back to many due to the uninviting feeling lingering in a room of usually, mostly Caucasian folks and the seemingly unavoidable awkward criticism. . . . Calling someone's culture, food, beliefs because they are unusual or out of the norm for you and especially using the word "exotic" or "foreign" IS offensive, derogatory and demeaning. . . .

[For example,] at a Vegan event there is a wide array of food. You spot a dish you have never tried before but you are all for trying it. You pick some out on to your plate. You sit down and chow down your food. You liked it and you state out loud to your neighbor, "Did you try that one potato-like thing? I'm not sure what it was but it was very exotic."

"Oh yeah. I love it when people bring foreign food," says your neighbor. The Vegan of color who brought the dish is sitting at the same table and perhaps says something or does not. But, they are offended.

Kram you asked, "Is that terrible? Is it demeaning (to the food or the person who made it)?" It is demeaning and offensive to the culture (where the food comes from) and to the person who made it. Once again, look at its history and its use.

Doris, a woman of color, responds to Kram's post as well:

Kram, as Carolina points out, the word "exotic" refers to something foreign. If you label a person or a food "exotic" you are implying that it doesn't belong here, wherever "here" is. For POC living in predominantly white cultures, we are too often marginalized and exoticized. We are constantly told in many ways that we do not belong where we are, and calling us or our food "exotic" is just one of those ways. It's particularly offensive in the United States, where white people are assumed to belong and other immigrant populations are exoticized.

Later on in the thread, Carolina continues explaining her experience:

The word "exotic" triggers many feelings of disgust as well. Take "Exotic" pornography for example. It has little to no white women/men. It is exclusively geared at demeaning women of color. It sells them as not being the norm, being unusual, alien, etc. To what is seen as regular porn (featuring white women/ men).

The word "exotic" is fueled with a demeaning sexual connotation as well.

Personally, I have been called an "exotic beauty" several times and it is not a complement. I am not the only woman who has been seen and labeled as such and was offended by it. . . .

Ignoring POC and other issues within the vegan community is what will hurt the cause of veganism. You have millions of POC who could be possible vegans but if vegan events are not seen as welcoming and there is ignorance flowing, you cannot further the cause.

Like the original posts, Carolina and Doris's responses associate being called exotic with the presumption, often made by whites, that what is culturally common to them is universal or normal, and what is not is foreign or other. Carolina's second post, like Johanna's original questions, also references the sexual exploitation that is often associated with the word *exotic*. All three women suggest that labeling food from non-white cultures as exotic invokes white/European food and culture as an unacknowledged norm. In white-settler nations such as the United States, the sociohistorical narrative of who looks like the "common citizen" is always expected to be white (Basson 2008; Dei and Calliste 2000; Clarke 1997). Kram's comment indicates a lack of awareness about the ways in which people of color have historically been erased as subjects (Hoagland 2007, 99).

Carolina, Doris, S, and Johanna's discomfort about the use of the word *exotic* is about more than food. It is about feeling and being out of place when they should not have to. It is about always being put on display for, and being judged (tasted) by a white gaze that appears to be ignorant of the entire colonial history of what it means to place a non-white person, food, culture, and so forth, into the categories of "exotic" and "foreign." Food is not separate from racially constructing a nation's "common citizen" into whiteness; food is nation (Pillsbury 1998; Watson and Caldwell 2005). Thus, for Carolina, Doris, S, and Johanna, the labeling of food from communities of color as exotic signals that they do not belong at vegan or animal rights events. Moreover, by identifying her reluctance to continue to attend vegan and animal rights events in which this terminology is used, Carolina exemplifies the ways that whiteness can have what Guthman (2008) calls a "chilling" effect people of color within the food movement.

Whiteness, Vegan Spaces, and Racialized Bodily Places

A second thread from the *Vegans of Color* blog also serves to exemplify the ways that whiteness manifests in the vegan and animal rights movements, and the effects of this whiteness on vegans of color. Many contributors to the discussion thread, "POCs Feeling 'Discomfort' at Predominantly White AR and Veg Events?" experience feeling out of place as nonwhite vegans. As their bodies enter white vegan spaces, their nonwhiteness is immediately visible and is constantly in the forefront of the gaze of white participants. Nassim, a vegan and woman of color living in Canada, writes:

I simply wanted to share an experience I've had as a vegan of colour—but first mention that although I am familiar with critical race theory/discourse. Identity politics confuse me . . . I am a middle eastern woman, but I feel as though I cannot say "I identify as . . ." because identity seems to be out of my hands sometimes . . . is this where the idea of agency comes in? Anyway, I'm trying to say that as a racialized woman it's hard to even discuss why I feel out of place in predominantly white environments, because those feelings are everyday realities—even fears of mine. Which brings me into my example. . . .

Not too long ago I attended the first and only AR/veggie workshop I've been to in my community. My ex-partner and I were two of only four people of colour in a crowd of at least 60–70 people. Not to mention we knew the two others, how funny is that, right? As soon as I walked in, I felt uncomfortable. I remember specifically signing in, and having issues with a woman because she "just can't seem to hear me" when I told her my name numerous times. . . . All of this crap was really getting to me, but I think what bothered me the most was the over-whelming assumption of privilege in the room. I mean the "issues" discussed were things such as "how to get vegan options at your favourite restaurant." I had never, before this day been exposed to this "other side of veganism" I suppose you could call it. Dare I say, the trendy vegans, the white vegans, that are vegan because veganism [is] becoming a status symbol?

I was so frustrated with the population, the cause, and felt like I could not call myself a vegan. As if "vegan" was a white word.

Nassim's experiences of feeling invisible as a Middle Eastern woman navigating through white spaces resonates with the experiences of many racialized minorities who live in white-settler nations. Her psychic experience of her place at this vegan workshop is deeply connected to the politics of mobility. Dr. Sara Ahmed, a scholar on racialized embodiment within white spaces, writes of similar experiences:

The politics of mobility, of who gets to move with ease across the lines that divide spaces, can be re-described as the politics of who gets to be at home, who gets to inhabit spaces, as spaces that are inhabitable for some bodies and not others, insofar as they extend the surfaces of some bodies and not others. Those who get stopped are moved in a different way. I have suggested that my name slows me down. A Muslim name. . . . For the body recognized as "could be Muslim," which translates into "could be terrorist" (Ahmed 2004), the experience begins with discomfort: spaces we occupy do not "extend" the surfaces of our bodies. (Ahmed 2007, 162–163)

I am not suggesting that white participants at the workshop that Nassim attended saw her as a "terrorist." However, what I am suggesting is that one of the reasons Nassim feels invisible comes from having to repeat her nonwhite and non-Anglo name within a white space of radical food politics in which vegan praxis manifests from a collective of white privileged experiences. While Nassim must constantly deal with the emotional stress of moving throughout white Canadian spaces as a racialized minority and a vegan, the white participants (in terms of racial politics and racialized mobility) worry only about moving through Canadian spaces as a vegan. Nassim's interpretation of this group's veganism as being rooted in whiteness is quite telling; it implies, on the one hand, that getting a vegan item on the menu of one's favorite restaurant is a privileged concern. Nassim, on the other hand, must also wonder if she can comfortably enter this restaurant as a nonwhite vegan with a nonwhite sounding name within a political climate that marks nonwhite people as foreign. This latter question would most likely not be part of the workshop that Nassim attended. In reading Nassim's experience, to be a white racially privileged vegan in Canada means to produce a space of vegan politics and praxis that unconsciously assumes the "taken for granted" mobility of whiteness/white privilege.

Supernovadiva, a black identified female vegan living in the United States also describes uncomfortable experiences in predominantly white vegan/animal rights movement settings:

i don't attend AR events maybe because of my own prejudice. AR people that always approached me were young, white, single, still in college, living on the parent's dime. . . .

i don't—sorry white readers above—like being singled out and bee lined because of my blackness. i don't like it because then the subject becomes my blackness. then the colorblind thing comes up and how that person don't see color BUT you bee lined straight to me to tell me you're colorblind. seriously.

not to sound like i have a chip on my shoulder, but those friendly, though well meaning, interrogations "why are you here?" is something i experienced in dept stores, malls, and any place that may be predom[inantly] white that i have a right, for my own reasons, to be. i can't explain it, but it's unwelcoming. your action can be saying 'you don't belong here.'

[i'm] not saying don't chat up POCs. if the oppo[r]tunity to chat comes around or you see a lone person (any race/sex/whatevs) not being talked to/ giving vibes of not feeling welcome—then chat. i'm just saying i don't bee line to the white girl in the reggae club asking questions.

For Nassim, being asked to repeat her name continuously makes her nonwhite identity hypervisible. Supernovadiva describes a similar feeling of hypervisibility when she is approached by well-intentioned white vegans. Both women demonstrate the ways that people of color must move through spaces that, while seemingly culturally neutral to whites, are to them, culturally exclusive.

In response to the earlier posts, Sue, a white vegan, commented in a way that placed responsibility for their exclusion on people of color:

I'm white, and have also noticed that the few vegan conferences I've attended are attended by mostly white people. I don't know what to think about that, except that you might feel less discomfort if you brought your friends along—or if you could find a way to connect with other people.

I suspect that many vegans feel alone. Where I feel most alone is at my church.

I can't speak to the racism issue. But it seems to me that the conferences tend to address all types of oppression, and have a definite "eco-feminist" element. Do you feel like the events are inaccessible? Or do you just feel like you're alone?

Like Kram did in his post, Sue clearly places responsibility for including vegans of color in predominantly white events on the people of color. S responds to Sue with the following statement:

Sue, so what you're saying is, if vegans of colour want to see some other vegans of colour they should make some friends who are vegans of colour and bring them along with them? So the issue doesn't have anything to do with the lack of inclusion in the vegan community to vegans of colour at all, it's *our* responsibility to connect with other people? It's that sort of attitude that precludes some vegans of colour from getting involved in AR/veg events.

Sue's perception of the discomfort expressed by the women of color on this blog is that it is an individual problem that can be cured through the individual action of bringing vegans of color to these events. She does not reflect on the structural and institutional factors that could have caused so many AR and vegan events to be predominantly white in the first place. This indicates that for her, the problem lies with the individual rather than a culture that relies on whiteness as the foundation of a nation's ideologies and spaces for its citizens.

Maureen, a white-identified vegan, then joined the discussion. Her experience is noteworthy in that she does not engage in the implications of being a "white," albeit "fat" vegan, at an all white-bodied vegan or animal rights event. She wrote:

I'm white, but my difference is my weight. I'm a fat vegetarian (mostly vegan). I feel self-conscious at animal rights and vegan events as well. I stopped eating flesh 35 years ago, but I'm not one of those people who lost weight by changing my diet. In fact, over the years, I've "blossomed" to the point where I'm obese. I eat healthy food, but obviously too much of it. And I can't wear most of the cool animal rights t-shirts because they don't come in my size! I'm sure people attending AR events look at me and think, "If she'd just stop eating double cheese-burgers, she'd be fine."

As for the color issue, I've noticed it and commented about it over the years that I've been involved. I've asked participants and movement leaders why more POC's aren't present. I've also told some of the POC's how happy I am to see people of color at the events, prefacing my words with a disclaimer that I tend to be outspoken and blunt at times. . . .

So yes—I'm white, but I feel the "subtle" discrimination that comes to all who are different. But I feel it's ok—because it's not about me personally. There's discrimination everywhere aimed at everyone who is different. *White (whatever that means)* vs color, old vs young, male vs female, straight vs gay, republican vs democrat, conservative vs liberal, rich vs poor . . . , and so it goes.

Bottom line—yes, it's intimidating to be obviously different from most of the other people at any event. But so what? If I let my differences keep me from attending, I miss out on something I want to do. And even more important, the rest of the people there miss out on seeing that fat people care about animals, about the environment, about vegetarianism. (emphasis added)

Maureen's situation reminds us that the connotations associated with veganism are not only "white," but also thin. However, instead of using her own difference as a standpoint from which to recognize the multiple ways that difference can manifest within the vegan/animal rights movement, Maureen avoids engaging with questions of racial privilege. By equating all differences, she attempts to assign her own meaning to the experiences articulated by Nasseem and Supernovadiva. This rhetorical move effectively erases the ability of vegans of color to interpret their own experiences and implies a lack of willingness to consider racial difference. Toward the end of her post, Maureen writes "white (whatever that means) vs color, old vs young, male vs female, straight vs gay, republican vs democrat, conservative vs liberal, rich vs poor . . . , and so it goes." Notice that she does not write "whatever that means" after "color," or "young," or any other category she has written. She has strategically detached herself from the racialization of "white" by claiming she does not even know what it really means, yet she feels comfortable making the claim that being "fat" is the same feeling as racial discrimination. Sue and Maureen fail to interrogate their own whiteness. They may know that they are at an event at which 95 percent of the participants are white, but critical thinking around the structure and institutions of whiteness does not go any deeper. They fail to notice all of the unconscious symbols through which white vegans are told they belong at vegan and animal rights events.

Jillian, however, does not want Maureen to detract from the initial topic of the dialogue and tries to recenter the dialogue back on the experience of racism and whiteness for people of color:

Maureen—Your experience is your experience, but you seek to minimize the experience of being a POC experiencing racism, and that is where I want to call you out. . . .

So, the most important thing about being a POC at a vegan event is that white people will see that POCs care about veganism? So . . . a vegan POC is really just supposed to show up at events in order to help white people with their racism? The post is talking about experiencing racism. Your experience is not the same as this. The fact that you feel the freedom to wax on at length in this forum seems to be evidence of your white privilege.

With this comment, Jillian reasserts her own subjectivity. She does not attempt to define Maureen's experience for her, but neither does she allow Maureen's ignorance of racial difference to stand. Moreover, she argues that people of color should attend vegan/animal rights events, when they choose to do so, for their own benefit, rather than to teach others that people of color can care about animals. Last, Jillian argues that the fact that Maureen has joined a listserv explicitly dedicated to the experiences of vegans of color, and yet writes authoritatively on the meaning of racial difference, demonstrates again the presumption of universality that accompanies a social position of normativity, and the general lack of regard for racialized histories and consciousnesses often espoused by "colorblind" whites.

Conclusion

This chapter provides examples from the Vegans of Color blog in the hope of helping "colorblind" white vegan activists to engage in a critical reflexivity around racially privileged oriented ways of being and knowing within food justice movements. Such a racialized consciousness can help white vegans to see speciesism and animal rights within a context of multiple oppressions including race, class, and nation. Examples of white antiracist consciousness can also be found on the *Vegans of Color* blog. Claire, for example, conveys her critical reflections on her whiteness:

My views of systemic racism, the process of racialization, white privilege, etc. have all developed a lot in the last two years . . . I simply feel an obligation to continue to listen to and learn from the voices and experiences of people of color, and to question other white people when they are doing or saying something that seems to come from a place of white privilege. . . .

I don't think it's impossible for people of color to feel comfortable in white-lead movements/spaces . . . Rather, I think there is a fundamental difference between asking "How can we get more POC involved here?" (which maintains the idea that whites are in control of the movement and that that is somehow natural or unproblematic and not really a big deal) and asking "What are we doing that makes this issue seem irrelevant or off-putting to POC? How can we change that?"

Claire provides a good example of what antiracist consciousness among white identified vegans can look like.

People involved in vegan food activism claim to encounter fear, denial, and defensiveness from humans benefiting from, and unwilling to interrogate, institutionalized speciesism. In the same manner that many nonvegans cannot easily see why they should be concerned with speciesism, many white vegan/animal rights activists cannot see how they benefit from institutionalized whiteness, or how this impacts their engagement with veganism. Food justice cannot be a reality, vegan or not, if the overwhelmingly white food movements, fail to engage in antiracism and critical whiteness-awareness activism.

Veganism is about harmlessness through consumption. Its fundamental core is to reflect on how human privilege and "speciesism as the norm" create a society in which nonhuman animals are placed in spaces of exploitation and objectification, and how this suffering is made invisible to the masses that consume them. However, when asked to extend this philosophy of harmlessness to reflect on how white privilege and "whiteness as the norm" creates a society in which nonwhite humans are placed in spaces of exploitation and objectification, and how this suffering is made invisible with vegan spaces, the plea is frequently ignored or is seen as the responsibility of vegans of color. The labeling of one's food as exotic or foreign serves as a constant reminder to people of color that they do not fit into the "colorblind" nation's model of what those who belong within these national borders should look like. Moreover, vegans of color acknowledge that their very bodies mark them as other, constraining their mobility in society in general and in predominantly white vegan and animal rights events in particular. Such constraints produce consistent feelings of discomfort.

Vegans of color blogging on this listserv respond to statements implicating unacknowledged whiteness with anger. They also acknowledge various ways that whiteness manifests in their interpersonal interactions with white vegans, and how such interactions dissuade them from attending vegan and animal rights events. Given these vegans of colors' accounts of how unacknowledged whiteness provokes anger toward, and lack of participation in, the vegan and animal rights movements, it seems likely that the spreading of a white antiracist consciousness, as exemplified by Claire's statement earlier, might create spaces in which the kind of racialized food politics advanced by vegans of color in this forum can thrive.

Note

1. *Speciesism* is the concept that human beings believe they have the right to exploit nonhuman animals because human beings are "superior" to animals.

References

Ahmed, Sara. 2004. *The Cultural Politics of Emotion*. Edinburgh: Edinburgh University Press.

Ahmed, Sara. 2007. A Phenomenology of Whiteness. *Feminist Theory* 8 (2): 149–168.

American Vegan Society. n.d. What Is Ahimsa? American Vegan Society. <http://www.americanvegan.org/ahimsa/htm> (accessed May 23, 2011).

Appel, Liz. 2003. White Supremacy in the Movement against the Prison-Industrial Complex. *Social Justice* 30 (2): 81–88.

Basson, L. L. 2008. *White Enough to Be American: Race Mixing, Indigenous People, and the Boundaries of State and Nation*. Chapel Hill: University of North Carolina Press.

Bonilla-Silva, Eduardo. 2006. *Racism without Racists: Color-Blind Racism and the Persistence of Racial Inequality in the United States*. 2nd ed. Lanham: Rowman & Littlefield Publishers.

Cherry, Elizabeth. 2006. Veganism as a Cultural Movement: A Relational Approach. *Social Movement Studies* 5 (2): 155–170.

Clark, Dylan. 2004. The Raw and the Rotten: Punk Cuisine. *Ethnology* 43 (1): 19–31.

Clarke, G. E. 1997. White Like Canada. *Transition* 73:98–109.

Dei, George J. Sefa, and Agnes Calliste. 2000. *Power, Knowledge and Anti-Racism Education: A Critical Reader*. Toronto: Fernwood.

Dwyer, Owen J., and John Paul Jones III. 2000. White Socio-Spatial Epistemology. *Social & Cultural Geography* 1 (2): 209–222.

Farr, Arnold. 2004. Whiteness Visible: Enlightenment Racism and the Structure of Racialized Consciousness. In *What White Looks Like: African-American Philosophers on the Whiteness Question*, ed. George Yancy, 143–157. New York: Routledge.

Flagg, B. J. 2005. Whiteness as Metaprivilege. *Washington University Journal of Law and Policy* 18 (1): 1–11.

Guthman, Julie. 2008. "If they only knew": Color Blindness and Universalism in California Alternative Food Institutions. *The Professional Geographer* 60 (3): 387–397.

Hamanaka, Sheila, and Tracy Basile. 2005. Racism and the Animal Rights Movement. *Satya Magazine* (June/July).

Harper, A. Breeze. 2009. *Sistah Vegan: Black Female Vegans Speak on Food, Identity, Health, and Society*. New York: Lantern Books.

Heldke, Lisa M. 2003. *Exotic Appetites: Ruminations of a Food Adventurer*. New York: Routledge.

Hoagland, Sarah Lucia. 2007. Denying Relationality. Race and Epistemologies of Ignorance, ed. Shannon Sullivan and Nancy Tuana, 95–118. Albany: SUNY Press.

Iacobbo, Karen, and Michael Iacobbo. 2006. *Vegetarians and Vegans in America Today*. Westport, CT: Praeger Publishers.

Kang, Jerry. 2000. Cyber-Race. *Harvard Law Review* 113 (5): 1131–1209.

Lander, C. 2009. Stuff White People Like. <http://stuffwhitepeoplelike. com/2008/01/27/32-veganvegetarianism/> (accessed August 25, 2009).

Larsen, Nella. 2006. *Quicksand*. Mineola, NY: Dover Publications, Inc.

Leary, Dr. Joy Degruy. 2005. *Post Traumatic Slave Syndrome: America's Legacy of Enduring Injury and Healing*. Milwaukie, OR: Uptone Press.

Lipsitz, George. 2006. *The Possessive Investment in Whiteness: How White People Profit from Identity Politics*. Rev. and expanded ed. Philadelphia, PA: Temple University Press.

Mahrouse, Gada. 2008. Race-Conscious Transnational Activists with Cameras: Mediators of Compassion. *International Journal of Cultural Studies* 11 (1): 87–105.

McKinney, Karen D. 2005. *Being White: Stories of Race and Racism*. New York: Routledge.

Mills, Charles W. 1997. *The Racial Contract*. Ithaca, NY: Cornell University Press.

Nagra, Narina. 2003. Whiteness in Seattle: Anti-Globalization Activists Examine Racism within the Movement. *Alternatives Journal* (Winter): 27–28.

Nakamura, Lisa. 2002. *Cybertypes: Race, Ethnicity, and Identity on the Internet*. New York: Routledge.

Patterson, Charles. 2002. *Eternal Treblinka: Our Treatment of Animals and the Holocaust*. New York: Lantern Books.

Pillsbury, R. 1998. *No Foreign Food: The American Diet in Time and Place*. Boulder, CO: Westview Press.

Poldervaart, Saskia. 2001. Utopian Aspects of Social Movements in Postmodern Times: Some Examples of DIY Politics in the Netherlands. *Utopian Studies* 12:143–164.

Said, Edward W. 2003. *Orientalism*. London: Penguin.

Salamon, Gayle. 2006. The Place Where Life Hides Away: Merleau-Ponty, Fanon, and the Location of Bodily Being. *Differences* 17 (2): 96–112.

Slocum, Rachel. 2006. Anti-Racist Practice and the Work of Community Food Organizations. *Antipode* 38 (2): 327–349.

Spiegel, Marjorie. 1996. *The Dreaded Comparison: Human and Animal Slavery*. Rev. and expanded ed. New York: Mirror Books.

Sullivan, Shannon, and Nancy Tuana. 2007. *Race and Epistemologies of Ignorance*. Suny Series, Philosophy and Race. Albany: SUNY Press.

Terry, Bryant. 2009. *Vegan Soul Kitchen: Fresh, Healthy and Creative African-American Cuisine*. New York: Da Capo Press.

Torres, Bob, and Jenna Torres. 2005. *Being Vegan in a Non-Vegan World*. Colton, NY: Tofu Hound Press.

VeganOutreach. 2007. Guide to Cruelty Free Eating. December 7. <http://www .veganoutreach.org/guide/gce.pdf>.

Watson, J. L., and M. L. Caldwell. 2005. *The Cultural Politics of Food and Eating: A Reader*. Malden, MA, and Oxford: Blackwell Publishers.

Yancy, George. 2004. *What White Looks Like: African-American Philosophers on the Whiteness Question*. New York: Routledge.

11

Realizing Rural Food Justice
Divergent Locals in the Northeastern United States

Jesse C. McEntee

The local food movement has experienced wide-scale buy-in across the United States with advocates promoting the social, economic, and environmental benefits of local food initiatives. Although these initiatives may have experienced some success in promoting sustainable farming methods and supporting local farmers, they have overwhelmingly disregarded the needs of low-income consumers, especially in rural areas. Some exceptions to this trend exist in urban settings, such as sliding-scale payment options for low-income customers, but the vast majority of local food initiatives operate on capitalist principles, stressing profit, growth, and efficiency. As long as food distribution, regardless of its place on either a local or global scale, is dictated by these principles, marginalized and particularly poor people will be denied access to the food required to live a healthy life. Though the majority of local food initiatives are trending in this direction, what I call *traditional localism* offers new hope for rural food justice.

As local food initiatives grow in number, food access investigations in the United States and the United Kingdom are undoubtedly increasing, the majority of which are urban focused. As a result, methods and outputs emerging from these investigations tend to have an urban slant, thus overlooking the particularities of the rural experience. Academics and policy makers are thus faced with a shortage of rurally focused policy and planning interventions.

Like the majority of food access investigations, food justice efforts typically occur in urban areas and target low-income minority populations as well (Wekerle 2004; Alkon and Norgaard 2009; Welsh and MacRae 1998). Employed by food security and food access advocates and increasingly by academics, the concept of food justice supports the notion that people should not be viewed as consumers, but as citizens (Levkoe 2006). These efforts attempt to link populations with alternative modes of food production and consumption, which typically occur at the local level and are grounded in democratic and social justice values (Welsh and MacRae 1998). The argument stands that when profit is prioritized above human well-being and the need for survival, the result is food injustice. The food justice movement represents what Wekerle describes as a movement away from a focus on emergency food sources, which took precedence in the 1980s, to one that focuses on "the right to food as a component of a more democratic and just society . . . and a theoretical framing of local initiatives as both the practice of democracy and as a means of de-linking from the corporate global food system" (Wekerle 2004, 378–379). This represents "more than a name change" departure from conventional food security concerns; rather, it is a systemic transformation that alters people's involvement in food production and consumption (Ibid., 379).

Rural food justice efforts are practically nonexistent and only a small number of references to rural food justice exist in the literature (Gottlieb 2009). Despite the fact that many, if not most, food justice principles (e.g., access to fresh and healthy food, community empowerment, social justice) could apply, in theory, to a rural context, their actual employment would need to first account for the nuances of rural culture, geography, and politics. Using Grafton County, New Hampshire, as a research site, this chapter aims to fill this gap by articulating how food justice efforts could address the food access concerns of a low-income rural community, primarily through an avenue that I have termed *traditional localism*.

Grafton County, New Hampshire, is located in the northeastern United States and has a population of 81,743 with the majority (65 percent) of people living in areas defined as rural by the United States Census Bureau (U.S. Census Bureau 2008a). Of the forty northern New England (Maine, New Hampshire, and Vermont) counties, Grafton County's population density ranks sixteenth (47.7 people per square mile of land) (Ibid.). Approximately 7.7 percent of New Hampshire's population is food insecure (Nord, Andrews, and Carlson 2008) while 8 percent lives in poverty (compared to 9.4 percent for Grafton County) (U.S. Census Bureau 2008b). At the time this research was conducted, the number of small farms in this area is growing, community supported agriculture (CSA) programs are increasing in popularity, and farmers markets are progressively more common (USDA AMS 2008b). A burgeoning local food movement, the presence of food insecurity, and the county's rural character combine to create a unique research opportunity.

Using primary data from qualitative fieldwork, I discuss the food justice nuances of two new terms that I have previously discussed (McEntee 2010): *traditional* and *contemporary* localisms. These represent two categories of participation in local food provisioning, distinguished from each other primarily by the actor's *intent*. Contemporary localism aligns with the increasingly popular local food movement; it is accessed by individuals who intend to impact their communities environmentally or socially through their food provisioning activity. Traditional localism, in contrast, attracts mostly those who intend to pursue means of obtaining *affordable* and fresh food.

While activities associated with both traditional and contemporary localism might manifest themselves in the same food provisioning experience (such as a farmers market), data presented here suggests that food justice advocates aiming to improve food insecurity and who operate among rural and poor populations would have greater efficacy working via the avenue of a traditional localism rather than a contemporary one.

The State of Food Access and Local Foods

An Urban Bias in Food Access Inquiries

The urban bias in food access examinations and food justice movements marginalizes the acute food access problems faced by rural poor populations. This could be due partly to the fact that rural areas have distinct geographies and sociopolitical landscapes that can make them difficult to penetrate as a researcher. A small number of rurally based food access studies do exist (Morris, Neuhauser, and Campbell 1992; Olson et al. 1996; Kaufman 1999; Skerratt 1999; Furey, Strugnell, and McIlveen 2001; Blanchard and Lyson 2006; Hendrickson, Smith, and Eikenberry 2006; Liese et al. 2007; Morton and Blanchard 2007; Burns and Inglis 2007; Morton et al. 2008; Sharkey 2009; Smith and Morton 2009; McEntee and Agyeman 2010), and these authors have distinguished some important challenges of a rural food environment, such as fewer supermarkets and grocery stores (Morris, Neuhauser, and Campbell 1992), higher food prices, and lower-quality produce and meat compared to suburban and urban stores (Morris, Neuhauser, and Campbell 1992; Kaufman 1999). Transportation barriers are particularly pronounced in rural areas and a well-defined mixture of reciprocal and redistributive economies are also present (Morton et al. 2008).[1] Areas of inadequate food access in a rural setting have been identified based on retail density

(Morris, Neuhauser, and Campbell 1992), distance to food retailers (McEntee and Agyeman 2010; Blanchard and Lyson 2006), and household food spending capability (Kaufman 1999). One thing that all food access examinations (urban or rural) have in common is that they tend to rely predominantly on quantitative measurements, such as distance to food sources and the price of food. These elements of access are certainly valuable in painting an initial picture of food access challenges facing a community, but informational access and perception play an equally important role in dictating one's food provisioning experience (Kirkup et al. 2004).

Food access is an important concept in addressing problems of food insecurity. Inadequate food access can lead to food insecurity, which can lead to higher rates of adverse health outcomes, such as obesity and diabetes, especially in rural areas (Olson and Bove 2006; Eberhardt et al. 2001). Households that struggle to buy enough food frequently stretch their budgets by purchasing calorically dense items that compromise nutrient content while simultaneously leading to overconsumption (Food Research and Action Center 2008). As such, an inverse relationship exists between energy or caloric density and food cost (Drewnowski and Specter 2004). It is clear, then, considering that nationwide rural areas have higher food insecurity rates (11.5 percent of nonmetropolitan households are food insecure, compared to 10.2 percent in metro areas, a statistically significant difference [Nord 2002]), they deserve equal attention from efforts to prevent such health disparities.

I use the term *food access* as a concrete indicator of one's ability to obtain food. This ability can be envisaged as a series of three choices that correspond to informational, economic, and physical access (see McEntee 2009 for additional discussion). Adequate access is therefore defined as having the ability to make *informed choices* throughout one's food provisioning experience, not only at the level of informational choice (adequate nutritional/cooking knowledge), but including economic and physical choices as well (having enough money and the physical ability to get and prepare food). That is, does the individual make her or his food provisioning decisions with full ability and knowledge of the consequences and advantages thereof? Later in this chapter I focus my discussion on the concepts of contemporary and traditional localisms, revealing that informational access, culture, and perception are vital gauges in assessing how one obtains food.

Localization: Contexts and Critiques

Local food efforts are typically grouped under the umbrella of alternative food initiatives; programs intended to counteract the ecological, social, and economical impacts of an increasingly globalized food system and which represent "a resistance to large producers and retailers" (Ilbery and Maye 2005, 825). Alternative food initiatives include a range of activities and provisioning agendas, including sustainable agriculture, fair trade, farm-to-school, as well as many others that have been reviewed extensively elsewhere (Kloppenburg et al. 2000; Allen et al. 2003). It is from this alternative context that local food initiatives emerge.

Local food initiatives have been coupled with a number of food justice campaigns with considerable success in urban and minority communities (Alkon and Norgaard 2009; Wallace 2008). In addition, many local food activities have emerged in both urban and rural communities that are not motivated by a food justice agenda, but by an aspiration to "localize" a food system environmentally, economically, and socially; such reasons to buy local are provided by New Hampshire Made (2009) and New Hampshire Farm to Restaurant Connection (2009). Yet it is not well established how emerging local food initiatives and the subsequent increased market visibility of local foods influence food access in rural, predominantly white communities in the United States. Previous literature offers some criticisms of alternative food efforts, from organics to sustainable agriculture to local food movements; namely their potential for exclusion due to a misalignment between programmatic goals and actual outcomes (Guthman 2008) or an inability to serve a diversity of income, education, and occupational backgrounds (Hinrichs and Kremer 2002). Patricia Allen and Claire Hinrichs have written about the social justice and equity elements of local foods by highlighting the implicit disregard of the needs of low-income consumers with a disproportionately weighted focus on farms and farmers (Allen and Hinrichs 2007) and how defensive localism stresses a homogeneity leading to nativist sentiments (Hinrichs 2003; DuPuis, Harrison, and Goodman, chapter 13, this volume).

Grafton County: Rural Aspirations for Food System Relocalization

Grafton County has a population of 81,743 people and a population density of 47.7 people per square mile of land (U.S. Census Bureau 2008a). Approximately half of the land area is located in the White

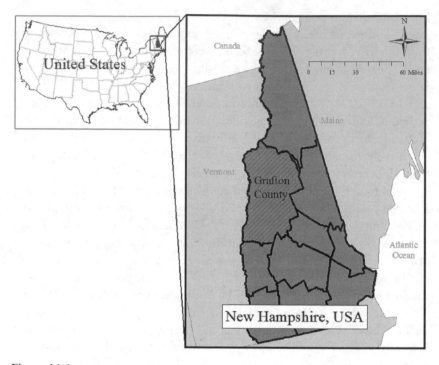

Figure 11.1
Grafton County, New Hampshire

Mountain National Forest. The 2008 U.S. Census records that 96 percent of Grafton County is classified as white (Ibid.). Unlike the other two primarily rural northern counties (see figure 11.1) of New Hampshire (Carroll County and Coos County), Grafton County contains two universities that serve as educational and cultural centers (Dartmouth University in Hanover and Plymouth State University in Plymouth), which attract residents with above-average educational attainment and incomes. These universities are both located in the southern sections of the county.

Grafton County faces many of the same challenges as the rest of rural America that are often overlooked by policy makers and researchers, including higher childhood poverty rates, transportation barriers, rising housing costs, higher food prices, and a growing population (Johnson 2006; Hanson and Andrews 2008). Simultaneous to these challenges, venues and opportunities for local food purchasing are on the rise nationally, in New Hampshire and in Grafton County. Statewide, the number of small farms is beginning to increase for the first time in decades (Langley 2009). New Hampshire has sixty-seven farm stands (New

Hampshire Department of Agriculture, Markets, and Food 2008), twenty-one farms that participate in community-supported agriculture (Adam 2006), and fifty-six farmers markets (USDA AMS 2008a). Nationwide, the number of farmers markets grew by nearly 3,000 between 1994 and August 2008 (USDA AMS 2008b).

Two prominent local food campaigns are the New Hampshire Farm to Restaurant Connection and New Hampshire Made. Both adopt a similar list of justifications and benefits to buying local, including: it is fresher, consumers know where the food comes from, it supports the local economy and farmers, and it is better for the environment (New Hampshire Made 2009; New Hampshire Farm to Restaurant Connection 2009). These programmatic goals are consistent with Allen and Hinrich's assessment of "Buy Local" campaigns, which determined that these promotional materials place greater emphasis on environmental and economic sustainability objectives, "comparatively subordinating social justice" (Allen and Hinrichs 2007, 263) by neglecting to include food access goals, especially for low-income populations. This growing local food interest in New Hampshire is in many ways similar to that of other regions of the country (Severson 2009).

Methods: Qualitative Inquiry

Findings presented in this chapter are derived from seventy-five semi-structured interviews I conducted in 2008 and 2009. Respondents included twenty-nine key stakeholders, such as state agency employees, farmers, nonprofit employees, and academics, as well as forty community respondents, such as food store patrons, food pantry clients, homeless shelter clients, senior center clients, and people participating in local food activities. Some key stakeholders worked at the state level, and all community respondents were from Grafton County. Key stakeholders were purposefully selected based on their knowledge of food-related activities and programming at the state and regional levels, and interviews with these individuals were prearranged and lasted from thirty to ninety minutes.

I approached community respondents during operating hours of the food retailer or nonprofit entity and interviews ranged in length from ten to ninety minutes. Guiding interview questions were used for both sets of respondents, therefore prompting a semi-structured, open-ended discussion. Access was gained to respondents through a number of avenues. For food store patrons, I contacted the store's manager, owner, and/or corporate headquarters to obtain permission to approach patrons within the store. Once I received permission, and depending on the conditions agreed upon (for example, Shaw's Supermarkets, Inc. would only allow one store visit and only allowed me to approach patrons directly outside of the store entrance), customers were approached for interviews.

After establishing a relationship with nonprofit employees and volunteers (from food pantries, homeless shelters, and the Plymouth Regional Senior Center), I asked them to serve as a liaison between their clients and me in order to arrange for interviews with the clients. Local food participants were identified through two venues, namely, the Plymouth Farmers Market and Local Foods Plymouth. Farmers market patrons were approached while shopping. Local Foods Plymouth is an online farmers market that operates year-round and is run by the Plymouth Area Renewable Energy Initiative. Each week customers place orders online and pick up orders three days later in downtown Plymouth, where interviews were conducted.

Participant observation was also conducted at a soup kitchen operating out of the basement of a local church, which serves weekly hot meals for free. During these meals I ate with clients and introduced the project and myself to them. No notes or recordings were taken during participant observation, in order to facilitate productive dialogue.

Divergent Locals: Traditional and Contemporary Localism

Traditional Localism: Tradition, Reciprocity, Affordability, and Freshness

Two types of approaches toward local food emerged from my primary data collection, which have distinct as well as intersecting characteristics. Although both localisms occur within a relatively short geographic proximity to consumers, *contemporary* localism and *traditional* localism are informed by two distinct sets of motivations. The majority of community respondents throughout Grafton County not only expressed some type of affinity for local food products, but also personally participated in food production or procurement, ranging from various forms of gardening to small-scale livestock production to hunting. However, respondents who participated in these local food activities did not frequently purchase "local" because they perceived it to be "too expensive"; in essence, people appeared to apply different standards to their purchasing habits than to their food production activities. They preferred local when they could grow it (or shoot it), but not when they had to pay for it. This behavior is part of what I call traditional localism.

Food activities within traditional localism might be more likely to improve food access, especially for those struggling to participate in the conventional global and contemporary local food systems. These activities, especially gardening, improve access to fresh produce—produce that would otherwise be perceived as too expensive (including both conventionally grown [see Dibsdall et al. 2003] and local produce). According to individual respondents, the reasoning behind their decision-making process can be attributed to the enjoyment derived from partaking in these activities, continuing a tradition, as well as from the fact that people view food production as a cheaper (and better-value) alternative to buying food from a retailer, especially in times of economic uncertainty (Pudup 2008; National Public Radio 2009). According to George Ball Jr. who owns Burpee Seed Company, seed sales are expected to rise 20 to 30 percent in 2009 (National Public Radio 2009); the company also is marketing "Burpee's Money Garden," a $10 package of seeds "capable of producing $650 worth of vegetables" (Burpee & Company 2009). The increase in seed sales suggests that people view gardening as an economical alternative to supermarket produce. Considering that most respondents viewed local produce as being too expensive to purchase regularly, having your own garden would seem like a prime opportunity to obtain fresh *and* affordable produce.

Contemporary Localism: Economic Development and Environmental Impact

These food-growing activities are situated next to a "new" local alternative food network of increasing popularity, which appears to be associated primarily with middle- and upper-income consumers (this phenomenon has been discussed at length by Hinrichs 2000). I call this local movement *contemporary localism*; this is represented by aspects of the local food movement that are explicitly recognized by advocates as representing holistic farming methods; presence of local food venues; commitment to social, economic, and environmental sustainability; and shorter distances between grower and consumer (Jarosz 2008). Some have cited the opportunity for local foods to increase accessibility (Feenstra 1997; Gottleib and Fisher 1996), yet my findings suggest that contemporary localism is largely inaccessible to low-income consumers. Instead, contemporary localism is an extension of the current globalized food system, dictated by the same core values of efficiency and profit, where the end consequence of both systems is inadequate access to food for marginalized populations. The large majority of respondents interviewed regarded contemporary local products as prohibitively expensive to purchase regularly. One homeless shelter resident described how the farmers market prices were "outrageous." In describing the local produce, he said, "Yeah, prices up there, corn and stuff was totally outrageous." And he later described his participation in a family garden plot: "Yeah, we did tomatoes and stuff like that so we had a little garden . . . I think we tried some peas and tomatoes and just a few things and it came out pretty good." The exclusivity due to both perceived price, and perhaps culture, of contemporary localism is countered by the approachability of traditional localism. Evidence suggests that perception of the high price of contemporary local products is not the only mechanism restricting access. One key stakeholder told me the following story about a local farmers market:

A year or two ago, I understood from other vendors that they attempted to bring in classical music. They "upgraded" their market. And the people that had [WIC] coupons—they don't feel comfortable because it was trying to cater to the higher income people, which I don't think [town name] is like that, but I understand [another town name] was. And I don't know if that was true this year, but I heard it in the past one of my vendors said people are not coming with those coupons because there is probably a stigma attached to it. Its like, "Oh, you're getting help."

In this instance price did not instigate a change in participation, but a general "upgrading" of the market, which contributed to the pressures for WIC recipients to stop attending. This indicates that lower-income people are trying to participate in local food activities, but rejecting the atmosphere that is often attached to contemporary localism.

As a result, only privileged consumers can purchase local products on a regular basis. For instance, the owner of a mid-sized independently owned supermarket located in a poor region of the county indicated that people do exhibit some preference for local, but can only afford to purchase the items for brief periods in the summer; "buying local is a treat" is how he put it. According to individual interview responses, affordability appears to be the primary obstacle associated with contemporary local food. This was true especially in lower-income areas where locally produced food may be widely viewed as healthier for one's physical well-being and for the community as a whole than nonlocal alternatives, but remain unaffordable for most individuals.

One food retailer donated its locally sourced food products, including produce, meat, and bread, to local pantries when the products did not conform to or meet the retailer's standards, for reasons such as reduced visual appeal and quality or expiration. This is no different from the power dynamic within the conventional globalized system; privileged people get higher quality food while poor people get the cheapest food or food that is left over (donated, expired, nonperishable, or a combination) from the primary market and is usually inferior in quality.

The contemporary localism that is emerging in Grafton County alleviates some industrialized food system concerns, but appears to still foster "socially exclusive niches" (Goodman 2004, 13) due to the perception of unaffordable prices that residents articulated. From a community food security standpoint, Allen's (1999) statements ring true regarding the potential incompatibility of local food production and prioritization of low-income people: "CFS [Community Food Security] has prioritized the needs of low-income people; in its focus on production, it emphasizes local and regional food systems. Although the CFS movement is working to integrate these objectives, it is also facing the question of where it should place its emphasis—on low-income access or on local food systems. These objectives are not necessarily compatible and may even be contradictory" (117).

Though some local farmers and retailers may occasionally donate local produce to food pantries, this transaction operates within the current redistributive system, and for that reason does little to alleviate the causes of inadequate physical access to food. From a food safety net perspective, rural residents tend to have less access to emergency food sources and entitlement programming than urban residents, which underscores the importance of Meert's (2000) "survival infrastructure" (236). The lack of institutional support compared to urban areas (Hofferth and Iceland 1998) is coupled with potentially less accessibility to contemporary local food resources such as community and school gardens (Garasky, Morton, and Greder 2004). Moreover, contemporary localism may reinforce conventional food injustices, reaffirming that a shorter commodity chain does not result in a better one (Allen and Hinrichs 2007).

Divergent Localism: Intentions and Motivations

Contemporary and traditional localisms describe motivations and actions regarding food procurement strategies at the local geographic scale. Traditional localism is effective in alleviating food insecurity among rural food-insecure populations because of its accessibility within a rural

	Contemporary local	Traditional local
Intent	Explicit desire to support local farmers, promote environmental sustainability, etc.	To obtain fresh and affordable food, reciprocal exchanges, tradition.
Access barriers	Culture, price (perception of).	Availability of soil, water, and land resources.

Figure 11.2
Motivation and barriers to access of traditional and contemporary locals

setting. At the same time, contemporary localism appears to promote exclusion through price and culture. Dissimilarities between contemporary and traditional localisms exist and a great deal of the difference between the two is found in the intent underlying participation and perceived barriers to access (see figure 11.2).

Respondents in our study area chose not to buy local products frequently because of the perception that prices were unaffordable. However, most people in Grafton County appear to be engaged in *traditional* localism; a form of localism that is not overtly guided by the tenets of the alternative food movement characteristic of contemporary localism, but by a desire to obtain fresh and affordable food. For example, one food stamp recipient raised broiler chickens in her garage because it was economical and something her father did when she was growing up on a farm. A farmers market patron also raised chickens because he believed it reduced his family's carbon footprint. The former respondent was motivated by a desire to obtain fresh and affordable food and continue a tradition; she would be classified as a traditional localist. The latter respondent is a contemporary localist; he is inspired by a desire to help the environment. This respondent continued his commitment to the "local" through his everyday purchasing habits by buying as many locally grown products as possible. This is in sharp contrast with the traditional localist, who obtains most of her food from a wholesale retailer and corporate supermarkets—entities that are often viewed as flagships of globalization and represent everything contemporary localism works against.

In a rural setting, there is a higher probability that people have the land, soil, and water resources necessary to be involved in a traditional local activity. Traditional local activities are more likely to improve food access, especially for those who struggle to participate in both contemporary localism and conventional global food systems. Growing one's own food—while at the same time benefiting from the efficiency-driven global food system, which ensures low prices and consistent quality—makes fiscal sense for low-income people. Traditional localists are very involved in local food activities, yet some of their actions are largely ignored by contemporary localism, possibly due to cultural barriers. For instance, we have yet to see a "buy local" campaign that supports hunting as an appropriate local food procurement mechanism. In Morton et al.'s (2008) examination of low-income urban and rural counties, the authors discovered that reciprocal exchanges (giving and getting "garden produce, fish, meat, and other food exchanges among people who know each other—friends, family, and neighbors" [110]) were more frequent in rural areas. These activities are in part a reaction to the failure of the predominant economic system to effectively integrate all members of society, especially financially limited households; consequently noneconomic systems develop to obtain food resources (Morton et al. 2008). Reciprocal activities contribute to traditional localism and are common among Grafton County respondents. In fact, the most common traditional local activity, gardening, was found by Morton and colleagues (Ibid.) to increase the likelihood of rural households meeting recommended vegetable servings. Many respondents who found locally produced products too expensive to purchase happily shared garden produce (from blueberries to squash) with neighbors, relatives, and even strangers through reciprocal exchanges. One food pantry employee described a how a local hunter donated moose meat:

Interviewer: What are the most popular items that you have here in the pantry?
Respondent 1: Meat. It's the most expensive . . .
Respondent 2: Oh was it last year we got the moose meat? We got 500 pounds. And we're thinking, what are we gonna do with all this moose meat? And it flew out of here. I mean, people were calling us and asking us for some.

Contemporary and traditional localisms do not represent concise rationales and actions. Instead, these mix and overlap, continually influenced by new information, time, and locality. The distinguishing motivations for participation in these two localisms can potentially manifest themselves in the same activity. Take, for instance, Hinrichs and Kremer's (2002) finding regarding CSA membership: "In every component of class those in the least privileged category joined the CSA for more basic reasons of accessing food, while those in other categories became members to obtain high quality food or to exercise their beliefs and demonstrate support" (85).

We also see tensions between these localisms and the conventional globalized food system. The example of the "traditional localist" who raises broiler chickens in her garage, but has no ideological problem with obtaining most of her food from a "big box" store or corporate supermarket, illustrates a tension between traditional localism and the conventional food system.

Consider another respondent's justification for shopping at big box stores:

Interviewer: And do you ever cut back on your food purchases to pay for other items?
Respondent: Not really. I would say that last winter when the price of gas and oil was going up, I started to get a lot smarter about my choices. I went back to clipping some coupons, you know. I'll use them here occasionally, I'll go to the big box stores and stock up on things that um, are bulk items. And so, I really did start to look at things differently. *If I get things like that cheaper, then I can afford to buy the local produce. Or the natural ground beef.* (italics added)

Taking advantage of the efficiency of, and therefore supporting, the globalized food system in order to be able to afford local foods could be viewed as a way to increase participation and support for local food producers and growers. Yet the contemporary local food movement, at least according to much of the programmatic literature, strives to change this globalized conventional system in a number of ways, such as bringing "you closer to the people who grow and produce your food" and supporting "the local economy and family farms" (New Hampshire Made 2009).

Future Directions in Rural Food Justice: The Role of Traditional Localism

As alternative food initiatives such as local food movements expand in popularity, they offer new hope for establishing a sustainable food system. Yet data presented here shows that caution should be taken in the support of local foods to ensure that disadvantaged consumers in Grafton County, and underprivileged rural consumers in general, are not adversely affected. This region could benefit from the presence of food justice efforts that stress equity for consumers. Food justice campaigns could accomplish this through an awareness of the difference between contemporary and traditional localisms, and subsequent support of a *traditional localism* that resonates with low-income people. In this chapter I have begun to outline the nuances of a traditional localism and the different motivations that make it distinct from the localism promoted by alternative food systems.

Gaining more knowledge about people's motivations for participating in a traditional localism could lead to the development of marketing strategies that speak to an audience that is more likely to include low-income people. Emphasizing the economic benefit to the individual not of *buying* locally through alternative agriculture systems—where the benefit goes to producers—but of *growing, producing, or bartering* locally in order to save money, would be more effective in garnering support for accessible and therefore socially just food systems. Having a garden or keeping livestock has economic benefits and speaks directly to those who are financially disadvantaged (a target population of antihunger campaigns) more so than contemporary localism's claims of reducing one's food miles, "building community," or other claims that tend to appeal to those with greater financial resources. Aspects of local food procurement that are currently completely ignored by contemporary localists, such as hunting, could be embraced by food justice advocates not only to promote local food procurement, but also to bridge cultural divides that currently exist between low-income, traditional, and affluent contemporary localists.

In rural areas like Grafton County there is an absence of local food programs such as farmers markets and CSAs that are run by and for low-income and homeless people. Though there are models in low-income areas, such as the Homeless Garden Project (2009) in Santa Cruz, California, or the reVision House Farm (2009) in Dorchester, Massachusetts, they tend to be more prevalent in urban areas. Establishing rural versions of these types of food justice programs would ensure that the needs of disadvantaged community members are acknowledged. From the grower/producer perspective, additional research is needed to understand how these actors (farmers, distributors, retailers) could shift market foci to accommodate the abilities and needs of low-income consumers, not just those with financial means.

Local food initiatives in urban areas have a greater food justice impact than identical initiatives in rural areas. For instance, a farmers market or community garden in Boston is accessible by pedestrians who reside in nearby neighborhoods and those who have access to public transportation. This is in contrast to a farmers market in a small community of Grafton County, which is accessible only by those who have a private automobile; that is, the geographic density of a city is more conducive to centralized food sources or distribution mechanisms, such as a farmers market, community garden, or supermarket, than a rural location where long distances mandate additional travel and opportunity costs. The dispersed land-use pattern and lack of public transportation options create significant barriers to food access for low-income populations. These types of rural factors are important considerations in the development of rural food justice initiatives.

Food justice aims to establish food as a right decoupled from the "corporate global food system" (Wekerle 2004, 378–379) and to enable democratic and socially just food production and consumption, especially for marginalized populations. Contemporary localism does not improve food access—it simply reinforces the well-known barriers to adequate food access that are part of our existent retail food system. Traditional localism, in contrast, contributes toward a *reflexive justice* (see chapter 13, this volume), in which the alternative economy of contemporary local food production serving the "homogenous few" is replaced by one that is reflexive in order to "work across difference" (see chapter 13, this volume). Those working to promote food justice in a rural setting would benefit from adopting the rhetoric of and acknowledging the concerns of those operating within a traditional localism. The modern, industrialized food system promotes "physical and psychological displacement of production from consumption" (Feagan 2007, 38), and contemporary localism appeals to people because they can reconnect with their food. This sense of belonging is important to community building. However, like other selective patronage campaigns (Hinrichs and Allen 2008), on the one hand, local food efforts emphasize social justice needs for some groups (such as local farms) and can potentially disregard the needs of others (low-income consumers), especially if a defensive localism is embraced. A traditional localism, on the other hand, might resonate more meaningfully with low-income rural people, and thus represents an important future direction for the food justice movement.

Traditional localism in rural areas engage participants through non-capitalist, decommodified means that are affordable and accessible. Food is grown/raised/hunted, not with the intention to gain profit, but to obtain fresh and affordable food. A traditional localism disengages from the profit-driven food system and illustrates grassroots food production where people have direct control over the quality of the food they consume—a principal goal of food justice.

Acknowledgments

The author gratefully recognizes the ESRC Centre for Business Relationships, Accountability, Sustainability and Society, Cardiff University; and Plymouth State University's Center for Rural Partnerships for providing funding for research described in this chapter.

Note

1. *Redistributive activities* are described as "the provision of human services as well as the direct collection and redistribution of foods"; *reciprocal exchanges* are informal strategies for obtaining food from friends and family that work toward "solving the problems of inadequate food and dietary quality" (Morton et al. 2008, 108–109).

References

Adam, K. 2006. Community Supported Agriculture. ATTRA Publication #IP289. <http://attra.ncat.org/attra-pub/csa.html#trends> (accessed October 1, 2008).

Alkon, A. H., and K. M. Norgaard. 2009. Breaking the Food Chains: An Investigation of Food Justice Activism. *Sociological Inquiry* 79 (3): 289–305.

Allen, P. 1999. Reweaving the Food Security Safety Net: Mediating Entitlement and Entrepreneurship. *Agriculture and Human Values* 16 (2): 117–129.

Allen, P., and C. Hinrichs. 2007. Buying into "Buy Local": Engagements of United States Local Food Initiatives. In *Alternative Food Geographies: Representation and Practice*, ed. L. Holloway, D. Maye, and M. Kneafsy, 255–272. Oxford, UK: Elsevier Press.

Allen, P., M. FitzSimmons, M. Goodman, and K. Warner. 2003. Shifting Plates in the Agrifood Landscape: The Tectonics of Alternative Agrifood Initiatives in California. *Journal of Rural Studies* 19 (1): 61–75.

Blanchard, T. C., and T. L. Lyson. 2006. *Food Availability and Food Deserts in the Nonmetropolitan South*. A publication of the Southern Rural Development Center, no. 12, April. <http://srdc.msstate.edu/publications/other/foodassist/2006_04_blanchard.pdf> (accessed January 1, 2011).

Burns, C. M., and A. D. Inglis. 2007. Measuring Food Access in Melbourne: Access to Healthy and Fast Foods by Car, Bus and Foot in an Urban Municipality in Melbourne. *Health & Place* 13 (4): 877–885.

Burpee & Company. 2009. Burpee's Money Garden. <http://www.burpee.com/gardening/pressrelease/families-can-cultivate-financial-savings-with-burpee-s-money-garden-seed-pack/10011.html > (accessed March 7, 2009).

Dibsdall, L. A., N. Lambert, R. F. Bobbin, and L. J. Frewer. 2003. Low-income Consumers' Attitudes and Behaviour towards Access, Availability and Motivation to Eat Fruit and Vegetables. *Public Health Nutrition* 6 (2): 159–168.

Drewnowski, A., and S. E. Specter. 2004. Poverty and Obesity: The Role of Energy Density and Energy Costs. *American Journal of Clinical Nutrition* 79 (1): 6–16.

Eberhardt, M. S., et al. 2001. *Urban and Rural Health Chartbook. Health, United States, 2001.* Hyattsville, MD: National Center for Health Statistics.

Feagan, R. 2007. The Place of Food: Mapping out the "Local" in Local Food Systems. *Progress in Human Geography* 31 (1): 23–42.

Feenstra, G. W. 1997. Local Food Systems and Sustainable Communities. *American Journal of Alternative Agriculture* 12 (1): 28–36.

Food Research and Action Center. 2008. <http://frac.org/> (accessed October 1, 2008).

Furey, S., C. Strugnell, and H. McIlveen. 2001. An Investigation of the Potential Existence of "Food Deserts" in Rural and Urban Areas of Northern Ireland. *Agriculture and Human Values* 18 (4): 447–457.

Garasky, S., L. W. Morton, and K. Greder. 2004. The Food Environment and Food Insecurity: Perceptions of Rural, Suburban, and Urban Food Pantry Clients in Iowa. *Family Economics and Nutrition Review* 16 (2): 41–48.

Goodman, D. 2004. Anti-racist Practice and the Work of Community Food Organizations. *Sociologia Ruralis* 44 (1): 3–16.

Gottlieb, R. 2009. Where We Live, Work, Play . . . and Eat: Expanding the Environmental Justice Agenda. *Environmental Justice* 2 (1): 7–8.

Gottlieb, R., and A. Fisher. 1996. Community Food Security and Environmental Justice: Searching for a Common Discourse. *Agriculture and Human Values* 3 (3): 23–32.

Guthman, J. 2008. Bringing Good Food to Others: Investigating the Subjects of Alternative Food Practice. *Cultural Geographies* 15 (4): 431–447.

Hanson, K., and M. Andrews. 2008. *Rising Food Prices Take a Bite Out of Food Stamp Benefits.* ERS Report Summary, Economic Research Service, U.S. Department of Agriculture, Washington, DC, December.

Hendrickson, D., C. Smith, and N. Eikenberry. 2006. Fruit and Vegetable Access in Four Low-income Food Deserts Communities in Minnesota. *Agriculture and Human Values* 23 (3): 371–383.

Hinrichs, C. C. 2000. Embeddedness and Local Food Systems: Notes on Two Types of Direct Agricultural Market. *Journal of Rural Studies* 16 (3): 295–303.

Hinrichs, C. C. 2003. The Practice and Politics of Food System Localization. *Journal of Rural Studies* 19 (1): 33–45.

Hinrichs, C. C., and P. Allen. 2008. Selective patronage and social justice: local food consumer campaigns in historical context. *Journal of Agricultural & Environmental Ethics* 21:329–352.

Hinrichs, C. C., and K. S. Kremer. 2002. Social Inclusion in a Midwest Local Food System Project. *Journal of Poverty* 6 (1): 65–90.

Hofferth, S. L., and J. Iceland. 1998. Social Capital in Rural and Urban Communities. *Rural Sociology* 63 (4): 574–598.

Homeless Garden Project. 2009. <http://www.homelessgardenproject.org/> (accessed April 2009).

Ilbery, B., and D. Maye. 2005. Alternative (Shorter) Food Supply Chains and Specialist Livestock Products in the Scottish–English Borders. *Environment and Planning* 37 (5): 823–844.

Jarosz, L. 2008. The City in the Country: Growing Alternative Food Networks in Metropolitan Areas. *Journal of Rural Studies* 24 (3): 231–244.

Johnson, K. 2006. *Demographic Trends in Rural and Small Town America.* Reports on Rural America, vol. 1, no. 1, Carsey Institute, University of New Hampshire, Durham. <http://www.carseyinstitute.unh.edu/publications/Report_Demographics.pdf> (accessed May 22, 2011).

Kaufman, P. R. 1999. Rural Poor Have Less Access to Supermarkets, Large Grocery Stores. *Rural Development Perspectives* 13 (3): 19–26.

Kirkup, M., R. D. Kervenoael, A. Hallsworth, I. Clarke, P. Jackson, and R. P. D. Aguila. 2004. Inequalities in Retail Choice: Exploring Consumer Experiences in Suburban Neighbourhoods. *International Journal of Retail & Distribution Management* 32 (11): 511–522.

Kloppenburg, J., Jr., S. Lezberg, K. De Master, G. W. Stevenson, and J. Hendrickson. 2000. Tasting Food, Tasting Sustainability: Defining the Attributes of an Alternative Food System with Competent, Ordinary People. *Human Organization* 59 (2): 177–186.

Langley, K. 2009. A Farm Fades Out. *Concord Monitor.* March 29. <http://www.concordmonitor.com/article/farm-fades-out> (accessed April 2009).

Levkoe, C. 2006. Learning Democracy through Food Justice Movements. *Agriculture and Human Values* 23 (1): 89–98.

Liese, A. D., K. E. Weis, D. Pluto, E. Smith, and A. Lawson. 2007. Food Store Types, Availability, and Cost of Foods in a Rural Environment. *Journal of the American Dietetic Association* 107 (11): 1916–1923.

McEntee, J. C. 2009. Highlighting Food Inadequacies: Does the Food Desert Metaphor Help this Cause? *British Food Journal* 111 (4): 349–363.

McEntee, J. 2010. Contemporary and Traditional Localism: A Conceptualisation of Rural Local Food. *Local Environment* 15 (9–10): 785–803.

McEntee, J., and J. Agyeman. 2010. Towards the Development of a GIS Method for Identifying Rural Food Deserts: Geographic Access in Vermont, USA. *Applied Geography* 30 (1): 165–176.

Meert, H. 2000. Importance of Reciprocal Survival Strategies. *Sociologia Ruralis* 40 (3): 319–338.

Morris, P. M., L. Neuhauser, and C. Campbell. 1992. Food Security in Rural America: A Study of the Availability and Costs of Food. *Journal of Nutrition Education* 24 (1): 52S–58S.

Morton, L. W., and T. C. Blanchard. 2007. Starved for Access: Life in Rural America's Food Deserts. *Rural Realities* 1 (4): 1–10.

Morton, L. W., E. A. Bitto, M. J. Oakland, and M. Sand. 2008. Accessing Food Resources: Rural and Urban Patterns of Giving and Getting Food. *Agriculture and Human Values* 25: 107–119.

National Public Radio. 2009. Vegetable Seed Sales Shoot Skyward. *All Things Considered*. Original air date Febuary 21. <http://www.npr.org/templates/story/story.php?storyId=100969802> (accessed March 7, 2009).

New Hampshire Department of Agriculture, Markets, and Food. 2008. *New Hampshire Farm Stand Directory*. <http://www.nh.gov/agric/publications/documents/farmstand.pdf > (accessed January 19, 2011).

New Hampshire Farm to Restaurant Connection. 2009. <http://www.nhfarmtorestaurant.com/> (accessed June 2009).

New Hampshire Made. 2009. <http://www.nhmade.com/> (accessed June 2009).

Nord, M. 2002. Food Security in Rural Households: Rates of Food Insecurity and Hunger Unchanged in Rural Households. *Rural America* 16 (4) (Winter): 42–47. Food Assistance and Rural Economy Branch, Food and Rural Economics Division, ERS. <http://www.ers.usda.gov/publications/ruralamerica/ra164/ra164g.pdf>.

Nord, M., M. Andrews, and S. Carlson. 2008. Measuring Food Security in the United States: Household Food Security in the United States, 2007. United States Department of Agriculture, Economic Research Service, Food Assistance and Nutrition Research Report Number 66. <http://www.americanveterannewspaper.org/images/foodsecurity07.pdf>

Olson, C., B. Rauschenbach, E. Frongillo, Jr., and A. Kendall. 1996. Factors Contributing to Household Food Insecurity in a Rural Upstate New York County. Institute for Research on Poverty Discussion Paper, no. 1107–96, Division of Nutritional Sciences. Cornell University, Ithaca, NY.

Olson, C. M., and C. F. Bove. 2006. Obesity in Rural Women: Emerging Risk Factors and Hypotheses. In *Rural Women's Health: Mental, Behavioral, and Physical Issues*, ed. R. T. Coward, L. A. Davis, C. H. Gold, H. Smiciklas-Wright, L. E. Thorndyke, and F. W. Vondracek, 63–81. New York: Springer Publishing Company.

Pudup, M. B. 2008. It Takes a Garden: Cultivating Citizen-Subjects in Organized Garden Projects. *Geoforum* 39 (3): 1228–1240.

reVision House Farm. 2009. <http://www.vpi.org/Re-VisionFarm/index.html> (accessed April 2009).

Severson, K. 2009. When "Local" Makes It Big. *New York Times*, May 12, 2009.

Sharkey, J. R. 2009. Measuring Potential Access to Food Stores and Food- Service Places in Rural Areas in the U.S. *American Journal of Preventive Medicine* 36 (4S): S151–S155.

Skerratt, S. 1999. Food Availability and Choice in Rural Scotland: The Impact of "Place." *British Food Journal* 101 (7): 537–544.

Smith, C., and L. W. Morton. 2009. Rural Food Deserts: Low-income Perspectives on Food Access in Minnesota and Iowa. *Journal of Nutrition Education and Behavior* 41 (3): 176–187.

U.S. Census Bureau. 2008a. Census 2000. American FactFinder. <http:// factfinder.census.gov/home/saff/main.html?_lang=en> (accessed August 2009).

U.S. Census Bureau. 2008b. Small Area Income and Poverty Estimates, Estimates for New Hampshire Counties, 2006. <http://www.census.gov/cgi-bin/saipe/ saipe.cgi?year=2006&type=county&table=county&submit=States%20%26 %20Counties&areas=all&display_data=Display%20Data&state=33> (accessed August 2009).

USDA AMS. 2008a. Farmers Market Database, USDA Agricultural Marketing Service. <http://apps.ams.usda.gov/FarmersMarkets/> (accessed October 2008).

USDA AMS. 2008b. Number of Farmers Markets Continues to Rise in U.S. Program Announcement, USDA Agricultural Marketing Service. <http://www .ams.usda.gov/AMSv1.0/getfile?dDocName=STELPRDC5072472&acct=frmrdi rmkt> (accessed October 2008).

Wallace, M. 2008. The Spirit of Environmental Justice: Resurrection Hope in Urban America. *World Views: Global Religions, Culture, and Ecology* 12 (2–3): 255–269.

Wekerle, G. R. 2004. Food Justice Movements: Policy, Planning, and Networks. *Journal of Planning Education and Research* 23 (4): 378–386.

Welsh, J., and R. MacRae. 1998. Food Citizenship and Community Food Security. *Canadian Journal of Development Studies* 19 (Special issue): 237–255.

IV
Future Directions

12

"If They Only Knew"

The Unbearable Whiteness of Alternative Food

Julie Guthman

"If people only knew where their food came from. . . ." This phrase resounds in alternative food movements. My students voice it in the classroom, and it is often the first sentence of papers they write. It undergirds many of the efforts of local food system activists, who focus a good deal of effort in encouraging more personalized relationships between producers and consumers. It is the end goal for contemporary muckraking led by the likes of Eric Schlosser and Michael Pollan whose writings on industrial food production practices often evoke a "yuck" reaction. It animates the long list of ingredients on upscale restaurant menus.

The phrase warrants additional parsing. Who is the speaker? How do we identify those who do not know the source of their food? What would they do if they only knew? Do they not know now? When pushed, those who employ this rhetoric will argue that such an unveiling of the American food supply would necessarily trigger desire for local, organic food, and people would be willing to pay for it. And then, so the logic goes, the food system would be magically transformed to one that is ecologically sustainable and socially just. To be sure, many alternative food advocates in the United States see lack of knowledge as the most proximate obstacle to a transformed food system, and in their elevated esteem for farmers—and chefs—relative to others who make their living in the provision of food, these advocates think that consumers should be willing to pay the "full cost" of food (Allen et al. 2003). This assertion is made in respect to the growing sense that food in the United States is artificially cheap due to both direct and indirect subsidies to agriculture, which include not only crop payments but also water, university research and extension, and even immigration policy. It follows that food produced in more ecologically sustainable and socially just ways would necessarily cost more. As stated by Michael Pollan (2008, 184), whose paeans to the local and organic have made him a hero of sorts of the alternative food movement: "Not everyone can afford to eat high-quality food in America, and that is shameful: however, those of us who can, should."

While there is much to say about the perverse ecological, social and health-related effects of the industrialized food system (Magdoff, Foster, and Buttel 2000; Kimbrell 2002; Pollan 2006), this chapter takes on the cultural politics of "if they only knew" and "paying the full cost" as it relates to alternative food. By alternative food, I refer to the broad range of practices and programs designed to bring producers and consumers into close proximity and to educate people of the value of local, sustainably grown, and seasonal food. Alternative food institutions thus include the relatively prolific (and thus decreasingly alternative) farmers markets, subscription farms that function as community supported agriculture (CSA), farm-to-school programs, community gardens, and a variety of hands-on educational programs and demonstration sites (Allen et al. 2003). Defined as such, my critique of alternative food will partially extend to some of the more race-conscious efforts toward those ends, specifically those that come under the banner of food justice—what I will later refer to as the alternatives to the alternatives. Following Stuart Hall, I define *cultural politics* as the relationship between signifying practices and power and am particularly concerned with how racialized representations and structural inequities are mutually reinforcing (Chen 1996, 395; Hall 1996). In this vein, I want to argue that many of the discourses of alternative food hail a white subject and thereby code the practices and spaces of alternative food as white. Insofar as this coding has a chilling effect on people of color, it not only works as an exclusionary practice, it also colors the character of food politics more broadly and may thus work against a more transformative politics. My objective in raising this issue is not to condemn, but to remark on the importance of a less messianic approach to food politics, and even the need to do something different than "invite others to the table," an increasingly common phrase in considering ways to address diversity in alternative food movements.

The empirical portions of this chapter are drawn from two sets of data. One set is the results of a study I led on the convergences and contradictions of food security and farm security in two kinds of alternative food institutions—farmers markets and CSAs. Specifically, I focus on managers' responses to survey questions and interviews regarding the dearth of participation of people of color in their markets (particularly CSA) to reveal the pervasiveness of rhetoric of "if they only knew" and its cognates. The other data set is evidence my undergraduate students have collected while on field study with various food justice and food security organizations. I will use these data to suggest that even some of the more race-conscious alternatives lack resonance in communities of color, particularly among African Americans, and then suggest why that may be so.

Coding Alternative Food as White

Thus far, existing research suggests that African Americans especially, do not participate in alternative food institutions such as farmers markets and CSAs proportionate to the population (e.g., Hinrichs and Kremer 2002; Payne 2002; Perez, Allen, and Brown 2003). While it may also be the case that working-class or, more likely, less formally educated whites do not participate equal to their numbers either, they have not been subject to the same sort of scrutiny regarding their food provisioning practices, including attempts to engage them in alternative food practice.[1] Of course there are exceptions, some of which are discussed in this volume, and, as I will discuss in this chapter, some people are working hard to alter perceptions of alternative food.

This point must not be construed as a claim that African Americans do not participate in these markets at all (which surely varies by region) or, worse, a blanket indictment of African American food provisioning practices. For example, a study of a farmers market in a working-class predominantly African American neighborhood in Chicago found that shoppers at that market were much happier with the food at the farmers market relative to nearby stores, although the study did not suggest much breadth in participation (Suarez-Balcazar et al. 2006). I myself have noted a loyal following of African American shoppers and activists who attend the markets in my home region, although through informal conversations I have learned that many are highly cognizant of the "whiteness" of these spaces in ways that they themselves have had to overcome.

In any case, the problem I am addressing is not negated by the presence of a few black bodies in these alternative food institutions. Indeed, to the extent that studies only count bodies as a way of determining if all phenotypes are adequately represented, they in certain ways contribute to the problem. For they not only reinforce notions of race that are based in stabilized categories (cf. Moore, Pandian, and Kosek 2003; Reardon 2005), they allow the conceit that racism is solved merely by attention to distributional outcomes. My concern is the extent to which these practices, institutions, and spaces are coded as "white"—or at least "not black"—not only through the bodies that tend to inhabit and participate in them but also the discourses that circulate through them. As I will show, many advocates and admirers of these institutions are "colorblind" to the racial content of these discourses. I posit that their representations contribute to the production of "white space" that in effect works as an exclusionary practice.

Another caveat is that whiteness is a messy and controversial concept to work with, variably referred to as the phenotype of pale bodies, an attribute of particular (privileged) people, a result of historical/social processes of racialization, a set of structural privileges, a standpoint of normalcy, or a particular set of cultural politics and practices (Frankenberg 1993; Kobayashi and Peake 2000). My interest in using whiteness is to make the invisible visible, to decenter white as "normal" or unmarked. I do so cognizant of the critique that the prominence given to whiteness scholarship has effectively recentered whiteness, as noted by McKinney (2005) and Sullivan (2006). Nevertheless, as Rachel Slocum (2006) first noted, community food movements have been slow to address issues of white privilege, which she attributes both to the persistent invisibility of whiteness as a racial category and to resistance within the movement to embrace an antiracist practice for fear of offending allies. So there is work to be done. For that matter, concerns about race and whiteness have been notably absent in agrifood scholarship until quite recently. This volume is an important effort to fill in that lacuna.

For some scholars of whiteness, the point is to encourage more reflexivity among whites regarding their privileged social position. Building on the work of Frankenberg, whose point was to bring into view the "social geography of race," several scholars have highlighted the presumptions and effects of those who inhabit white bodies. So, according to McKinney (2005), the purpose of an engagement with whiteness is not to determine who is racist or not, but to uncover what whites think about being white and what effects that has on a racial system. Her position is that one can be nominally nonracist and still contribute to a racial society. Further augmenting this line of argument, Sullivan (2006) makes the point that the unconscious habits of white privilege are in some respects more pernicious than the explicit racism of white supremacy because it is not examined. She draws particular attention to how nonrecognition of being the beneficiaries of privilege allows whites to retain a sense of being morally good. These insights, along with those drawn from a growing body of literature, particularly sociological, as to how whites experience their whiteness are important, but they also tend to personalize whiteness (Perry 2002; Bettie 2003).

In this argument, I am drawing, then, from geographers of whiteness who as a whole seem less concerned with white personhood and instead focus on the work that whiteness as an unmarked category does in shaping social relations and, hence, space (Holloway 2000; Dwyer and Jones 2003; Schein 2006; Shaw 2006). Kobayashi and Peake (2000, 394) capture the subtle distinctions in this possibly more geographic approach when they state that "whiteness is indicated less by its explicit racism than by the fact that it ignores, or even denies, racist implications." So it is not only a matter of making white people, those inhabiting pale-skinned bodies, accountable as to their effects on others (Frankenberg 1993; Lipsitz 1998; McKinney 2005)—although that is an important project in its own right. Rather, it is also a matter of showing how discourses associated with whiteness are implicated practically and spatially. No space is race neutral; there is an iterative coding of race and space (Thomas 2005; Saldanha 2006; Schein 2006). In addition, focusing on the discursive aspects of whiteness opens the door to understanding how the doctrine of colorblindness, for example, can be embraced by all kinds of people, whites and nonwhites alike.

With this concern in mind, two related manifestations of whiteness are particularly important for how they define alternative food practice and space. One is colorblindness. For many, *colorblindness* or the absence of racial identifiers in language, or both, are seen as nonracist (Frankenberg 1993; McKinney 2005). Refusing to see (or refusing to admit) race difference for fear of being deemed racist has its origins in liberal thought, yet as many have remarked, the doctrine of colorblindness does its own violence by erasing the violence that the social construct of race has wrought in the form of racism (Holloway 2000; Brown et al. 2003). Inversely, colorblindness erases the privilege that whiteness creates. This is the point made by various scholars who have considered how whiteness acts as property—a set of expectations and institutional benefits historically derived from white supremacy that in their contemporary invisibility work to naturalize inequalities (Roediger 1991; Harris 1993; Lipsitz 1998).

The second manifestation of whiteness is universalism: the assumption that values held primarily by whites are normal and widely shared. Sometimes this takes the form of an aesthetic ideal that is not obviously racialized but is predicated on whitened cultural practices (Kobayashi and Peake 2000, 394). This move erases difference in another way by refusing to acknowledge the experience, aesthetics, and ideals of others, with the pernicious effect that those who do not conform to white ideals are justifiably marginalized (Moore, Pandian, and Kosek 2003). In other words, when particular, seemingly universal ideals do not resonate, it is assumed that those for whom they do not resonate must be educated about these ideals or be forever marked as different. It is in this classic missionary impulse that universalism works to reinscribe difference (Hall 1992; Stoler 1995). As the following sections show, both colorblindness and universalism are salient in alternative food movements.

Evidence of Colorblindness and Universalism in Alternative Food Institutions

In 2004–2005, I led a study of farmers markets and CSAs in California. The purpose of the study was to examine to what extent these market forms meet the twin goals of farm security and food security—goals that have been championed as synergistic by the community food security movement. In addition to surveys sent to all California CSAs and farmers market managers, for whom we received about a 35 percent response rate, we conducted interviews with a directed sample of CSAs and farmers market managers to explore some of these issues in more depth. Overall, we found that managers of these institutions generally support the idea of improving the affordability of the food they provide, and most have made an effort to do so, although these efforts vary with institutional capacity. Still, some hedged their interest in supporting food security goals with countervailing concerns such as the need to support farmers first (Guthman, Morris, and Allen 2006).

The research also provided additional evidence that these institutions disproportionately serve white and middle- to upper-income populations, although it almost goes without saying that farmers markets are more racially and class diverse than CSAs. Most CSA managers reported that the vast majority of their customers were white. In response to a survey question as to why European Americans appeared to be the dominant ethnic group, one CSA manager wrote, "cause unfortunately we are in honky heaven! And the only people who seem to be able to afford to live here are people of this race." Farmers market managers reported having more ethnic diversity at their markets, mainly because farmers markets more closely mirror the demographics of the area in which they are located, as many managers noted. Still, few farmers markets are located in communities of color, especially those that are primarily

African American, and those that exist in African American neighborhoods tend to be very small. As one farmers market manager noted in his/her survey form: "Farmers' markets are good for everyone, but many of them are being located in 'high-end' areas. The farmers may make more money there, and the higher income communities are 'entertained' by outdoor markets." To be fair, since the primary purpose of such markets is to serve farmers by providing a regular source of income, most markets are set up in areas where palpable demand exists for them (which, again, also includes many Asian immigrant communities), unless market charters require otherwise.

Putting these important demographic issues aside, it is worth considering why these institutions tend to be disproportionately white even in communities with a more racially mixed population. To this end, our attempts to explore how managers regarded the importance of racial inclusion and what they had done to encourage participation are telling. Following the suggestions of various scholars of whiteness, paying attention to the subjects rather than the objects of racializing discourses is a compelling way to understand the work that representational practices do (Morrison 1992; Frankenberg 1993; McKinney 2005). In this case, they illustrate the whitened cultural politics that operate in these institutions and specifically the ways in which themes of colorblindness and universality recur.

At one level, most respondents were sympathetic to a project that would make their markets more inclusive. Seventy-four percent of farmers market managers and 69 percent of CSA managers thought it important to address the ethnic diversity of their markets, although the enthusiasm among CSA managers dropped to 59 percent when asked if they would consider strategies that increase the ethnic diversity of their customers. The ideas expressed in open-ended written comments and interviews were more revealing.

Most of the managers surveyed and interviewed in this study believed their market spaces are universal spaces and speak to universal values. As one CSA manager stated, the purpose of CSAs is to "have people eat real food and understand where it comes from." For some, that entails rejecting the very idea of having strategies to reach out to particular communities of color. When asked how to improve diversity at the market, one manager responded, "We always hope for more people and do not focus on ethnic—what we present attracts all!" Likewise, a CSA manager said, "Targeting those in our communities that are ethnic or low income would show a prejudice we don't work within. We do outreach programs to reach everyone interested in eating locally, healthily, and organically."

Some managers explicitly invoked the language of colorblindness. Aversion to questions regarding the ethnicity of customers was founded on the presumption that the questions themselves were racist. As one farmers market manager put it, "Some of your questions are pretty intrusive—I also found some to be racist. I left these questions blank. This was intentional, not accidental." Echoed the CSA respondent quoted earlier, "Difference is wrong; it is better to try to become color blind in how we do things . . . your questioning has a slant of political correctness. . . . We are set up for our community." Yet another CSA manager said: "I think it is an admirable goal to try to get our customers to be more diverse, but I feel a bit troubled by all of this. I sometimes feel pressure to be perfectly politically correct . . . I wish we could elevate the farmers first, then it might be easier to bring the rest of the world along."

While in one register managers rejected the idea of difference, in another they invoked it. Importantly, this last comment was followed immediately by one in which she said "the [CSA] concept needs to be taken on by low income and ethnic folks." Indeed, another recurring theme throughout the responses was that healthy, local, sustainable eating is a "lifestyle choice" and one that people of color apparently do not adhere to. For example, in responding to the question "What do you think are some of the reasons that it is primarily European-American people who seem to participate in CSAs?" respondents consistently imputed personal characteristics and motives rather than problems with access and affordability. In the qualitative analysis, phrases such as "better education," "more concern about food quality," "more health consciousness," and even "more time" were mentioned repeatedly. One manager portrayed white people as "more aware and willing to do something with food for sociopolitical reasons rather than other reasons and involved in the social component of CSAs and what they represent." Another simply said, "Hispanics aren't into fresh, local, and organic products."

Farmers market managers named some of these same issues, but also tended to include additional factors regarding neighborhood demographics and location, and cost, as obstacles to participation. Even attributing behavior to cost, though, makes presumptions about differences in values. For example, in reference to a question regarding expanding entitlement programs to make farmers markets more affordable to all, one manager responded, "I'm not sure that I agree that subsidy is the best route. In my experience, the subsidy customers are the least committed and reliable. I believe that the food is affordable to all; it's just a matter of different values and priorities. Education and outreach are the only hope I have of interesting more low income people." One respondent who characterized his market as one that "caters to high income consumers seeking quality and freshness" said that "low income people shop elsewhere unless they are given freebies like WIC" and that he would not want to use strategies to attract low-income consumers because those strategies "may discourage the high end consumers that we cater to."

In short, these responses represent various ways in which lack of knowledge or the "right" values or both are seen as barriers to broader participation in alternative food institutions. As Nash (2007) discusses regarding farmworkers' horrific pesticide exposures, it is an old trope to attribute structural inequalities to cultural differences or lack of education. What I hope I have shown in addition is that this position involves certain significations: specifically, managers portray their own values and aesthetics to be so obviously universal that those who do not share them are marked as other. These sorts of sensibilities are hallmarks of whiteness. So, in assuming the universal goodness of fresh, local, and organic food, those who value this food ask those who appear to reject this food to either be subject to conversion efforts or simply be deemed as other. If they only knew.

Evidence of Lack of Resonance

Thus far this chapter has focused on a particular set of subjects in alternative food, namely those in the business of managing what are clearly market-driven institutions. As such, it is certainly reasonable that they would take for granted the superior qualities of alternative food and be less than wholly reflective of the effects of alternative food discourses. Some institutions in alternative food, however, are not so market driven and as a social service or social change activity work to provide people with or educate them about alternative food. To be sure, many of what might be called the alternatives to the alternatives have been created in the name of food justice or food security or both. Particularly in light of the dearth of healthy food venues in many low-income African American neighborhoods (the so-called food desert problem), these programs are premised on the idea that low-income African Americans lack access, or perhaps exposure, to this food. As such, they are more race conscious in their practices and rhetoric and seek, as Slocum (2007) puts it, "to bring this good food to others."

I have learned something about these too, primarily through my students who as part of the Community Studies undergraduate major at UCSC do six- month field studies with food- and agriculture-related social justice organizations. Many choose to work with food justice or food security projects. During the field study, they are expected to write field notes and updates, and upon their return they are required to analyze their notes and prepare a capstone essay or thesis. These notes and formal papers are the basis of the evidence I present in this section. Admittedly, some of these data should be taken with a grain of salt, given that they reflect student disappointments relative to their pre-field study expectations (see Guthman 2008a for an extended discussion of this). Nevertheless, the consistency of observations across different projects tells us something.

For example, many students note that organic or farm fresh food does not always resonate in the communities and projects they study. In a required faculty update letter, one student working at a gardening and environmental education project reported, "Often times the girls show up with Jack in the Box for breakfast, eating it while working. The newly hired market manager . . . is openly opposed to eating vegetables." Another student wrote of a field trip where she accompanied African American youth to a nearby intentional community and organic demonstration farm. As she reports it in her thesis most of the youth were repulsed by the food. "Eww! This lasagna is vegetarian?" inquired one girl. "This shit is organic," stated another. Later the youth were asked to say what they thought organic means. Many used the terms "disgusting," "gross," or "dirty" (Tattenham 2006).

One student worked with a job program for homeless people and recovering drug addicts in a midwestern city. The program is designed to teach people how to farm as a means of employment and empowerment. It also encourages them to change the way they eat. As my student wrote, many trainees did not take home as much produce as they could carry, despite encouragement to do so (Goy 2007). Another working at a garden project in New Orleans after Hurricane Katrina had a similar experience. She wrote of an exchange she had had with a local black activist whom she had encouraged to stop by and pick vegetables. The person "laughed and said she did not know how to cook any of the things we planted" (Goelz 2007). Garden projects faced similar indifference—and sometimes hostility. A student cited earlier wrote of another field trip to a nearby organic farm. The director of the youth program had said that it would be a good idea for the youth to "get their hands dirty" and pick fruit. As my student described it, the African American chaperone, as well as the youth, had scowls on their faces as they left for the field trip. In talking to the youth later, she learned that they resented the expectation to work not only for free, but also for white farmers (Tattenham 2006). In a somewhat similar vein, the letter from the student at the environmental education project went on to say:

I am also working at a community garden that is run by [the same organization]. It is approximately one acre and is located within public housing. It is far from a productive garden. It is under funded and run by a seventy year old woman with no gardening experience. She is assisted by teenagers [who] drag their feet and have little interest in being there, perhaps because their tasks mainly consist of weeding. They have little access to the growing or harvesting [of] crops and are unable to see the whole process through. They are used as laborers instead of gardeners. There is little outside expertise brought in and no community involvement.

The student working in New Orleans noted, "few people ever acknowledged our presence or accompanied us in the garden" (Goelz 2007, 31). The student working in the job training program found that those working in the garden could not identify crops they had seeded, transplanted, cultivated, and harvested all season. Several students who have worked in school gardening programs have reported back to me that the kids do not like to garden—they don't like getting their shoes or hands dirty. They have also corroborated the finding that that many of the youth of color participating in garden projects see their efforts more as donated labor than therapy.

Some of my students observe that such projects lack resonance precisely because they are alternatives. In a conversation with her African American neighbor one day, my student mentioned that she worked for the organization that brought a truck load of organic fruits and vegetables to their neighborhood. Her neighbor's response to why she did not shop from the truck (which was both convenient and sold at below market prices) was "Because they don't sell no food! All they got is birdseed." She went on to exclaim, "Who are they to tell me how to eat? I don't want that stuff. It's not food. I need to be able to feed my family." When my student asked her what she would like the truck to offer, the neighbor said, "You know, what normal grocery stores have" (Tattenham 2006). Another student working at an umbrella organization for several different projects noted in her thesis how very few people either worked in the nearby garden or attended the fledgling farmers market. She also worked on a liquor store conversion project where, as she writes, it was difficult to get a steady following. One month they had some success, selling about twenty bunches of collard and mustard greens in one week. Most of the food, though, went into the compost (Biddle 2005). In a verbal report to me she noted that organizational staff had heard from community members on several occasions that what they really wanted was a Safeway supermarket in their neighborhood.

Of course the problems I describe here have not gone unnoticed by movement activists and project leaders. I have attended many public meetings of the sustainable agriculture/alternative food movements where people of color or whites working in communities of color insist that the messages of these movements are, simply put, "too white." Accordingly, many of these projects are making explicit efforts to attract African Americans, as names such as The Peoples' Grocery, Mo' Betta Foods, Food from the Hood, Mandela's Farmers' Market, Black to Our Roots, and Growing Power amply demonstrate. These organizations endeavor to rework some of the whitened idioms associated with alternative food practice, specifically as a way to enroll African Americans in their projects.

For example, as described by Alkon (2008), the West Oakland farmers market (Mandela market) frames its work in terms of black identity. It publicizes the plight of African American farmers (who are few and far between in California), and encourages community support of these farmers in the name of racial justice. She describes how vendors see their work in terms of providing alternatives to the supermarkets that have abandoned the cities and, as one put it, "sell poison." The market also emphasizes black cuisine and culture. And unlike, say, the nearby Berkeley farmers market, food is sold substantially below supermarket prices (some vendors have been subsidized at times). Mo' Better Foods similarly sees its work in terms of rectifying a history where African Americans have been consistently stripped of landholding possibilities. They also work to establish a more positive relationship to agrarian production among African American consumers.

Still, leaders and staff of these organizations constantly have to struggle to create and maintain an African American presence, sometimes at the expense of keeping out white people who want to do good, including interns. Another one of my students, who worked in a community garden in a largely African American community, wrote in her field notes (not publicly available) of the discussions she often had with her supervisor. "She often tells me and other people who come into [the name of the garden] wanting to help out, that 'we are not trying to serve the anarchist/hippy/crust sector' of [that city]. This makes sense because she wants to serve the African American/Latino families and not groups of privileged white people who come around 'wearing ripped up clothes and generally insulting everyone.'" The student went on to note that the "struggle for inclusivity in the garden was complicated by the inadvertent creation of white spaces which put off a majority of the low income people."

In sum, the evidence my students provide suggests that even these more race-conscious projects tend to get coded as white. They are coded so not only because of the prevalence of white bodies, as salient as that is. As some of their field notes illustrate, it also appears that the associations of the food, the modes of educating people to its qualities, and the ways of delivering it lack appeal to the people such programs are designed to entice. It is worth considering why this seems to be the case. Is it a lack of knowledge?

"If They Only Knew"?

In her dissertation entitled "Black Faces, White Spaces: African Americans and the Great Outdoors" Carolyn Finney (2006) found a tendency among whites to attribute the lack of participation of African Americans in the U.S. national parks to such things as different values, lack of interest, or the costs of getting there. When she queried African Americans on the same issue, many rejected those sorts of prompts and responded to an open-ended prompt of "exclusionary practices." Not all respondents specified these practices, but those who did pointed to issues such as cultural competency, white privilege, and varying levels of commitment by environmental groups.

I want to argue for a similar phenomenon with these spaces of alternative food provision. And the exclusionary practices I want to point to are a pervasive set of idioms in alternative food practice that are either insensitive or ignorant (or both) of the ways in which they reflect whitened cultural histories and practices (Kobayashi and Peake 2000). "Getting your hands dirty in the soil," "if they only knew," and "looking the farmer in the eye" all point to an agrarian past that is far more easily romanticized by whites than others (Guthman 2004). It is important to recall that U.S. agricultural land and labor relations are fundamentally predicated on white privilege. As elucidated by Romm (2001), land was virtually given away to whites at the same time that reconstruction failed in the South, Native American lands were appropriated, Chinese and Japanese were precluded from landownership, and the Spanish-speaking *Californios* were disenfranchised of their ranches. So, when I teach Wendell Berry, a poet of agrarianism much beloved by the sustainable agriculture movement, I do so not to depict Berry as a racist because of his skin color, but instead to show how a romanticized American agrarian imaginary erases the explicitly racist ways in which, historically, American land has been distributed and labor has been organized, erasures that ramify today in more subtle cultural coding of small-scale farming. For African Americans, especially, putting your hands in the soil is more likely to invoke images of slave labor than nostalgia. Such rhetoric thus illustrates a lack of cultural competency that might be deemed an exclusionary practice.

Yet, it is the rhetoric of paying the full cost that best demonstrates what Lipsitz (1998) has called "the possessive investment in whiteness," referring to the relationship between whiteness and wealth accumulation and other structural privileges in U.S. society. To return to this quote from Michael Pollan: "Not everyone can afford to eat high-quality food in America, and that is shameful: however, those of us who can, should." What at first appears to be a compassionate statement that acknowledges class disparities is, in effect, a punt. It assumes the persistence of inequality and, for that matter, ignores that the racialized land and labor relationships embedded in the U.S. food system continue to contribute to structural inequality. As such, it puts people who cannot afford to eat high-quality food beyond consideration—the very definition of an exclusionary practice. As Barack Obama stated, when he was a presidential candidate running against John McCain, "it's not that he doesn't care, it's just that he doesn't get it."

If Who Only Knew?

The data presented in this chapter say much more about those who champion alternative food than those who are the objects of conversion—or dismissal—in the context of alternative food efforts. Still, I have also tried to suggest some of the reasons that these institutions do not seem to resonate among African Americans, especially, as much as they do for whites. Clearly more research is needed to understand how and to what degree people of color experience exclusionary practices in the spaces of alternative food provision. It is research I hope to pursue. By the same token, we also need to pay attention to the openings. To that end, Jessica Hayes-Conroy's (2009) long-term observations of a middle school cooking and gardening program in Berkeley, California, is an important antidote to what I have reported here. While her findings confirm some of mine in terms of students' distaste for the gardens and tendencies for the African-American students to write off the food they experienced as "white food," she also witnessed their agency in making alternative food their own—more so when opportunities to eat were not scripted in a certain way. In general we need to understand much more about food practices and not assume that people who shop for "industrial food" are any less knowledgeable or ethically driven. In that vein, we need to understand more about class differences as well as race differences in foodways, especially in light of the Obamas' championing of organic food, which is likely to reorganize the idioms of the local and organic.

In terms of activism, we need to think a lot more about the ethics of "bringing good food to others" in alternative food. My underlying concern is that because alternative food tends to attract whites more than others, whites continue to define the rhetoric, spaces, and broader projects of agrifood transformation. As I have argued elsewhere (Guthman 2008b), the current menu reflects a fairly delimited conception of the politics of the possible, with a tremendous emphasis on market-driven alternatives, which often take root in the most well-resourced localities. This is an enormous problem given that race intersects with agriculture and food in myriad ways, yet many substantial health and livelihood inequalities are barely addressed through existing social movement activity. Insofar as people of color die younger due to such lack of attention, the problem in its totality surely meets Gilmore's (2002) definition of racism. In other words, the implications of these perhaps minor exclusions are far reaching.

In that vein, it is worth taking note that those African Americans who do participate in alternative food have tended to become involved because they have been sickened (literally or figuratively) by industrial food provisioning practices. Still, with a persistent dearth of people of color in alternative food spaces, and perhaps in the absence of explanations that might render their indifference to alternative food practice understandable, the rhetoric of "if they only knew" appears to be vindicated. The goodness of the food continues to go without saying. This is the hallmark of whiteness and its presumption of normativity; it goes to the deeper way in which colorblindness and acts of doing good can work to separate and scold others.

My point, however, is not to disable activists and advocates who have good intentions, out them for being overtly racist, or even to claim that without whiteness food activism would take a substantially different course and be wildly successful. My immediate goal for this chapter is to encourage much deeper reflection on the cultural politics of food activism. Saldanha (2006, 11) is surely right that "the embodiment of race . . . encompasses certain ethical stances and political choices. It informs what one can do, what one should do, in certain spaces and situations." Following Sullivan (2006), whites need to think about how to use the privileges of whiteness in an antiracist practice. In the realm of food politics, this might mean turning away from proselytizing based on universal assumptions about good food. Perhaps a place to start would be for us whites to state how much we do not know to open up the space that might allow for others to define the spaces and projects that will help spurn the transformation to a more just and ecological way of providing food.

Note

1. In contrast, community gardens seem to have significant participation of Latino immigrants (Terwilliger 2006) and farmers markets are frequently peopled by both Asian Americans and South, Southeast, and East Asian immigrants. Such observations would seem to support the case for a cultural politics of these spaces.

References

Alkon, Alison. 2008. Paradise or Pavement: The Social Construction of the Environment in Two Urban Farmers Markets and their Implications for Environmental Justice and Sustainability. *Local Environment: The International Journal of Justice and Sustainability* 13 (3): 271–289.

Allen, Patricia, Margaret FitzSimmons, Michael Goodman, and Keith Warner. 2003. Shifting Plates in the Agrifood Landscape: The Tectonics of Alternative Agrifood Initiatives in California. *Journal of Rural Studies* 19 (1): 61–75.

Bettie, Julie. 2003. *Women without Class: Girls, Race, and Identity.* Berkeley: University of California.

Biddle, Christie. 2005. Sowing Alternatives: A Critical Look at the Community Food Security Movement. Senior thesis, Community Studies, University of California, Santa Cruz.

Brown, Michael K., Martin Carnoy, Elliott Currie, Troy Duster, David B. Oppenheimer, Marjorie M. Shultz, and David Wellman. 2003. Of Fish and Water: Perspectives on Racism and Privilege. In *Whitewashing Race: the Myth of a Color-Blind Society*, ed. M. K. Brown, M. Carnoy, E. Currie, T. Duster, D. B. Oppenheimer, M. M. Shultz, and D. Wellman, 34–65. Berkeley: University of California.

Chen, Kuan-Hsing. 1996. Cultural Studies and the Politics of Internationalization: an Interview with Stuart Hall. In *Stuart Hall: Critical Dialogues in Cultural Studies*, ed. D. Morley and K.-H. Chen, 392–408. London: Routledge.

Dwyer, Owen J., and John Paul Jones. 2003. White Socio-spatial Epistemology. *Social & Cultural Geography* 1 (2): 209–222.

Finney, Carolyn. 2006. Black Faces, White Spaces: African Americans and the Great Outdoors. PhD diss., Department of Geography, Clark University, Worcester, MA.

Frankenberg, Ruth. 1993. *White Women, Race Matters: The Social Construction of Whiteness*. Minneapolis: University of Minnesota.

Gilmore, Ruth Wilson. 2002. Fatal Couplings of Power and Difference: Notes on Racism and Geography. *Professional Geographer* 54 (1): 15–24.

Goelz, Mariah. 2007. Ideologies Aside: The Need for a Reformation in the Community Food Security Movement. Senior thesis, Community Studies, Santa Cruz, University of California.

Goy, Christina. 2007. Oversights of the Community Food Security Initiatives: Rural and Low-Income Access to Food. Senior thesis, Community Studies, Santa Cruz, University of California.

Guthman, Julie. 2004. *Agrarian Dreams? The Paradox of Organic Farming in California*. Berkeley: University of California Press.

Guthman, Julie. 2008a. Bringing Good Food to Others: Investigating the Subjects of Alternative Food Practice. *Cultural Geographies* 15 (4): 425–441.

Guthman, Julie. 2008b. Neoliberalism and the Making of Food Politics in California. *Geoforum* 39 (3): 1171–1183.

Guthman, Julie, Amy W. Morris, and Patricia Allen. 2006. Squaring Farm Security and Food Security in Two Types of Alternative Food Institutions. *Rural Sociology* 71 (4): 662–684.

Hall, Stuart. 1992. The West and the Rest: Discourse and Power. In *Formations of Modernity*, ed. S. Hall and B. Gieben, 275–320. Cambridge, UK: Polity Press.

Hall, Stuart. 1996. New Ethnicities. In *Stuart Hall: Critical Dialogues in Cultural Studies*, ed. D. Morley and K.-H. Chen, 441–449. London: Routledge.

Harris, Cheryl I. 1993. Whiteness as Property. *Harvard Law Review* 106 (8): 1709–1791.

Hayes-Conroy, Jessica. 2009. Visceral Reactions: Alternative Food and Social Difference in American and Canadian Schools. PhD diss., Geography and Women's Studies, Pennsylvania State University, State College.

Hinrichs, Claire G., and K. S. Kremer. 2002. Social Inclusion in a Midwest Local Food System. *Journal of Poverty* 6 (1): 65–90.

Holloway, S. R. 2000. Identity, Contingency and the Urban Geography of "Race." *Social & Cultural Geography* 1 (2): 197–208.

Kimbrell, Andrew, ed. 2002. *The Fatal Harvest Reader*. Washington, DC: Island Press.

Kobayashi, Audrey, and Linda Peake. 2000. Racism Out of Place: Thoughts on Whiteness and an Anti-racist Geography in the New Millennium. *Annals of the Association of American Geographers* 90 (2): 392–403.

Lipsitz, George. 1998. *The Possessive Investment in Whiteness*. Philadelphia: Temple University Press.

Magdoff, Fred, John Bellamy Foster, and Frederick H. Buttel, eds. 2000. *Hungry for Profit*. New York: Monthly Review Press.

McKinney, Karyn D. 2005. *Being White: Stories of Race and Racism*. New York: Routledge.

Moore, Donald S., Anand Pandian, and Jake Kosek. 2003. Introduction: The Cultural Politics of Race and Nature: Terrains of Power and Practice. In *Race, Nature, and the Politics of Difference*, ed. D. S. Moore, J. Kosek, and A. Pandian, 1–70. Durham, NC: Duke University Press.

Morrison, Toni. 1992. *Playing in the Dark: Whiteness and the Literary Imagination*. Cambridge, MA: Harvard University Press.

Nash, Linda. 2007. *Inescapable Ecologies: A History of Environment, Disease, and Knowledge*. Berkeley: University of California Press.

Payne, Tim. 2002. U.S. Farmers' Markets 2000: A Study of Emerging Trends. *Journal of Food Distribution Research, Food Distribution Research Society* 33 (1) (March). <http://agmarketing.extension.psu.edu/ComFarmMkt/PDFs/emerg_trend_frm_mrkt.pdf> (accessed July 19, 2007).

Perez, Jan, Patricia Allen, and Martha Brown. 2003. Community Supported Agriculture on the Central Coast: The CSA Member Experience. Center Research Brief #1, Center for Agroecology and Sustainable Food Systems, University of California, Santa Cruz.

Perry, Pamela. 2002. *Shades of White: White Kids and Racial Identity in High School*. Durham, NC: Duke University Press.

Pollan, Michael. 2006. *The Omnivore's Dilemma: a natural history of four meals*. New York: Penguin.

Pollan, Michael. 2008. *In Defense of Food: An Eater's Manifesto*. New York: Penguin.

Reardon, Jenny. 2005. *Race to the Finish: Identity and Governance in an Age of Genomics*. Princeton, NJ: Princeton University Press.

Roediger, David R. 1991. *The Wages of Whiteness: Race and the Making of the American Working Class*. New York: Verso.

Romm, Jeff. 2001. The Coincidental Order of Environmental Injustice. In *Justice and Natural Resources: Concepts, Strategies, and Applications*, ed. K. M. Mutz, G. C. Bryner, and D. S. Kennedy, 117–137. Covelo, CA: Island Press.

Saldanha, Arun. 2006. Reontologising Race: The Machinic Geography of Phenotype. *Environment and Planning D: Society & Space* 24 (1): 9–24.

Schein, Richard H. 2006. Race and Landscape in the United States. In *Landscape and Race in the United States*, ed. R. H. Schein, 1–21. New York: Routledge.

Shaw, Wendy S. 2006. Decolonizing Geographies of Whiteness. *Antipode* 38 (4): 851–869.

Slocum, Rachel. 2006. Anti-racist Practice and the Work of Community Food Organizations. *Antipode* 38 (2): 327–349.

Slocum, Rachel. 2007. Whiteness, Space, and Alternative Food Practice. *Geoforum* 38 (3): 520–533.

Stoler, Ann Laura. 1995. *Race and the Education of Desire*. Durham, NC: Duke University.

Suarez-Balcazar, Yolanda, Louise I. Martinez, Ginnefer Cox, and Anita Jayraj. 2006. African-Americans' Views on Access to Healthy Foods: What a Farmers' Market Provides. *Journal of Extension* 44 (2). <http://www.joe.org/joe/2006april/a2.php> (accessed July 19, 2007).

Sullivan, Shannon. 2006. *Revealing Whiteness: The Unconscious Habits of Racial Privilege*. Indianapolis: Indiana University.

Tattenham, Katrina. 2006. Food Politics and the Food Justice Movement. Senior thesis, Community Studies, University of California, Santa Cruz.

Terwilliger, Kaley. 2006. Sowing the Seeds of Self-Determination: Why People Are Community Gardening in Los Angeles County. Senior thesis, Community Studies, University of California, Santa Cruz.

Thomas, Mary. 2005. "I Think It's Just Natural": The Spatiality of Racial Segregation at a US High School. *Environment & Planning A* 37:1233–1248.

13

Just Food?

E. Melanie DuPuis, Jill Lindsey Harrison, and David Goodman

In 2002, the title for the Community Food Security Coalition Conference in Seattle was "Think Globally, Act Locally." In 2010, the title for the New Orleans conference is "Food, Culture and Justice: The Gumbo that Unites Us All." As the titles and talks for this year's conference indicate, local food movement activists have discovered the word "justice" and are rethinking localist strategies to link with more global food movements. This change reflects the fact that food movement groups are increasingly incorporating the word *justice* on their Web pages, in their nonprofit names, and in the names of the programs they initiate, such as the California Food and Justice Coalition or Growing Power's The Growing Food and Justice for All Initiative. Yet, food movement activists continue to talk about "healthy communities," about increased social connectedness between eaters and producers, about neighbors becoming more caring of one another, about face-to-face relationships, and about trust, transparency, and access. Emotional words like *caring* and *trust* become signposts on which the notion of food justice hangs. Therefore, while one seldom sees the statement "Local food systems are more just," ideas about "local food" and "just food" tend to get connected in food activist talk.

Many scholars who study alternative food systems often make similar arguments about the links between local food systems and social justice, with case studies showing that relocalization is leading to more equitable and more democratic food systems and, in general, to healthier civil societies (Holloway and Kneafsey 2004; Hassanein 2003; DeLind 2002; Lyson 2005; Miele and Murdoch 2002; Kloppenburg, Hendrickson, and Stevenson 1996). Jack Kloppenburg and his colleagues' engaging essay "Coming into the Foodshed," widely regarded as a seminal work in local food scholarship, exemplifies this connection between local food and a more just world. In it, the authors describe the foodshed as "one vehicle through which we assemble our fragmented identities, reestablish community and become native not only to a place but to each other" (Ibid., 34). The Drake University Agricultural Law Center's State and Local Food Policy Project Web site describes food policy councils as localizing institutions "playing a role in building a better food system—strengthening food democracy" (State and Local Food Policy Project 2011). Scholars see food system localization as part of progressive participatory democracy movements around the world (e.g., see Hassanein 2003).

From the perspective of these observers, a local food system fosters sustainability, democracy, and justice. Just societies and local healthy ecologies—of both nature and the bodies of consumers—are framed as inextricably linked. Yet, the question remains: is "the local" intrinsically just? (Born and Purcell 2006).

Critics have argued against the intrinsic justice (or sustainability, or healthiness) of local food systems (Born and Purcell 2006), noting that *the local* is not an innocent term (Harvey 1996). Yet, while food activists and scholars have repeatedly examined the idea of *the local*, they seldom explain what they mean by the term *just*. In fact, their work has ignored another, larger conversation about the meaning of justice itself. It is time to bring these ideas together, to more closely examine the meaning of the term *social justice* in local food politics.

In this chapter we will look specifically at the current theories of justice to argue that the alternative food movement would be more effective if it worked with a more "reflexive" notion of justice. A reflexive notion of justice is one with a clear understanding of the complexities of justice in terms of its various and contradictory meanings. Without a reflexive approach, work toward a more just food system will remain problematic and incomplete. We will lay out the foundation of a reflexive approach to local food activism by interrogating the notions of justice that are part of the current alternative food movement, and then showing how a multifaceted definition of justice can improve the movement's effectiveness in terms of making food-society relationships more socially just.

Understanding the diverse and sometimes contradictory definitions of justice, we argue, is essential to the remaking of food systems that we can all agree are "just" no matter how we define this term. Specifically, we will look at the pervasive but unexamined conceptions of justice currently underlying much food movement work—communitarian, anticorporate, and liberal egalitarian—as well as the current embrace of a more cultural notion of justice in food movement agendas. The second part of the chapter will, therefore, articulate a more reflexive agenda as steps toward a reflexively just food localism.

It is not within the scope of this chapter to cover the political theory conversations about justice in total. Instead, we will focus on how these theories deal with the tension between the universal and the particular, and how localizing food movements have ignored these tensions. In the final section, we will argue that food movement activists will need to work reflexively within these tensions in order to achieve imperfect, yet more socially just goals.

Theories of Justice

To understand theories of justice, we need to ask some of the basic questions that have pervaded political philosophy since its inception. Is justice a matter of determining "the good"—such as "good food"—and then attempting to move society toward this ideal? Or is the creation of a single set of universal values intrinsically nonegalitarian, creating a situation where some determine a singular set of ideals not necessarily shared by others, values that others resist as a violation of their freedom? Can even those with good intentions create particular visions of "the good life" that unconsciously exclude some people? These tensions between the democratic ideals of individual autonomy and sovereign notions of "the common good" as a particular vision of "the good society" constitute a central debate in modern political theory. Understanding the political theory debate over the meaning of justice can help us clarify questions of local food and social justice as well.

The classical philosopher Plato devoted himself to discovering the ideals that would lead to a perfect—just—society. In the Middle Ages, religious and regal authorities determined these ideals for society as a whole. Beginning in the seventeenth century, however, Enlightenment political philosophers, such as Thomas Hobbes and John Locke, rejected religious authority and instead saw the ideals of justice in God and nature, as human rights for democratic participation in the creation of a social contract. In *The Social Contract,* Jean-Jacques Rousseau argued that individuals, through democratic processes, could define just ideals of "the common good" and "the good life" for a particular society. Jefferson used the ideas of natural rights and universal truths in the Declaration of Independence, translating democratic ideals of a just society through his understanding of Locke and Rousseau (Kymlicka 2002; Bailyn 1967).

But some political philosophers responded skeptically to this idea of a social contract. Utilitarian political philosophers Jeremy Bentham and John Stewart Mill rejected social contract idealism as injurious to an individual's personal freedom to choose his or her own idea of the good, just life. For utilitarians, the perfectionist vision of the good life was inevitably authoritarian, since it imposed a particular idea of good living on everyone and did not admit alternative views. In other words, perfectionist idealism is undemocratic—unjust—denying the individual's own rationality and ability to determine "right living." Bentham and Mill's utilitarian political philosophy was specifically aimed at overcoming this platonic, social contract-based, perfectionist view of justice. Utilitarian Adam Smith took this a step further, arguing that involvement in market exchange was the way in which individuals pursued their interests and fulfilled their individual notions of the good life and greater wealth for all. From this perspective, the market facilitates social justice by enabling each person to choose what life to lead, specifically by offering in the market various "goods" a person can choose, and thereby maximizing these choices and increasing the wealth necessary to give people more choices.

In response to the utilitarian critique, social-contract political philosophers have argued that a world based solely on individual interests leads to a war of all against all, a state of social chaos as first laid out by Thomas Hobbes as the main reason for the creation of a social contract. Hobbes argued that, in a society without traditional hierarchies like priests and kings, a social contract was necessary to avoid this chaos, a state of society often depicted nowadays in post-apocalyptic movies like *Bladerunner*. Ever since Hobbes first laid out the idea of a social contract—and the Hobbesian nightmare of social chaos—tensions between individual liberty and social order have pervaded the political theory debate about social justice.

John Rawls's seminal work, *A Theory of Justice* (Rawls 1971), has represented the most prominent attempt to reconcile these tensions. Rawls's liberal egalitarian theory of justice revived the Enlightenment ideals of equal rights and individual freedom while taking seriously both social contract ideas of fairness and utilitarian critiques of perfectionism. Rawls formulated a universal political philosophy of right action that did not rely on a preformed, perfectionist set of ideals or values, except for a basic social commitment to equality (Rawls 1971, 1993). In other words, Rawls's theory of justice took seriously the utilitarian notions of liberty from authority while maintaining the social contract idea that there is one universal idea of justice as "fairness." He did this by balancing liberal ideas of equality with utilitarian ideas of freedom through a thought experiment: what would the social contract look like if people agreed to an idea of justice under a "veil of ignorance" about their own economic and social position in society? He argued that, under those (theoretical) circumstances, people would agree to the maximum possible liberty and equality—including equality of access to political institutions—but that they would accept an unequal distribution of wealth if that inequality benefited the least well-off members of society. In other words, you deserved more if your efforts and talents were dedicated to improving the lives of those who have the least.

Rawls's work initiated an intense debate over the nature of justice in contemporary society. Critiques of his ideas tend to come from three alternative perspectives: political economy, communitarian, and cultural.

Political economy perspectives place the onus for injustice on the structure of capitalism, particularly global industrial capitalism and its market-based individualism, as celebrated by libertarians. The political economy perspective on capitalism draws on Marx's critique of capitalism as drawing a veil over the real inequalities in a society. For Marx, economic goods—what utilitarians see as freedom—in fact were a "fetish," an illusory form of freedom. For Marx, market relationships between production and consumption, and capital and labor, formed the basis for social injustice. From this perspective, one of the goals of a social justice movement is to uncover this "truth" by unveiling this exploitative relationship, thereby making people more aware. Awareness, from this perspective, leads to social action. In today's society, critiques of global corporate control and monopoly power tend to follow this line of argument.

Yet, both Marx and Rawls critiqued perfectionist ideas of social justice, agreeing with utilitarianism insofar as the solution to social problems was not the reestablishment of particular normative values, but the establishment of better political processes (although they disagreed about the nature of those processes). Marx advocated a radical restructuring of society, but he did not paint a vision of the "perfect society" that would result from that process. Both Marx and Rawls agreed with Smith that perfectionist visions of social ideals were inimical to freedom. Marx critiqued the communitarians, especially Proudhon, and other utopian visionaries for their attempts to imagine and create such alternative, ideal societies. Unlike Rawls, however, Marx railed against liberal egalitarian ideas as part of the fetishization—the veiling—of politics under capitalism.

Communitarian theorists of justice begin by reembracing ideas of the good life as community-shared values. Communitarian notions of justice are based in the particular, although the definition of this particularity tends to be spatial (as in a particular neighborhood) or voluntary (as in a particular group joined together for a common goal and with common values). Communitarians seek to reestablish moral economies (Scott 1976) with an attachment to a larger set of social and cultural values (Etzioni 1993), a move both utilitarians and liberal egalitarians criticize as perfectionist and authoritarian, therefore inimical to personal freedom.

Communitarian concepts of the good life are embedded in particular relationships of trust, often based on the physical proximity of living in the same place. Communitarians "are united by the belief that political philosophy must pay more attention to the shared practices and under-standings within each society" (Kymlicka 2002, 209). Proponents of communitarian forms of social justice, such as Michael Walzer (1990) and Michael Sandel (1984), argue in favor of community-based auton-omy, in which people join together to articulate decisions about the good life and how to go about making a good society. Their stance counters the individualism predominant in egalitarian liberal notions of social justice, which defines the good life as equal political representation within a "neutral state." For communitarians, in contrast, social justice "is conceived of as a substantive conception of the good life which defines the community's 'way of life.' This common good, rather than adjusting itself to the pattern of people's preferences, provides a standard by which those preferences are evaluated. The community's way of life forms the basis for a public ranking of conceptions of the good, and the weight given to an individual's preferences depends on how much she conforms or contributes to this common good" (Kymlicka 2002, 220).

Cultural perspectives on justice add to the complexity of the debate on perfectionism. Perfectionist ideals, culturalist theorists argue, are masked under false notions of rational, scientific objectivity underlying liberal notions of rights. Instead, feminists and postcolonial and critical race theorists argue that universalist notions of justice are exclusionary, as are particularist notions of social justice advocated by communitari-ans. Both universal and particularist ideas of justice tend to make a particular type of person, generally the Western white male citizen, into a universal category (Lipsitz 2006; Omi and Winant 1986). Cultural theorists, however, agree with the communitarians that justice must include considerations of group autonomy, yet the cultural perspective on justice includes racial, religious, and ethnic "group differentiated rights" within communities. Cultural theorists and communitarians, therefore, agree that justice requires shared norms, although in different ways.

Cultural perspectives of justice stem from identity-based group definitions of the good life, whether from a gender, race, or religious perspective. These more particularistic and contingent notions of justice emphasize differences in group epistemological standpoints derived from histories of oppression and exclusion (Collins 2000). Culturalists criticize liberal egalitarian forms of justice as imposing an idea of the individual, autonomous subject either as intrinsically white (Lipsitz 2006), Western civilized (Said and Sjöström 1995), or predominantly male (Butler 1999; Scott 1999).

Yet, cultural theorists also agree with utilitarians that a single, universal idea of "the good life" is authoritarian and does not admit the diversity of viewpoints. Identity-based theories of justice begin by uncovering the racialized aspect—"whiteness"—of seemingly universal ideals of the good life, and argue that all universal ideals are, in fact, embedded in particular forms of racial dominance.

From a culturalist perspective, critical race scholars have looked at the history of local control politics and have shown how localism has been used as a tool for exclusion rather than inclusion. For example, Lizabeth Cohen's history of U.S. consumerism (Cohen 2004) shows how localism has been implicated in nonegalitarian local agendas in many different arenas, particularly housing, access to financial capital, and education. Her study of the rise of consumerism after World War II shows that localism became the main tool in the creation of "stratified communities with mass suburbanization" (230). Cohen shows how the rise of local "home rule" increased segregation and an inequitable distribution of state resources toward white neighborhoods.

Cohen's historical focus is the state of New Jersey, but her analysis echoes work on California by Carey McWilliams (1973), Don Mitchell (2003), Mike Davis (1991), and other scholars who have shown that maintaining local control over land use planning, housing, and agricultural practices has enabled large-scale farmers, developers, and white middle-class suburbanites to monopolize control over the state's natural resources. In California and New Jersey, suburban supporters of home rule and local control supported the practice of "upzoning": "a strategy of requiring substantial plots for home construction to preserve high property values but also to cap the municipality's population and thereby control the cost and demand for social services" (Cohen 2004, 231). Upzoning in New Jersey has resulted in "two societies: a sparser, wealthier and better serviced New Jersey, on the one hand, and a denser, poorer and overburdened New Jersey on the other" (Ibid.). Critical race scholars have also shown how moral notions of health and rightness serve as instruments of power and exclusion. For example, Shah (2001) shows that moral notions about community public health represented particular ways of life and particular groups practicing those ways of life as "healthy," which set other groups apart as vice ridden and unsanitary.

Communitarianism as group autonomy is not always defensive, however. Sometimes it is linked with tolerance of other cultures and practices in a community, as with the Amish communities. Yet, even the most well-intentioned communitarians sometimes clash with cultural theorists, especially when definitions of "community values" exclude definitions of "the good life" embraced by other groups in a locality (Guthman 2008; Slocum 2007).

Some scholars have sought to envision a multicultural "transcommunal" ideal (Childs 2003) that takes Bentham's perfectionist critique into account, yet accepts a coexistence of different group-based ideals. This reflexive approach tends to focus on the discovery of good *processes* rather than *visions* of good life. From this perspective, processes of collaboration can emerge from groups that have different conceptions of the good life, or even of life itself. Some scholars have studied how certain civil rights groups that identify as autonomous communities also work within the tension of practicing egalitarian decision making within the group, combining liberal and communitarian notions (Polletta 2002). The reflexive approach we describe in later sections draws much of its inspiration from this approach.

Food Justice

In their justice rhetoric and activism, food movement activists tend to draw upon one or another of these perspectives without reflexively recognizing the tensions between them. Instead, they tend to work at the nexus of communitarian, political economy, and egalitarian perspectives, without acknowledging different views of justice. Many popular authors, such as Michael Pollan, *The Omnivore's Dilemma* (2006), Barbara Kingsolver, *Animal, Vegetable, Miracle* (2007), Francis Moore Lappé and Anna Lappé, *Hope's Edge* (2002), Brian Halweil, *Eat Here* (2004), and

Gary Nabhan, *Coming Home to Eat* (2002), call for the rebuilding of local food systems as a way to fight the global industrialization of the food system. Joining them are the "New Agrarians" such as Wendell Berry and Wes Jackson who envision a just society as coming out of remaking community connections between consumers and farmers. The agrarian perspective "offers useful guiding images of human living and working on land in ways that can last. In related reform movements, it can supply ideas to help rebuild communities and foster greater virtue" (Freyfogle 2001, xviii). Agrarians see virtue as coming out of people working in small, local, economic structures that are closely linked to nature.

Food movement groups also work from a political economy perspective, envisioning localization as an alternative to the injustices of globalization. Movement activists often cite anti-globalization critics such as David Korten, whose work both decries the global corporation and celebrates countermovements that seek to fight the global through the local:

Initiatives throughout America are seeking to counter the trend toward corporate control and ownership. Some 3,000 community development corporations across the country support local business development. More than a thousand family farms in the U.S. and Canada have contracts with local residents to provide fresh produce. . . . These and countless related initiatives are the proactive side of the living democracy movement, demonstrating the possibilities of local democratic control within a framework of commitment to creating healthy, ecologically sound communities that work for all. (Korten 2001, 319)

From this antiglobalist perspective, food movements tend to embrace grassroots democratic processes as a way to (re-)create community values and resist the universal, instrumentalist juggernaut of industrial agriculture.

The maxim "Think Globally, Act Locally" brings together the anti-corporate critique with a communitarian preference for value-based food systems that are based on group interconnections in local moral economies. "Think Globally," in the food politics context, means a critique of the industrial food system that begins with a Marxian pulling off of the veil that hides the true system of food production, showing in particular "where your food comes from." Food activist discourses then tend to "Act Locally" through the communitarian "coming together" of people in local food networks.

The documentary *Food, Inc.* illustrates the approach typically taken in alternative food movement visions of social justice. It begins by uncovering the true unhealthiness, cruelty, and exploitation behind our current industrial food system and then includes "Buy foods that are grown locally" in its list of solutions at the end of the movie. The current anticorporate, leftist critique of capitalism, therefore, embraces this part of Marx's critique of capitalism—that commodities such as industrial foods hide the relationships behind their production—at least in the global mode of capitalist production. Yet, the "Act Locally" part of the equation tends to lead antiglobal-food activists away from Marx's revolutionary ideas of radical social restructuring and toward communitarian relocalization through consumer support for local farmers—and the utilitarian solution of the market—as the ideal.

The political alliance between Marxian anticorporatists and communitarians is therefore subtle: the communitarians see the local as the remaking of community connections while the anticorporatists see localism as fostering greater transparency and smaller production structures. For the anticorporatists, food is better if you get to see how it is made. Once you know the truth, you will opt for buying your food from a local system of smaller farms. For the communitarians, food is better if you create a personal connection of trust and shared values between you and the person who makes your food. Either way, local food solves the "justice problem" as defined from both perspectives.

However, it is important to note that the pursuit of local food as justice also allows for key alliances with more conservative groups whose idea of community integrity may come at the exclusion of particular groups. Some embrace local protection from a "politics of fear" (Davis 1998) in which references to community are heavily related to anticrime, antigang conservative rhetoric and calls for more police to protect against "outsiders." While most food localists do not ascribe to these politics, there has been little discussion in the food movement of the differences among food localism, nativist "defensive localism" described by Winter (2003) and Hinrichs (2003), and "crunchy conservatives" as described by popular social commentator Rod Dreher (2006). In many ways, this lack of reflexivity about the meaning of local food justice enables a political alliance among communitarians, Marxians, and crunchy conservatives that, from a pragmatic perspective, may seem better to keep pushed under the rug. The "moral economy" of local food systems embedded in communities with shared values may therefore be more of a pragmatic alliance between substantially different political interests with substantially different ideas about the meaning of social justice.

In fact, this pragmatic politics violates liberal, Marxist, and cultural notions of justice. A pragmatic local food politics that ignores exclusionary politics and local elite control violates the liberal egalitarianism notion of individual equality, that is, the right of everyone to pursue their notions of the good life. If, as Hassanein (2000) argues, work on local food systems is a type of "participatory democracy," then this participation needs to take Rawls's liberal democratic ideas of equal access to political voice into account. It ignores Marxian critiques of market exchange as intrinsically exploitative of labor. It also violates cultural notions of justice as inclusivity and tolerance toward different cultural viewpoints of "the good life."

However, rather than decrying localism as a "trap" (Born and Purcell 2006), a *reflexive* approach to localist food politics takes seriously Kloppenburg and colleagues' claim that localized institutions and networks—even beyond food—are a powerful strategy for resisting the alienating, disabling, and unsustainable features of increasingly globalized food systems. Reflexive localism examines the potential for localist strategies to create more just forms of governance, seeing local forms of control as "experiments in sub-national regional governance that are themselves a response to wider problems in managing global capitalism" (Lawrence 2005, 3). Hess (2009), for example, acknowledges that localist politics may be problematic but argues nonetheless that support for local businesses can move toward a more progressive mode of politics, due to alliances between local business groups and local consumers concerned with issues of justice and equality. Relocalization politics and business initiatives have grown alongside the restructuring of government toward local "governance": the devolution of decision making to local networks and self-governing actors, coordinated through multilayered institutional structures.

In local food system discourse, communitarian ideas of justice often combine with feminist notions of justice as care, embedding local farmers and consumers in relations of regard, empathy, and trust. In many of these communitarian schemes, "tradition" is redeemed from the Enlightenment rejection of authority to become a part of a social contract based on production of quality, artisan goods by local businesses. While communitarianism and localism are not entirely congruent ideas, communitarians, like localists, often tie their notions of a good society to a good place. Additionally, personal connection in both perspectives is intrinsically place based. In other words, for both communitarians and localists, justice, territory, and a sense of place tend to go hand in hand.

Critics see the relocalization of food systems as certainly providing benefits to some, but not necessarily providing greater democratic participation. It can also provide the ideological foundations for reactionary exclusionist politics and nativist sentiment (Allen et al. 2003; Winter 2003; Born and Purcell 2006). Political sociologists who study power elites (Dahl and Rae 2005; Davis 1991; Domhoff 1967; Logan and Molotch 1987; Molotch 1976) have described in detail the power of local elites and the defeat of democratic participation in U.S. cities and localities. Increasing the power of local elites can deepen material inequalities such as wealth and health (Harrison 2006), exacerbate inequalities between regions (Gray and Lawrence 2001; Bauman 2004), and further marginalize certain groups through new exclusionary practices (Morrissey and Lawrence 1997). Furthermore, some anticorporate scholars have shown how localist governance can be part of a neoliberal global regime that extends rather than resists the dominance of global corporations (Harvey 2005).

One example of localism as reactionary and local elite political control is Harrison's (2006) research on conflicts in California over pesticide drift (the off-site airborne movements of pesticides away from their intended target). This study provides one example of the power of local elites to further their own interests at the price of social justice. Harrison shows that devolved regulations and the local interactions that drive devolved politics can deeply exacerbate existing material inequalities. In California and elsewhere in the United States, Progressive Era "direct democracy" reforms of the late nineteenth and early twentieth centuries took power away from centralized agencies and legislative processes. It favored decision making through local county agricultural boards, commodity commissions territorialized to particular production areas, and water boards covering single watersheds. As a result, water, land use, and agricultural management have been in local hands for over a century (Starr 1985; Pincetl 2003). The devolved, local regime of control enables powerful agricultural groups in the state to continue to pollute resources and to play down the worker and community health effects of pesticide use, thereby also reinforcing industrial agriculture.

Although California's devolved pesticide regulatory structure prioritizes "local" needs, these have historically been defined solely in terms of crop protection (Baker 1988; Nash 2004). The associated regulatory abandonment of ecological and human health concerns is particularly easy when agricultural regions are populated largely by politically disenfranchised immigrant farmworker communities, which characterizes

California's major agricultural regions. Agriculture continues to be an economic entry point for new, undocumented immigrants, and researchers estimate that approximately 50 percent of California's 800,000 farmworkers lack documentation (Aguirre International 2005, 15). In general, being "illegal" means that individuals are rarely likely to assert their rights or seek damages in terms of fair wages, benefits, working conditions, on-the-job injuries, or exposures to pesticides. Consequently, the local politics that drive pesticide regulation and other social services disproportionately reflect the economic interests of the dominant agricultural elites, while sidelining and rendering invisible the concerns of the farmworker communities. Thus, because of the inequalities of localist politics, the highly devolved pesticide regulatory structure exacerbates the problem of human exposure to pesticides while maintaining growers' power to pollute in counties dominated by agricultural production. This case shows that, under certain circumstances, empowering the local may increase local material inequalities.

Regulatory protection of local growers' interests has deep historical roots in California. Consequently, students of the California food system tend to be less sanguine about making connections with local farmers, especially when those farmers are the industrial producers other parts of the country seek to resist. Yet, it is worth asking whether local politics in agricultural regions in the rest of the country function as "growth machines" (Molotch 1976) where local politics prioritize the interests of local landowning elites.

Such exclusionary and disempowering forms of localism are clearly counter to the stated food security goals of local food movements explicitly concerned with the equitable distribution of healthy, nutritious food. However, some scholars point out that not all localization projects are concerned with equality. Potential allies in the local food movement include those pursuing individual health goals, those interested in maintaining the power of local small businesses over "chain" stores, faith-based groups interested in pursuing their moral agendas, and local elites interested in maintaining control over local labor (Gaventa 1982). These are powerful local actors and potential allies in the fulfillment of local food agendas, yet egalitarian forms of justice may not be high on their agendas. As the history of the Civil Rights movement indicates, local subordinated groups may need to call on extralocal powers in their pursuit of equity.

The irony of embracing communitarian justice by anticorporate critics is that it belies Marx's own class-based, antiperfectionist, anti-utopian

critique. Allen and colleagues draw out the implications of this discon-
nect, showing how local food organizations put justice issues such as
farmworker rights in the background (Allen et al. 2003, 73). Rather,
these organizations frame their ideas of social justice around notions of
consumer choice through "increasing the diversity of alternative markets"
(72). Critical race scholars who have examined the culture of whiteness
in local food movements also argue that the white middle-class health
concerns of the "worried well" tend to erase deeper issues of food justice
and equity related to food diseases of poverty (Slocum 2007; Guthman
2008).

The history of food policy is fraught with examples of food reformers
and nutrition experts insisting on the moral superiority of particular
"good food" imaginaries with exclusionary consequences (Coveney and
Staff 2006; DuPuis 2007). The conflation of territory with equality has
been questioned in the agrifood literature, particularly in critical decon-
structions of "defensive localism" (see, e.g., Hinrichs 2003).

The problem of perfectionist politics has been a part of U.S. food
reform movements ever since Benjamin Rush, signer of the Declaration
of Independence, created a hierarchy of beverages with water at the top
and alcohol at the bottom. DuPuis's (2002) social history of the rise of
American milk consumption shows how food reformers created a perfectionist vision of eating based on a Northern European food practice:
drinking fresh, fluid milk, the "perfect food." This vision of good food
reinforced the privilege of the Northern European middle class as the
practitioners of a superior middle-class lifestyle, or "habitus" (Bourdieu
1984). These practices entered U.S. food reform politics as middle-class
white families lobbied for "pure" and "sanitary" milk to feed infants,
replacing a mother or a wet-nurse. Pure fluid milk was therefore the
product of an alliance between Northern European middle-class families,
emerging large-scale industrial processors, a new class of health professionals, and an expanding group of state officials, who all benefited from
food sanitation reform movements. The rise of the sanitary milk regime
was therefore the product of political alliances.

Are food relocalization movements therefore intrinsically unjust? No,
but food politics and politics of the body have in common a number of
serious potential pitfalls. The first problem is the notion of "health" as
a universal category. This prevents "good food" movements from being
more tolerant forms of communitarian justice, since it is difficult to be
tolerant of food practices not considered "good" if they are seen as
debilitative for the health of these groups. This problem is particularly

relevant to issues of diabetes and obesity (Guthman and DuPuis 2006). Second, the politics of the body tends to be racialized. The cultural links between food practices and both universalist and racialized notions of the body makes food politics particularly entangled in social injustices. Even local food groups with the best of intentions can sometimes arouse suspicion in groups that have been historically marginalized in these "local control" discourses; "local" for some people with these histories can mean "not us" (Guthman 2008). Slocum (2007) and Guthman (2008) have pointed out the tendency toward an unintentional politics of whiteness in local food movements.

However, a reflexively localist approach sees these pitfalls not as a reason to abandon local food activism, but instead as an argument for a new kind of localist practice, one that includes more careful "reflexive" interrogation of the relationships between "the local" and "justice." What kinds of justice can a localist food movement strategy pursue, and what tends to be left out of such strategies? Can food relocalization movements overcome these shortfalls, and if so, how?

Toward Reflexive Food Justice

The answers to these questions are bound to be incremental, messy, and bound in imperfect politics (DuPuis 2002). In this section, we will explore how to rebuild localism along more just lines, through a more "reflexive localism" that takes into account different visions of justice, community, and good food (Staeheli 2008). A reflexive perspective works within an awareness of the tensions between different definitions of justice, environmental and bodily health, and good food, while admitting that localist strategies are imperfect and contradictory. Reflexivity is not a set of values, but a process by which people pursue goals while acknowledging the imperfection of their actions. It is also not a particular, fixed process, but one that responds to changing circumstances, imperfectly, but with an awareness of the contradictions of the moment. Here is an imperfect start to a description of reflexive localism as a practice.

Reflexivity Begins by Admitting the Contradictions and Complexity of Everyday Life

A number of contemporary scholars have been attempting to formulate a theory of justice that both takes into account reflexive notions of equality while maintaining rights of group or community autonomy. Several influential political theorists, for example, are exploring the idea of "reflexive" or "dialectical" equality (Benhabib 1996, 2002; Young 1990; Beck, Giddens, and Lash 1994). These theorists see reflexivity as a way to escape a politics of perfection that both hides and perpetuates hegemony. The challenge, therefore, becomes the discovery of practices that make society "better" without reinforcing inequalities.

Reflexive Approaches Emphasize Process Rather than Vision

While it would be difficult to predict what a reflexive localism would look like, it is possible to speculate on some possible practices and processes (as opposed to ideals) that might lead to better (as opposed to perfect) local food systems. First, along with an awareness of ways in which a local food system can help improve the health of community members should be an awareness of how ideas about "good food" and "community values" have been used to marginalize certain groups and that these groups are aware of these practices, as recognized in numerous studies of sociology of deviance and social control (Becker 1997; Merton 1972). Urban consumers could practice an "economy of regard" with farmers that included an awareness of the history of rural–urban inequalities and consumer-producer conflicts. Local food activists could also practice with an awareness of the extent that "defensive localism" (Winter 2003) is tied to larger urban politics around territorial boundaries (in terms of restricting services, housing, shopping, schooling), particularly between more and less wealthy neighborhoods. Relocalization activists, whether focusing on local food systems or other "buy local" initiatives (Hess 2009), need to recognize the ways in which this type of boundary setting between places of the poor and places of the well-off contribute to, rather than resolve, injustices. Local food movement activists must be cautious about making alliances with defensive localists, or else their efforts will only serve to perpetuate this trend.

Reflexivity Does Not Favor Any One Scale of Political Practice

Born and Purcell (2006) have argued that localism is a "trap" that causes political movements to rely on a particular scale to further the goals of environmental sustainability, human health, and social justice. They argue that localism should be seen as a possible strategy and not as an intrinsic solution to the problems of the global food system. Reflexive approaches to food justice assess the value of localism as a strategy on a case-by-case basis. Furthermore, a just and reflexive food localism needs to be aware of the possible inequalities and unfairnesses caused by the creation of economic boundaries between a particular "here" and the global "there."

For example, what is the more "fair" banana: the one grown by smaller farmers in cooperatives in the Caribbean or those grown on Central American plantations by unionized and politically empowered banana workers? (Frank 2005). Food justice activists are beginning to learn from New Regionalist and Smart Growth initiatives that attempt to reunite inner-city and surrounding suburban and ex-urban interests (Clancy and Ruhf 2010; Pastor 2000; Swanstrom et al. 2003) as a way to put reflexive social and environmental justice into practice. New Regionalist movements emphasize political inclusion and region-wide decision making. This political approach also goes beyond the idea that city-country connections are only along the commodity chain. A just and inclusive city-country politics would give voice to political interests that go beyond consumers' interest in local farmers. Those we sell and buy from are not the only country people worthy of our thoughts.

Reflexivity Works within Multiple Notions of Privilege and Economy

A reflexive food politics would do more than just respond to environmental degradation and loss of livelihood experienced within industrial economies. It would address the ways in which racial notions of purity and privilege helped to usher in both our spatial and our dietary inequalities. In the same way that we need to realize that white flight and localist "upzoning" contributed to urban sprawl and that localist education policies contributed to the maintenance of an undereducated underclass in the United States, we need to understand the ways in which privilege, class, and status struggles contributed to the rise of the industrial food system that now threatens the health of the entire population (DuPuis 2002).

Reflexivity Distinguishes between Equality and Charity

From this political perspective, activists could forge local food systems that were products of political relationships that cut across categories of economy and identity, making for a more inclusive metropolitan regionalism promoting equitable distribution of resources and services, across the board. Food system relocalization could also contribute to New Regionalist political agendas by expanding beyond the suburban "rings" around cities. Ideas about "smart growth" and intelligent planning then become part of larger housing, nutrition, and economic development policies, which would include active partnership with rural hinterlands. For example, policies to reduce childhood obesity, such as school lunch reform, will be ineffective without thinking about how it fits into the overall problem of school funding (Guthman and DuPuis 2006; Allen and Guthman 2006). This more inclusive policy would make more evident how local food inequalities are tied to inequalities at higher geographic scales, such as federal school lunch policies.

Reflexivity Works within a Strong Memory of Past Inequalities

Those interested in pursing equitable relationships in local foodsheds have also ignored inequitable power relationships in the history of urban–rural politics in the United States. The political relationships of dairy "milkshed" politics at the beginning of the twentieth century provide another example of unequal rural–urban power relationships, as cities transformed their view of farmers as vital agrarian support for urban life to a view of the countryside as simply existing "in service" to the city (Block and DuPuis 2001). The split between ideas about production and consumption in food studies also exacerbates this blindness (Goodman and DuPuis 2002). Ever since eaters moved out of the countryside, food politics—whether the "farmer labor" politics at the turn of the twentieth century, the "urban-rural food alliances" of the 1970s and 80s (Belasco 1993; McLeod 1976) or today's "food policy councils"—have been based substantially in urban social movements. Most local food policy councils are named after the cities where the consumers reside, not the countryside that contains the producers. The ambiguous and mythic history of American agrarianism in opposition to American urbanism is also a disservice to localist food politics (Hofstadter 1955; Jacobs 1961). It entangles impersonal "gesellshaft" views of the city and value-filled "gemienshaft" views of the country competing with less idyllic views of the countryside as degraded crackerdom (Bell 1997) or spaces of racial exclusion (Agyeman 2002) and cities as the realm of tolerance and progressive thought (Jacobs 1961). These perspectives influence policy, as does the question of urban consumer vs. rural farmer interests in farmers market policy (McPhee and Zimmerman 1979). Farmers markets also reflect urban political relationships (Alkon 2008). To pretend that foodsheds avoid such entanglements is a potentially fatal blindness in the local food movement. Again, this suggests that we need to draw on urban political sociology if we are to understand local food systems.

Reflexivity Does Not Insist on Shared Values or Even Shared Views of the World

Of course, it is just such alliances—between groups with different interests but some overlapping agendas—that makes politics work, especially in the United States where people must work out political differences within a two-party system. As early as the 1830s, an observer of U.S. politics, Alexis de Tocqueville, noted that the U.S. system requires people to work together in order to further their personal agendas. In a point of view more recently reflecting de Tocqueville's views, the alliances necessary to get things accomplished in the U.S. system bring people together in ways that enable them to better understand each other, leading to what de Tocqueville called "self-interest rightly understood"—a kind of reflexive approach to political alliances. Local politics will always require people with different, sometimes countervailing, interests working together (Hassanein 2003). A reflexive local politics works within, and not against, the awareness of these differences in political viewpoints, what "wisdom of crowds" expert Paola De Maio refers to as "open ontologies" and critical race scholar John Brown Childs refers to as "transcommunality" (Childs 2003; Di Maio 2007). Conversely, in accepting localism as the main tool of antiglobal resistance without reflexivity, local food activists may be playing into the hands of groups with less egalitarian goals.

Conclusion: Embracing Imperfect Politics

In this chapter, we have interrogated the local food movement through the lens of political theories of justice, emphasizing that local food system discourse is wrapped up in particular and problematic notions of justice that conflict with other, more cultural notions of what justice might look like. We therefore argue that the food movement would be more effective if it worked with a more "reflexive" notion of justice, one that takes into account different visions of community, good food, and environmental and bodily health, and that pursues change well beyond the realm of the "local."

From this more reflexive, nonperfectionist viewpoint, true reform of our food system requires that we muck ourselves up in the imperfection of political contestation over food. To start, we need to recognize the diverse and "situated" (Haraway 1988) productions of knowledge about "good" food. This will enable a more complex and inclusive discussion about what a just food system should look like, rather than reflecting certain "community values" as if all communities were defined by a monolithic set of values. This could bring us closer to looking at inequalities beyond the local that affect the choices we have closer to home, such as food industry monopolies, USDA agency capture, nutrition policy, agricultural subsidy policies, food dumping, and food deserts, as well as extraordinary inequalities in wages and access to health care. A reflexive food politics can include relocalization in a larger struggle over the global food system, combining a "not in my body" politics of boycotting and a "yes in my body" politics of "buycotting" within a more realist perspective of local and global power. Reflexive justice brings activism back to the imperfect politics of process and away from the perfect and privileged politics of standard setting. Rather than creating an alternative economy for the homogenous few, reflexive localism could work across difference, and thereby make a difference, for everyone.

References

Agyeman, J. 2002. Constructing Environmental (in) Justice: Transatlantic Tales. *Environmental Politics* 11 (3): 31–53.

Aguirre International. 2005. The California Farm Labor Force: Overview and Trends from the National Agricultural Workers Survey, June. <http://agcenter.ucdavis.edu/AgDoc/Didyouknow.php> (accessed January 18, 2011).

Alkon, A. 2008. Paradise or Pavement: The Social Constructions of the Environment in Two Urban Farmers' Markets and Their Implications for Environmental Justice and Sustainability. *Local Environment* 13 (3): 271–289.

Allen, P., M. FitzSimmons, M. Goodman, and K. Warner. 2003. Shifting Plates in the Agrifood Landscape: The Tectonics of Alternative Agrifood Initiatives in California. *Journal of Rural Studies* 19 (1): 61–75.

Allen, P., and J. Guthman. 2006. "From "Old School" to "Farm-to-School": Neoliberalization from the Ground Up. *Agriculture and Human Values* 23 (4): 401–415.

Bailyn, Bernard. 1967. *The Ideological Origins of the American Revolution.* Cambridge, MA: Belknap Press.

Baker, B. P. 1988. Pest Control in the Public Interest: Crop Protection in California. *UCLA Journal of Environmental Law & Policy* 8 (1): 31–72.

Bauman, Z. 2004. *Wasted Lives: Modernity and Its Outcasts.* Oxford: Polity.

Beck, U., A. Giddens, and S. Lash. 1994. *Reflexive Modernization: Politics, Tradition, and Aesthetics in the Modern Social Order.* Stanford, CA: Stanford University Press.

Becker, H. S. 1997. *Outsiders: Studies in the Sociology of Deviance.* Glencoe, IL: Free Press.

Belasco, W. J. 1993. *Appetite for Change.* Ithaca, NY: Cornell University Press.

Bell, D. 1997. Anti-idyll: Rural Horror. In *Contested Countryside Cultures: Otherness, Marginalisation, and Rurality,* ed. P. Cloke and J. Little, 91–104. London: Routledge.

Benhabib, S. 1996. *Democracy and Difference: Contesting the Boundaries of the Political.* Princeton, NJ: Princeton University Press.

Benhabib, S. 2002. *The Claims of Culture: Equality and Diversity in the Global Era.* Princeton, NJ: Princeton University Press.

Block, D., and E. M. DuPuis. 2001. Making the Country Work for the City: von Thünen's Ideas in Geography, Agricultural Economics and the Sociology of Agriculture. *American Journal of Economics and Sociology* 60 (1): 79–98.

Born, B., and M. Purcell. 2006. Avoiding the Local Trap: Scale and Food Systems in Planning Research. *Journal of Planning Education and Research* 26 (2): 195–207.

Bourdieu, P. 1984. *Distinction: A Social Critique of the Judgment of Taste.* Cambridge, MA: Harvard University Press.

Butler, J. 1999. *Gender Trouble: Feminism and the Subversion of Identity.* Philadelphia, PA: Routledge.

Childs, J. B. 2003. *Transcommunality: From the Politics of Conversion to the Ethics of Respect.* Philadelphia, PA: Temple University Press.

Clancy, Kate, and Kathryn Ruhf. 2010. Is Local Enough? Some Arguments for Regional Food Systems. *Choices* 25 (1). <http://www.choicesmagazine.org/magazine/article.php?article=114> (accessed August 17, 2010).

Cohen, Lizabeth. 2004. *A Consumer's Republic: The Politics of Mass Consumption in Postwar America.* New York: Random House, Inc.

Collins, P. H. 2000. *Black Feminist Thought: Knowledge, Consciousness, and the Politics of Empowerment.* New York: Routledge.

Coveney, J., and C. J. Staff. 2006. *Food Morals and Meaning: The Pleasure and Anxiety of Eating.* New York: Routledge.

Dahl, R. A., and D. W. Rae. 2005. *Who Governs? Democracy and Power in an American City.* New Haven, CT: Yale University Press.

Davis, Mike. 1991. *City of Quartz: Excavating the Future in Los Angeles.* London: Verso.

Davis, Mike. 1998. *Ecology of Fear: Los Angeles and the Imagination of Disaster.* New York: Vintage Books.

DeLind, L. B. 2002. Place, Work, and Civic Agriculture: Common Fields for Cultivation. *Agriculture and Human Values* 19 (3): 217–224.

Di Maio, P. 2007. An Open Ontology for Open Source Emergency Response System. Open Source Research Community. <http://citeseerx.ist.psu.edu/viewdoc/summary?doi=10.1.1.93.1829> (accessed January 18, 2011).

Domhoff, G. W. 1967. *Who Rules America?* Englewood Cliffs, NJ: Prentice-Hall.

Dreher, Rod. 2006. *Crunchy Cons: The New Conservative Counterculture and Its Return to Roots.* New York: Three Rivers Press.

DuPuis, E. M. 2002. *Nature's Perfect Food: How Milk Became America's Drink.* New York: NYU Press.

DuPuis, E. Melanie. 2007. Angels and Vegetables: A Brief History of Food Advice in America. *Gastronomica* 7 (2): 34–44.

Etzioni, A. 1993. *The Spirit of Community: Rights, Responsibilities, and the Communitarian Agenda*. New York: Crown.

Frank, Dana. 2005. *Bananeras: Transforming the Banana Unions of Latin America*. Cambridge, MA: South End Press.

Freyfogle, E. T. 2001. *The New Agrarianism: Land, Culture, and the Community of Life*. Washington, DC: Island Press.

Gaventa, J. 1982. *Power and Powerlessness: Quiescence and Rebellion in an Appalachian Valley*. Urbana: University of Illinois Press.

Goodman, D., and E. M. DuPuis. 2002. Knowing Food and Growing Food: Beyond the Production-Consumption Debate in the Sociology of Agriculture. *Sociologia Ruralis* 42:5–22.

Gray, I., and G. Lawrence. 2001. *A Future for Regional Australia: Escaping Global Misfortune*. New York: Cambridge University Press.

Guthman, J. 2008. "If They Only Knew": Color Blindness and Universalism in California Alternative Food Institutions. *Professional Geographer* 60 (3): 387–397.

Guthman, J., and M. DuPuis. 2006. Embodying Neoliberalism: Economy, Culture, and the Politics of Fat. *Environment and Planning D: Society and Space* 24 (3): 427–448.

Halweil, B. 2004. *Eat Here: Reclaiming Homegrown Pleasures in a Global Supermarket*. W. W. Norton & Company.

Haraway, D. 1988. Situated Knowledges: The Science Question in Feminism and the Privilege of Partial Perspective. *Feminist Studies* 14 (3): 575–599.

Harrison, Jill. 2006. "Accidents" and Invisibilities: Scaled Discourse and the Naturalization of Regulatory Neglect in California's Pesticide Drift Conflict. *Political Geography* 25 (5):506–529.

Harvey, D. 2005. *A Brief History of Neoliberalism*. Oxford: Oxford University Press.

Harvey, D. 1996. *Justice, Nature and the Geography of Difference*. Malden, MA: Blackwell Publishers.

Hassanein, N. 2000. Democratizing Agricultural Knowledge through Sustainable Farming Networks. In *Science, Technology, and Democracy*, ed. Daniel Lee Kleinman, 49–66. Albany: SUNY Press.

Hassanein, N. 2003. Practicing Food Democracy: A Pragmatic Politics of Transformation. *Journal of Rural Studies* 19 (1): 77–86.

Hess, D. J. 2009. *Localist Movements in a Global Economy: Sustainability, Justice, and Urban Development in the United States*. Cambridge, MA: MIT Press.

Hinrichs, C. C. 2003. The Practice and Politics of Food System Localization. *Journal of Rural Studies* 19 (1): 33–45.

Hofstadter, R. 1955. *The Age of Reform: From Bryan to FDR*. New York: Knopf.

Holloway, L., and M. Kneafsey. 2004. Producing–Consuming Food: Closeness, Connectedness and Rurality. In *Geographies of Rural Cultures and Societies*, ed. L. Holloway and M. Kneafsey, 262–281. London: Ashgate.

Jacobs, J. 1961. *The Death and Life of Great American Cities*. New York: Random House.

Kingsolver, B., with S. L. Hopp and C. Kingsolver. 2007. *Animal, Vegetable, Miracle: A Year of Food Life*. New York City: Harper Collins.

Kloppenburg, J., J. Hendrickson, and G. W. Stevenson. 1996. Coming into the Foodshed. *Agriculture and Human Values* 13 (3): 33–42.

Korten, D. C. 2001. *When Corporations Rule the World*. San Francisco: Berrett-Koehler Publishers.

Kymlicka, W. 2002. *Contemporary Political Philosophy: An Introduction*. 2nd ed. Oxford: Oxford University Press.

Lappé, F. M., and A. Lappé. 2002. *Hope's Edge: The Next Diet for a Small Planet*. New York: Jeremy P. Tarcher/Putnam.

Lawrence, Geoffrey. 2005. Promoting Sustainable Development: The Question of Governance. In *New Directions in the Sociology of Global Development*, ed. F. Buttel and P. McMichael, 145–174. New York: Elsevier.

Lipsitz, G. 2006. *The Possessive Investment in Whiteness: How White People Profit from Identity Politics*. Philadelphia: Temple University Press.

Logan, J. R., and H. L. Molotch. 1987. *Urban Fortunes: The Political Economy of Place*. Berkeley: University of California Press.

Lyson, Thomas. 2005. *Civic Agriculture: Reconnecting Farm, Food and Community*. Lebanon, NH: University Press of New England.

McLeod, D. 1976. Urban-Rural Food Alliances: A Perspective on Recent Community Food Organizing. In *Radicalizing Agriculture*, ed. R. Merrill, 188–211. New York: Harper and Row.

McPhee, J., and W. Zimmerman. 1979. *Giving Good Weight*. New York: Farrar Straus Giroux.

McWilliams, C. 1973. *Southern California: An Island on the Land*. Santa Barbara, CA: Peregrine Smith.

Merton, R. K. 1972. Insiders and Outsiders: A Chapter in the Sociology of Knowledge. *American Journal of Sociology* 78 (1): 9–47.

Miele, M., and J. Murdoch. 2002. The Practical Aesthetics of Traditional Cuisines: Slow Food in Tuscany. *Sociologia Ruralis* 42 (4): 312–328.

Mitchell, D. 2003. *The Right to the City: Social Justice and the Fight for Public Space*. New York: Guilford Press.

Molotch, H. 1976. The City as a Growth Machine: Toward a Political Economy of Place. *American Journal of Sociology* 82 (2): 309–332.

Morrissey, P., and G. Lawrence. 1997. A Critical Assessment of Landcare: Evidence from Central Queensland. In *Critical Landcare*, ed. S. Lockie and F. Vanclay, 217–226. Wagga Wagga, NSW: Centre for Rural Social Research.

Nabhan, Gary P. 2002. *Coming Home to Eat: The Pleasures and Politics of Local Foods.* New York: W. W. Norton & Co.

Nash, L. 2004. The Fruits of Ill-Health: Pesticides and Workers' Bodies in Post-World War II California. *Osiris* 19:203–219.

Omi, M., and H. Winant. 1986. *Racial Formation in the United States: From the 1960s to the 1980s.* New York: Routledge.

Pastor, M. 2000. *Regions that Work: How Cities and Suburbs Can Grow Together.* Minneapolis: University of Minnesota Press.

Pincetl, S. S. 2003. *Transforming California: A Political History of Land Use and Development.* Baltimore: Johns Hopkins University Press.

Pollan, M. 2006. *The Omnivore's Dilemma: A Natural History of Four Meals.* New York: Penguin.

Polletta, F. 2002. *Freedom Is an Endless Meeting: Democracy in American Social Movements.* Chicago: University of Chicago Press.

Rawls, J. 1971. *A Theory of Justice.* Cambridge, MA: Harvard University Press.

Rawls, J. 1993. *Political Liberalism.* New York: Columbia University Press.

Said, E. W., and H. O. Sjöström. 1995. *Orientalism.* London: Penguin.

Sandel, M. J. 1984. The Procedural Republic and the Unencumbered Self. *Political Theory* 12 (1): 81–96.

Scott, J. C. 1976. *The Moral Economy of the Peasant: Rebellion and Subsistence in Southeast Asia.* New Haven, CT: Yale University Press.

Scott, J. W. 1999. *Gender and the Politics of History.* New York: Columbia University Press.

Shah, N. 2001. *Contagious Divides: Epidemics and Race in San Francisco's Chinatown.* Berkeley: University of California Press.

Slocum, R. 2007. Whiteness, Space and Alternative Food Practice. *Geoforum* 38 (3): 520–533.

Staeheli, L. A. 2008. Citizenship and the Problem of Community. *Political Geography* 27 (1): 5–21.

Starr, K. 1985. *Inventing the Dream: California through the Progressive Era.* New York: Oxford University Press.

State and Local Food Policy Project. 2011. <http://www.Statefoodpolicy.org> (accessed January 18, 2011).

Swanstrom, T., P. Dreier, C. Casey, and R. Flack. 2003 Pulling Apart: Economic Segregation in Suburbs and Central Cities in Major Metropolitan Areas, 1980–2000. In *Redefining Urban and Suburban America: Evidence from Census 2000*, ed. Bruce Katz and Robert E. Lang, 143–165. Washington, DC: Brookings Institution Press.

Walzer, M. 1990. The Communitarian Critique of Liberalism. *Political Theory* 18 (1): 6–23.

Winter, M. 2003. Embeddedness, the New Food Economy and Defensive Localism. *Journal of Rural Studies* 19 (1): 23–32.

Young, I. M. 1990. *Justice and the Politics of Difference*. Princeton, NJ: Princeton University Press.

14

Food Security, Food Justice, or Food Sovereignty?

Crises, Food Movements, and Regime Change

Eric Holt-Giménez

Hunger, Harvests, and Profits: The Tragic Records of the Global Food Crisis

2008 saw record levels of hunger for the world's poor at a time of record harvests and record profits for the world's major agrifoods corporations. The contradiction of increasing hunger in the midst of wealth and abundance unleashed a flurry of worldwide "food riots" not seen for many decades. These protests were sparked by skyrocketing food prices. In June of 2008, the World Bank reported that global food prices had risen 83 percent in three years and the United Nation's Food and Agriculture Organization (FAO) cited a 45 percent increase in their world food price index in just nine months (Wiggins and Levy 2008). *The Economist*'s comparable index stood at its highest point since it was originally formulated (USDA 2008). In March 2008, average world wheat prices were 130 percent above their level a year earlier, soy prices were 87 percent higher, rice had climbed 74 percent, and maize was up 31 percent (BBC 2008). The United States Department of Agriculture (USDA) predicted that at least 90 percent of the increase in grain prices would persist during the next decade (USDA 2008). Thus far, their dire predictions are proving to be true.

Not until the dramatic displacement of food crops by fuel crops began in 2006 did the FAO start to warn of impending food shortages. But in the winter of 2007, instead of shortages, *food price inflation* exploded on world markets—despite that year's record harvests. The overnight reversal of a thirty-year global trend in cheap food in 2007–2008 quickly became known as the "global food crisis."

When the FAO sounded the alarm on the current food crisis in 2007, a flurry of official solutions to global hunger were proposed by the G-8 governments and the world's multilateral institutions. Most aimed to reinforce the failing global food system, others sought to reform it. Unfortunately, but perhaps unsurprisingly, none of these "solutions" considered that these institutions themselves might be part of the problem, and there were no proposals to significantly transform the global order of food.

Encouragingly, people's movements for sustainable agriculture, food justice, and food sovereignty (the democratization of the food system in favor of the poor) have worked for decades to substantively change what is seen as a socially dysfunctional and environmentally destructive global food system. The current crisis brings both momentum and urgency to the broad array of grassroots initiatives around the world that make up the global food movement. But to understand where these initiatives fit into the current debate on hunger and what potential they have to bring about substantive and lasting change, we need to examine not only these movements, but also the power structures that have shaped the global food system—the corporate food regime. This analysis can help us visualize how different currents within the food movement might either transform, reform, or inadvertently even reinforce the ways we produce, process, trade, and consume our food.

Understanding the Crisis

Because they spend as much as 70–80 percent of their income on food, the world's poor—particularly women—were hit the hardest. Not surprisingly, people took to the streets in Mexico, Morocco, Mauritania, Senegal, Indonesia, Burkina Faso, Cameroon, Yemen, Egypt, Haiti, and over twenty other countries. Scores of people were killed and hundreds injured and jailed. In Haiti, the poorest country in the Western hemisphere, food prices increased by 50–100 percent, driving the poor to eat biscuits made of mud and vegetable oil. Angry protestors forced the Haitian prime minister out of office. Street rebellions have continued to flare in Haiti as successive hurricanes pummel the island, destroying livelihoods and putting food farther and farther out of people's reach.

The protests were not simply crazed "riots" by hungry masses. Rather, they were angry demonstrations against high food prices in countries that formerly had food surpluses, and where government and industry were unresponsive to people's plight. Painfully prophetic, they signaled the onset of the global financial crisis and economic recession now gripping the world economy.

The food crisis appeared to explode overnight, reinforcing fears that there are just too many people in the world. But according to the FAO, with record grain harvests in 2007, there was more than enough food in the world to feed everyone—at least 1.5 times current demand (Holt-Giménez and Peabody 2008). In fact, over the last twenty years, food production has risen steadily at over 2.0 percent a year, while the rate of population growth has dropped to 1.14 percent a year (Hansen-Kuhn 2007; Rossi and Lambrou 2008). Globally, population is not outstripping food supply. According to the World Food Program (WFP) over 90 percent of the world's hungry are too poor to buy enough food. This is also true for the United States, where one in nine Americans is now on food stamps (Reuters 2009).

Despite the oft-cited production gains of the Green Revolution (a campaign which injected high-yielding varieties of wheat, maize and rice coupled with the heavy use of subsidized fertilizers, pesticides, irrigation and machinery into the agricultural economies of the global South) and despite repeated development campaigns over the last half-century— most recently, the UN's Millennium Development Goals—the number of desperately hungry people on the planet has grown steadily from 700 million in 1986 to 800 million in 1998, to over a billion today (Lappé and Collins 1986; Lappé, Collins, and Rosset 1998; Blas and Walls 2009).

While seen as a serious problem, this situation was not generally referred to as a "global food crisis" by governments, the international aid institutions, or the mainstream media. For decades, thanks to market oversupply from northern grain-producing countries, food prices had been on a steady downward trend. It was widely assumed that eventually—as the promised benefits from global trade began to kick in—the poor would be able to buy the food they needed.

The immediate reasons for food price inflation were easily identified: droughts in major wheat-producing countries in 2005–2006; less than fifty-four days of global grain reserves; high oil prices; the diversion of 5 percent of the world's cereals to agrofuels and 70 percent to grain-fed beef; and, as prices began to rise, financial speculation. Though grain futures and oil prices have dropped and agricultural growth increased throughout 2009, retail food prices remain high in much of the global South, and in many cases have not come down at all.

But, if we focus only on food and oil prices, we are a long way from understanding—much less solving—the global food crisis. Why? Because drought, low reserves, agrofuels, and oil prices and speculation are only the *proximate* causes of the food crisis. These proximate factors tipped us over the precipice, but to understand how and why we drove our food system to the edge in the first place, we need to address the *root* causes of the global food crisis, as well. These reside in a skewed global food system that has made Southern countries and poor people everywhere highly vulnerable to economic and environmental shock. This vulnerability springs from the risks, inequities, and externalities inherent in food systems that are dominated by a globalized, highly centralized, *industrial agrifoods complex*. Built over the past half-century—with the help of public funds for grain subsidies, foreign aid, and international agricultural research—the industrial agrifoods complex is made up of multinational grain traders; giant seed, chemical, and fertilizer corporations; global processors; and supermarket chains. These global companies dominate markets and shipping, and increasingly control the world's food-producing resources: land, labor, water, inputs, genetic material, and investments.

While many activists assert that the global food system is "broken," for these companies it works extraordinarily well. Today two companies, Archer Daniels Midland (ADM) and Cargill, capture three-quarters of the world grain trade (Vorley 2003). Chemical giant Monsanto controls 41 percent of seed production (Grain 2007). Monopolization of the world's food provides these companies with unprecedented market power. This translates into profits in the midst of crisis. In the last quarter of 2007, as the world food crisis was breaking, Archer Daniels Midland's earnings jumped 42 percent, Monsanto's by 45 percent, and Cargill's by 86 percent. Mosaic Fertilizer, a subsidiary of Cargill, saw profits rise by 1200 percent (Lean 2008a, 2008b). Even the livestock sector in the United States—supposedly hard hit by soaring grain prices—saw profit increases in the first and second quarters of 2008 which were up 429 percent compared to a year earlier.

The trend toward monopolistic control over our food systems is particularly visible within the United States, where a handful of transnational corporations in the industrial agrifoods complex mediate the relationship between three million farm operators and 300 million consumers, gobbling up the lion's share of the food dollar. Over the last sixty years, the companies that buy, sell, and process farm products, and the chains that distribute and sell food, have steadily eroded farmers' profits. The monopolistic concentration of power in the global food system characterizes what some analysts refer to as the *corporate food regime*.

The Corporate Food Regime

A food regime is a "rule-governed structure of production and consumption of food on a world scale" (Friedman 1993, 31; qtd. in McMichael 2009). The first global food regime spanned the late 1800s through the Great Depression and linked food imports from Southern hemisphere and American colonies to European industrial expansion. The second food regime reversed the flow of food from north to south to fuel Cold War industrialization in the Third World. Today's corporate food regime based on fossil fuels and dominated by global monopolies is characterized, inter alia, by global concentration in the input, processing, and food retail sectors; global supply chains for meat, animal feed, and biofuels; and the rise of patented genetically modified seeds (Ibid.).

Virtually every food system in the world is tied to the global food system in one way or another, which is itself shaped by the corporate food regime (Friedman 1987; McMichael 2009). This regime is financially dominated by the monopolies of the industrial agrifoods complex and politically managed by the national governments and multilateral organizations that make (and enforce) the free trade, labor, and property rules that make it possible to create and enforce a globalized food regime. This political-economic partnership is complemented by the myriad institutions, both public and private, that carry out supportive, mitigating, and adaptive functions within the regime—from the World Bank/ International Monetary Fund (IMF), to the World Food Program, U.S. Agency for International Development (USAID), and big philanthropy.

The corporate food regime has proven to be as resilient and protean as capitalism itself. It constantly expands, squeezing profits from the food system by destroying existing forms of production and consumption and replacing them with new ones. No obstacle, crisis, or disaster in the food system is too large or too small that it can't be turned into *some* kind of opportunity for corporate profit. Granted, some "opportunities" (like agro-export and agrofuels), require strong state intervention and massive taxpayer subsidies. Food aid—a long-standing example—has been used by donating countries to destroy foreign agriculture and break open markets for northern agribusiness's surplus.[1] Other "opportunities"— like the current land grabs and wholesale colonization of seed markets in the developing world through foreign aid tied to the development and expansion of proprietary and genetically modified seeds—are reflective of capitalism's systematic process of accumulation through dispossession (Harvey 2003).

Like capitalism, the corporate food regime goes through periods of *liberalization* characterized by unregulated markets and breathtaking capital expansion, followed by devastating busts. These are followed by *reformist* periods in which markets, supply, and consumption are regulated in an effort to restabilize the regime. While these phases may appear politically distinct, they are actually two sides of the same system. As Polanyi (1944) pointed out, if capitalist markets were allowed to run rampant, they would eventually destroy both society and their own natural resource base. For this reason, as liberal markets begin to undermine society and the environment, capitalism periodically implements government reforms to reign in markets in crisis. As calls for reform ring from nearly every corner of the globe, it is important to remember that some reforms serve to merely prop up the existing food regime.

The last sixty years of global capitalist development was ushered in by a reformist period (following the Great Depression) of Keynesian deficit spending and supply regulation in the industrial North. Throughout the postwar expansion period of the 1950s and into the1970s, domestic policies built up surpluses. These were exported using food aid as an entry mechanism to open foreign markets. To keep developing countries from aligning with the Soviet Union, the United States (and later European countries) supported state-led economic development in the global South. The U.S. farm crisis and the Third World debt crisis of the late 1970s, followed by the fall of the Soviet Union, led to the dismantling of the state-led economic development model through the implementation of the now-infamous structural adjustment policies applied by the World Bank and the IMF. The North set about privatizing state-owned enterprises and destroying supply-management and regulatory frameworks in the global South to make way for the capital expansion of transnational corporations. The 1980s ushered in thirty years of liberalization, deregulation, privatization, and the consolidation of corporate monopoly power around the globe. (We commonly refer to the latter, neoliberal phase of capitalist expansion as "globalization." In fact, for thirty years prior to this, northern capital was using state-led, regulatory frameworks to expand global corporate control over the world's food systems.)

The present agrifoods industry has been an integral part of global capitalist expansion throughout its reformist and liberalization periods. The Green Revolution, introduced with public money during the 1960–1970s, spread a model of industrial agriculture to the "underdeveloped" world. Global food production increased dramatically. Despite population growth, the world produced over 1.5 times enough food for every man, woman, and child on the planet. However, at any given time, around a sixth of the world's population (most of whom were peasant farmers) was too poor to buy this food, resulting in chronic overproduction and a steady decline in global food prices. The balance, after sixty years of capitalist expansion, was the establishment of the corporate food regime, based on the northern production and consumption model dominated by the industrial agrifoods complex (box 14.1).

While the cheerleaders of the corporate food regime—in particular the champions of the Green Revolution—credit themselves for "saving a billion people from hunger," on balance, the picture is less rosy. In 1970 the developing world still produced food surpluses of $1 billion a year. By 2001 developing countries were running a yearly food deficit of $11 billion (FAO 2004). The environmental externalities of the corporate food regime are massive: Industrial agriculture and the global transport of food produce 20 percent of the world's greenhouse gases and use up 80 percent of the planet's water. The shift to industrial hybrid seeds has led to a loss of 75 percent of the world's crop diversity. While the quantity of food per capita did indeed increase by 11 percent in the Western world, the number of hungry people also increased by 11 percent (Lappé and Collins 1986). The neoliberal phase of the corporate food regime has been especially painful for the world's poor. The cost of twenty years of trade liberalization in sub-Saharan Africa is estimated at US$272 billion—an amount equal to what the region received in foreign aid (Christian Aid 2005). Millions upon millions of smallholder farmers around the developing world went bankrupt during this phase because they were unable to compete with subsidized grains that were being dumped in their markets (sold at under the cost of production) from the industrial North. In Mexico alone, following the North American Free Trade Agreement, over 1.3 million Mexican smallholders went bankrupt and a million immigrants poured across the border to the United States (Stone 2009). As the North destroyed the food systems of the global South, tremendous consolidation took place. Grain giants ADM, Cargill, and Bunge took control of 80 percent of the world's grain (Vorley 2003). Chemical corporations Monsanto and DuPont together appropriated 65 percent of the global maize seed market (Action Aid International 2005), four companies—Tyson, Cargill, Swift, and National Beef Packing Company—controlled 83.5 percent of the U.S. beef supply (Hendrickson and Heffernan 2007), and Tyson assumed a substantial portion of the world's meat production.

Box 14.1
Building Blocks of the Neoliberal Corporate Food Regime (1960–1990)

The Green Revolution was a campaign originally financed by the Ford and Rockefeller Foundations for the industrialization of agriculture in the global South.[2] Though credited for saving the world from hunger in the 1960s and 1970s, the Green Revolution led to the monopolization of seed and chemical inputs by northern companies, the loss of 90 percent of the global South's agricultural biodiversity, the global shift to an oil-based agricultural economy, and the displacement of millions of peasants to fragile hillsides, shrinking forests, and urban slums. Contrary to popular belief, the Green Revolution also *produced* as many hungry people as it saved (Lappé et al. 1998).

Northern Subsidies and Food Aid were instrumental after World War II, when the United States expanded its reach overseas through a combination of domestic and foreign polices designed to pry open foreign markets for the sale of the massive amounts of grain produced in the Midwestern breadbasket. In the United States and Europe direct farm payments, environmental subsidies, price supports, protective tariffs, and export assistance total more than $300 billion a year—almost the GDP of the entire African continent and *six times* the amount of OECD development assistance to the global South (OECD 2001). While rich counties in Europe and the United States can protect their farmers from low prices, the global South was prevented from doing the same. Over time, policies that encouraged overproduction in the global North systematically undermined food production in the South.

Structural Adjustment Programs (SAPs) imposed by the World Bank and the International Monetary Fund in the 1980s and 1990s broke down tariffs, dismantled national marketing boards, eliminated price guarantees, and destroyed research and extension systems in the global South. By deregulating agricultural markets, the SAPs cleared the way for the "dumping" of agricultural commodities by multinational grain companies into local markets with subsidized grain from the United States and Europe. This tied Southern food security to global markets dominated by multinational agribusinesses from the industrial North instead of encouraging developing countries to increase self-sufficiency through local farm production.

Regional Free Trade Agreements and World Trade Organization rules then cemented the SAPs into international treaties that overrode national labor and environmental laws and legally prevented countries from protecting their food systems from foreign dominance. While these policies were sold under the banner of "free trade," under WTO rules, the United States and EU can heavily subsidize their agribusinesses, but other countries are prohibited from doing the same.

Corporate Concentration resulted from the neoliberal food regime; it paved the way for higher and higher levels of corporate concentration across agricultural and food retail sectors. Five firms (WalMart, Kroger, Albertson's, Safeway, and Ahold) control 48 percent of U.S. food retailing (Hendrickson and Heffernan 2007). Over 80 percent of all beef packing in the United States is in the hands of four firms and 90 percent of the global grain trade is controlled by just three corporations.

During the first phase of the corporate food regime, the expansion of capitalist agriculture met with strong resistance from the world's peasantry. The Green Revolution—originally a state-led campaign to *avoid* land reform and defuse growing agrarian movements—eventually led to the massive appropriation of peasant seeds by industry and the concentration of peasant land in the hands of large, better-capitalized farmers. This resulted in the exodus of millions of smallholders to urban slums and to the fragile hillside and tropical forest perimeters of the agricultural frontier. In hopes of obtaining land, and in a desperate effort to protect themselves from the reactionary violence of large landholders, millions of peasants joined revolutionary movements in Indochina, India, the Philippines, and Central and South America. They were met by U.S.-financed counterinsurgency aid programs that included food aid, Green Revolution "packages" (seed, fertilizer, and agrochemicals) and militarily strategic land reforms.

Nevertheless, throughout the 1960s, 1970s, and mid-1980s, peasant movements around the globe consistently struggled for redistributive land reform, production credit, fair markets, and the right to dignified rural livelihoods (Borras 2007).

Similarly, in the industrial north, agro-industry benefited from public subsidies and supply-management policies that ensured a cheap surplus for both domestic and foreign markets. Overproduction forced down the prices of basic grains, leading to an explosion in grain-fed meat, processed food, and big supermarket chains. In the inevitable "bust" of the 1970s, the United States lost nearly half its farmers. This opened the way to greater corporate consolidation of U.S. farmland, and accelerated the growth of agrifoods monopolies. Aside from a marginal back-to-the-land movement and an embryonic organic foods movement, the burgeoning agrifoods industry encountered little resistance in the global North, even as it set in motion the social and environmental externalities that would usher in the next crisis.

During what is commonly referred to as the "neoliberal" phase of the corporate agrifoods regime (1980s to the present), smallholders struggled in the face of the rapacious programs of liberalization, deregulation, and privatization forced on the developing world by the World Bank and the International Monetary Fund, and later cemented into law by Free Trade Agreements (FTAs) and the World Trade Organization (WTO). The expansion of extractive industries (e.g., mining, oil, hydropower, etc.) on peasant and indigenous lands, as well as the metastasis of soy, sugar cane, and palm oil plantations on the marginal tropical lands where peasants had taken refuge, introduced even higher levels of violence into the corporate food regime's ongoing process of forcing smallholders off the land through policies for "de-peasantization" (Bryceson 2002).

During the neoliberal period, the food regime brought high costs. Hunger, malnutrition, and diet-related diseases rose dramatically in the underserved "food deserts" of the otherwise affluent countries of the industrial north. Many of the fifty million people now suffering from "food insecurity" in the United States are people of color and displaced immigrant farmers working in poorly paid jobs—frequently in the agri-foods sector. Pollution from confined animal feedlot operations (CAFOs) fouled the air and contaminated aquifers. Agricultural runoff choked rivers and estuaries with nitrates, producing a "dead zone" in the Gulf of Mexico the size of New Jersey (Science Daily 2009). Agriculture became the primary emitter of greenhouse gases. Ancient water tables on the Ganges plains of India and the Ogallala aquifer of the U.S. Midwest were drawn down past recharging levels.

The increasing social, economic, and environmental contradictions of the corporate food regime fueled the growth of the organic food sector and gave rise to grassroots movements for food justice, sustainable agriculture, urban gardening, localization, and health and lifestyle-based trends for quality food and food "authenticity." Many family farmers in the industrial North began to see their fates intertwined with the struggling peasants of the global South. Via Campesina, an international federation for smallholder-peasant-fisher-pastoralist organizations launched a cry for "food sovereignty" that quickly spread to both northern and southern small-scale producers. Food sovereignty is essentially a political demand, one that cuts through reformist proposals to address the power structures at the root of the corporate food regime.

With the food and financial crises, rumblings of reform have appeared within the food regime. The World Bank has been forced to quietly admit that the turn toward liberal markets did not result in economic development for the poor and has revived its moribund loan portfolio for agricultural development. The G-8 countries have made repeated statements invoking the importance of investment in agriculture and rural safety nets. The EU now allows some local procurement for food aid. However, the neoliberal focus on "free" markets remains intact and, as yet, there are no official proposals for corporate, financial, or supply regulation in the corporate food regime.

Food Enterprise, Food Security, Food Justice, Food Sovereignty

Combating the steady increase in global hunger and environmental degradation has produced an ever-expanding array of institutions, programs, initiatives, and campaigns, resulting in a wide diversity of
antihunger approaches from government, industry, and civil society.
Most address the proximate causes of hunger, though increasingly many
are beginning to focus on the root causes as well. Some programs treat
hunger and poverty as a business opportunity and call for solutions based
on public-private partnerships. Others address hunger normatively and
insist government and industry should be held accountable if they
advance policies or enterprises that undermine the human right to food.
In the United States, "food insecurity" is framed quite differently than
global hunger (usually understood as severe malnutrition and starvation). Food systems approaches address hunger differently than do agricultural systems approaches. Post-scarcity perspectives see food and
health as issues of informed consumer choices, and advocate "voting
with our forks." Food justice activists from underserved communities
ground their hunger work in critiques of structural racism and classism.
Some efforts are highly institutionalized and bureaucratized. Others take
the form of broad-based movements for social change.

What can be seen as a rich diversity of approaches also reflects important systemic divides. Where one stands on hunger depends on where
one sits. While strategic and tactical overlap exist, food and hunger
efforts tend to split ideologically between those who seek to stabilize the
corporate food regime, and those who seek to change it. This split is
further characterized by different tendencies, each with its own set of
discourses, institutions, models, and approaches. Making substantive
sense of the similarities and differences within these different approaches
is essential for charting equitable and sustainable ways forward through
the multiple crises plaguing our food systems.

The Food Regime and the Food Movement

Some actors within the growing food movement have an overtly *radical*
critique of the corporate food regime. These groups call for structural,
redistributive reforms of entitlements, for example land, water, markets.
Others advance a *progressive* agenda on the basis of rights of access for
marginalized groups defined by race and gender, or based on the desire

for pleasure, quality, and authenticity in the food system. While progressive groups are rich in practice, and radical organizations tend to engage more in political advocacy, both trends overlap significantly in their approaches. Together, they make up the global food movement that seeks to change food systems in favor of the poor and underserved, and strives for more sustainable and healthy environments. While the progressive trend tends to focus on localizing production and improving the service and delivery aspects of the food system, the radical trend directs its energy at the structural changes and enabling conditions for more equitable and sustainable food systems. The point is, both trends attempt to change food systems locally and globally. In this regard they are two sides of the same food movement.

A comparison of trends and tendencies within the corporate food regime and food movements can help us understand the dynamics at play in our food systems in the face of the current food crisis (table 14.1).

The Neoliberal Trend
The political trend of the food regime over the last three decades has been indisputably neoliberal. It is held firmly in place by the North's major international finance and development institutions, as well as the major agrifood corporations, the U.S. government, and important players in big philanthropy. Loyal to the belief in market-led economic development, the neoliberal approach employs a *food enterprise* discourse for ending hunger. In this view, hunger can be eradicated by expanding global markets and by increasing output through technological innovation. This trend reinforces a model of chronic overproduction and the corporate monopolization of the food system. Through this model, the solutions to hunger presently being advanced by northern governments, the World Bank, WTO, IMF, USDA, and USAID all call for more of the same neoliberal measures that created the present food regime and unleashed the recent food crisis. These include further liberalization of global markets, public financing of proprietary, technological "fixes," and "land mobility," that is, the continued disenfranchisement of the rural poor from food-producing resources to make way for "more efficient" producers.

The Reformist Trend
Even before the present food crisis, the social and environmental externalities unleashed by neoliberal globalization provoked a timid reformist response on the part of many UN agencies, aid organizations, socially

Table 14.1
Politics, production models, and approaches

	Corporate food regime		Food movements	
Politics	Neoliberal	Reformist	Progressive	Radical
Discourse	*Food enterprise*	*Food security*	*Food justice*	*Food sovereignty*
Main institutions	International Finance Corporation (World Bank); IMF, WTO: USDA; Global Food Security Bill; Green Revolution; Millennium Challenge; Heritage Foundation; Chicago Global Council; Bill and Melinda Gates Foundation; ONE Campaign	International Bank for Reconstruction and Development (World Bank); FAO; UN Commission on Sustainable Development; International Federation of Agricultural Producers; mainstream fair trade; Slow Food Movement; some Food Policy Councils; most food banks and food aid programs	Alternative fair trade and many Slow Food chapters; many organizations in the Community Food Security Movement; CSAs; many Food Policy Councils and youth food and justice movements; many farmworker and labor organizations	Via Campesina, International Planning Committee on Food Sovereignty; Global March for Women; many food justice and rights-based movements

Table 14.1
(continued)

Orientation	Corporate	Development	Empowerment	Entitlement
Model	Overproduction; corporate concentration; unregulated markets and monopolies; monocultures (including organic); GMOs; agrofuels; mass global consumption of industrial food; phasing out of peasant and family agriculture and local retail	Mainstreaming/ certification of niche markets (e.g. organic, fair, local, sustainable); maintaining northern agricultural subsidies; "sustainable" roundtables for agrofuels, soy, forest products, etc.; market-led land reform	Agroecologically produced local food; investment in underserved communities; new business models and community benefit packages for production, processing, and retail; better wages for ag. workers; solidarity economies; land access; regulated markets and supply	Dismantle corporate agrifoods monopoly power; parity; redistributive land reform; community rights to water and seed; regionally based food systems; democratization of food system; sustainable livelihoods; protection from dumping/ overproduction; revival of agroecologically managed peasant agriculture to distribute wealth and cool the planet
Approach to the food crisis	Increased industrial production; unregulated corporate monopolies; land grabs; expansion of GMOs; public-private partnerships; liberal markets; international sourced food aid Guiding document: World Bank 2009 Development Report	Same as neoliberal but with increased medium farmer production and some locally sourced food aid; more agricultural aid but tied to GMOs and "bio-fortified/climate-resistant" crops Guiding document: World Bank 2009 Development Report	Right to food; better safety nets; sustainably produced, locally-sourced food; agroecologically based agricultural development Guiding document: International Assessment on Agriculture Science Technology and Development	Human right to food sovereignty; locally sourced, sustainably produced, culturally appropriate, democratically controlled focus on UN/ FAO negotiations Guiding document: Peoples' Comprehensive Framework for Action to Eradicate Hunger

conscious entrepreneurs from the private sector, and some politicians. International institutions employ a *food security discourse* and seek to mainstream less inequitable and less environmentally damaging alternatives into existing market structures. Some advocate incentive-based certification and corporate self-regulation. These approaches aim to modify industrial behavior through the power of persuasion and consumer choice. The supporting notion is that by dint of a good example or "voting with our forks," less damaging trade and production alternatives will some day transcend their market niches (frequently high-end specialty products) and set new industrial standards. Common projects and institutions in this trend include the corporate mainstreaming faction of Fair Trade; the various industry-dominated "roundtables" for sustainable soy, sustainable biofuels, and sustainable agricultural standards; and corporate sectors of the organic foods industry. Many humanitarian, environmental and social service organizations like Bread for the World, Oxfam-USA, CARE, and World Vision are wholly or partially rooted in the reformist trend because their main sources of funding come from government, major corporations, or neoliberal philanthropic institutions such as the Bill and Melinda Gates Foundation. Because of their economic dependence on government agricultural surplus, many food banks are rooted in this trend as well. Rather than call for structural change, most work to increase and improve existing social safety nets (food stamps, food banks, food aid, food-for-work, etc.).

The Progressive Trend
This tendency—primarily based in northern countries—is possibly the fastest growing grassroots expression of the food movement. It is grounded in the notion of citizen empowerment and employs a *food justice discourse*. This discourse, coming from the traditions of environmental justice, denounces the ways people of color and underserved communities in rural and urban areas are abused by the present food regime, and invokes the notion of a gradual, grassroots-driven transition, or passage, to a more equitable and sustainable food system. Institutions and groups in this trend promote local production, processing, and consumption of food, and are busy inventing new business models to better serve economically disadvantaged communities. The approach to the food crisis is based on the right to food, better safety nets, and more citizen involvement in decisions regarding community food systems. In the United States, farm organizations in this trend represent smallholders seeking support for organic agriculture and family farming

over corporate agriculture and genetically modified organisms (GMOs). A host of locally based initiatives linking access to healthy food together with sustainable production has come from this trend, including farm-to-school programs, urban gardens, corner-store conversions, community markets, community-supported agriculture (CSA), and the spread of farmers markets into underserved communities. One strong expression of the progressive trend is the emergence of Food Policy Councils throughout the United States, Canada, and many parts of the industrial North. Food Policy Councils convene actors from local or state governments or both, local businesses, and civil society in an effort to better manage food systems within the existing food regime. The councils are characterized by strong citizen participation and a commitment to equity and sustainability. The involvement of youth in food system assessments, school and community garden programs, and farm-to-school, farmers markets, and food justice campaigns brings tremendous dynamism to this trend. While groups in this trend are often well aware of the global framework girding the corporate food regime, they are primarily active in local, state, and national political arenas.

The Radical Trend

The international trend seeking the radical transformation of food systems addresses the root causes of poverty and hunger in the food system, based on the notion of entitlement to food-producing resources and the redistribution of wealth. The discourse is framed by the concept of *food sovereignty* and the democratization of the food system in favor of the poor and underserved. The model advanced is one of local and international engagement that proposes dismantling the monopoly power of corporations in the food system and redistributing land and the rights to water, seed, and food-producing resources. Organizations leading this trend come from historical agrarian and labor struggles in the global South. The Brazilian Landless Workers Movement (MST) and Via Campesina, the international peasant-fisher-pastoralist federation, are emblematic of this trend. Activists in this trend frequently engage international spaces for advocacy, such as the UN and FAO. Following the momentum of the antiglobalization movement (the massive protests at the WTO's ministerial meetings in Seattle, Cancun, etc.), they have engaged in direct action across the globe to stop the WTO, protect peasant land from contamination by GMOs, and denounce appropriation of peasant and indigenous lands for extractive industries.

Food movement organizations, however, are fluid and have different and changing positions on key food system issues like GMOs, domestic hunger programs and foreign aid, biofuels, subsidies, supply management, land reform, and trade. Depending on their ideology, political awareness, support base, and funding, food movement organizations will adopt a range of stances, and will consciously (or unconsciously) form alliances with institutions across regime and movement trends. While some organizations are solidly neoliberal, reformist, progressive, or radical, others are much harder to categorize because they adopt politically different positions on the different issues (for example, a transformational position on trade, but a reformist position on GMOs). Some organizations say one thing, yet do another. Rather than ascribing fixity to the organizations, an appreciation of their heterogeneous and fluid political nature, coupled with an analysis of their positions on specific issues, might better help identify opportunities for alliance building within the food movement.

Solving the Food Crisis: The Imperative of Regime Change

The current food crisis is a reflection of the environmentally vulnerable, socially inequitable, and economically volatile corporate food regime. Unless there are profound changes to this regime, it will repeat its cycles of liberalization and reform, plunging the world's food systems into ever graver crises. While food system reforms—like localizing food assistance, increasing aid to agriculture in the global South, increasing funding for food stamp programs or for research in organic agriculture—are certainly needed and long overdue, they will not alter the fundamental balance of power within the food system, and in some cases may even reinforce existing, inequitable power relations. To put an end to hunger at home and abroad, the practices, rules, and institutions determining the world's food systems must change. This implies *regime change*.

The challenge for food movements is to address the immediate problems of hunger, malnutrition, food insecurity, and environmental degradation while working steadily toward the structural changes needed for a sustainable and equitable food system. The first task has been undertaken widely, and is reflected in the rich diversity of experiences, projects, and organizations that, in the words of activist-academic Harriet Friedman, "Appear everywhere like plants breaking through the cracks in the asphalt!" The second task—structural change—is a much more difficult project that requires countering the privatization of government and public spheres by big philanthropy and transnational agrifoods corporations and reversing the trends of monopolization. This in turn requires social movements bold, strong, and imaginative enough to wrest political will from the vise grip of the industrial agrifoods complex. To build this kind of social force, activists within food movements need to build strong alliances and must distinguish superficial change from structural change. This not only requires a vision and practice of the desired change, it also means making strategic and tactical sense of the matrix of actors, institutions, and projects at work within local, national, and international food arenas. The existing trends within the world's food systems are somewhat self-defining. Actors and institutions within the reformist, progressive, and transformational trends tend to associate among themselves, reaching out when specific opportunities for legislation, campaigns, or direct action arise. The challenge for movement building is to reach beyond the naturally occurring and tactical relationships, to form strategic alliances across the progressive and radical trends.

As the world's fuel, financial, and climate crises exacerbate the food crisis, the systemic differences between the food regime and food movements will likely deepen. However, unlike the symbiotic relationship between liberal and reformist trends in the food regime, in which the latter helps to stabilize the food regime following a crisis caused by the former, there is nothing intrinsically stable about the relationship between the progressive and radical trends that will keep them from splitting under pressure. The natural fragmentation of a broad-based food movement "growing up through the cracks in the sidewalk" already cedes strategic political ground to the corporate food regime, whose reformist trends are busy co-opting "organic," "local," and "fair." Both the discourse and practice of food movements can blur the lines between food security and food enterprise (e.g., with calls for a new Green Revolution, for example), which consolidates and amplifies the monopolistic narrative of the corporate food regime.

The systemic divergence within the food movement is visible in the different constituencies and strategies of organizations working for food aid and those struggling for structural change; between those working with the underserved and those working with "overserved" communities; between those working on national hunger issues with those working on international hunger; and between advocacy groups in Washington, DC, New York, and Rome with those actually implementing practices on the ground. Further, a profound area of silence commonly found across all trends in the food movement is the issue of labor in the food system. Under pressure, these fractures could lead to more divergence and fragmentation within the food movement, undermining the possibility of regime change.

However, when seen through the lens of regime change, these structurally induced divides can also be seen as opportunities for alliances, convergence, and the possibility of increased influence and impact on food systems. In fact, there are already indications than many groups are busy addressing these divides. Food Policy Councils are an excellent example of citizens working together to span the divide between advocacy and practice. The Slow Food Movement has made strong statements in favor of food sovereignty and some chapters are actively reaching out to food justice movements and underserved communities. Family farmers in the U.S. National Family Farm Coalition also belong to Via Campesina, and thus address their national issues within the context of the international struggle for food sovereignty. In Brazil, the Landless Workers Movement (MST) integrates the agroecological practice taught in their peasant university in Paraná with the agrarian advocacy and direct action of their land occupations. Increasingly, the positions and campaigns of labor organizations like the Coalition of Immokalee Workers, the United Food and Commercial Workers International Union, the Food Chain Workers Alliance, and others are finding their way into the agendas of food justice coalitions.

The food crisis has opened up new opportunities for transition and transformation. It has also resulted in a retrenchment of liberalization and mild calls for reform. This suggests that substantive changes to the food regime will originate outside the regime's institutions—from the food movement.

Changing the food regime will depend in large part on the ability of food activists to link issues and build bridges between the progressive and radical trends of the growing food movement. Will the political efforts of convergence among the diverse organizations of the food movement overcome their systemic fragmentation? This will likely depend on the extent to which organizations within the movement reach beyond their immediate organizational agendas to embrace regime change.

Because of their social weight, food movement organizations operating primarily within the progressive trend will ultimately determine not only the possibility of convergence, but also the political balance for or against regime change. If the balance of strategic relationships within the progressive trend tilts toward reform, the corporate food regime will be reinforced. If these relationships tend toward transformation, the possibilities for regime change will increase. To solve the food crisis, we hope the latter scenario prevails.

Notes

1. The world's food aid is dominated by the U.S. model, initiated in 1954 with the passing of Public Law 480. The United States' objective with PL 480 was "to lay the basis for a permanent expansion of our exports of agricultural products with lasting benefits to ourselves and peoples of other lands" By law, 75 percent of food aid from the United States must be purchased, processed, transported, and distributed by U.S. companies. In 2002, just two U.S. companies—ADM and Cargill—controlled 75 percent of global grain trade with U.S. government contracts to manage and distribute 30 percent of food aid grains. Only four companies control 84 percent of the transport and delivery of food aid worldwide. Bilateral trade agreements control 50–90 percent of global food aid. For example, U.S. aid requires recipient countries to accept genetically modified grains. In 2007, 99.3 percent of U.S. food aid was "in-kind," that is, food procured in the United States and shipped to recipient countries on corporate ships, rather than purchased with cash or coupons closer to recipients (see Holt-Giménez, Patel, and Shattuck, 2009, 88).

2. The term *Green Revolution* comes from a meeting of the Society for International Development in Washington, DC, in 1968. Referring to record yields in Pakistan, India, Philippines, and Turkey, William Gaud, director of USAID announced, "These and other developments in the field of agriculture contain the makings of a new revolution. It is not a violent Red Revolution like that of the Soviets, nor is it a White Revolution like that of the Shah of Iran. I call it the Green Revolution." A perfect Cold War sound bite, the term quickly spread worldwide (<http://www.agbioworld.org/biotech-info/topics/borlaug/borlaug-green.html> [accessed May 1, 2010]).

References

Action Aid International. 2005. Power Hungry: Six Reasons to Regulate Global Food Corporations. Action Aid International. <www.actionaid.org.uk/_content/documents/power_hungry.pdf> (accessed July 12, 2009).

BBC. 2008. The Cost of Food: Facts and Figures: Explore the Facts and Figures behind the Fluctuating Price of Food across the Globe. *BBC News.* <http://news.bbc.co.uk/2/hi/7284196.stm> (accessed April 8, 2008).

Blas, Javier, and William Walls. 2009. US Investor Buys Sudanese Warlord's Land. *Financial Times* (North American edition) (January): 9. <http://www.ft.com/cms/s/0/a4cbe81e-de84-11dd-9464-000077b07658.html#axzz17RmipoW3> (accessed January 14, 2009).

Borras, Saturnino. 2007. *Pro-Poor Land Reform.* Ottawa: University of Ottawa Press.

Bryceson, Deborah Fahy. 2002. The Scramble in Africa: Reorienting Rural Livelihoods. *World Development* 30 (5): 725–739.

Christian Aid. 2005. The Economics of Failure: The Real Cost of "Free" Trade for Poor Countries. <http://www.christianaid.org.uk/Images/economics_of _failure.pdf> (July 12, 2009).

Conner, Heidi, Juliana Mandell, Meera Velu and Annie Shattuck. 2008. The Food Crisis Comes Home. *Food First Backgrounder* 14 (3).

FAO. 2004. The State of Agricultural Commodity Markets 2004. Food and Agriculture Organization. <http://ftp.fao.org/docrep/fao/007/y5419e/y5419e00. pdf> (September 25, 2008).

Friedman, Harriet. 1987. International Regimes of Food and Agriculture Since 1870. In *Peasants and Peasant Societies*, ed. T. Shanin, 258–276. Oxford: Basil Blackwell.

Friedman, Harriet. 1993. The Political Economy of Food: A Global Crisis. *New Left Review* 197:29–31.

Grain. 2007. Corporate Power—Agrofuels and the Expansion of Agribusiness. GRAIN. <http://www.grain.org/seedling/?id=478> (October 24, 2008).

Hansen-Kuhn, Karen. 2007. *Women and Food Crises: How US Food Aid Policies Can Better Support Their Struggles—A Discussion Paper*. Action Aid USA, Washington, DC.

Harvey, David. 2003. *The New Imperialism*. Oxford: Oxford University Press.

Hendrickson, Mary and William Heffernan. 2007. Concentration of Agricultural Markets: National Farmers' Union. <http://www.nfu.org/wp-content/2007 -heffernanreport.pdf> (accessed June 29, 2010).

Holt-Giménez, Eric, and Loren Peabody. 2008. *From Food Rebellions to Food Sovereignty: Urgent Call to Fix a Broken Food System*. Food First. <http:// www.foodfirst.org/files/shared_staff/audio/From_Food_Rebellions_to_Food _Sovereignty.pdf> (July 15, 2008).

Holt-Giménez, Eric, Raj Patel, and Annie Shattuck. 2009. *Food Rebellions! Crisis and the Hunger for Justice*. Oakland, CA: Food First Books.

Lappé, Frances Moore, and Joseph Collins. 1986. *World Hunger: 12 Myths*. New York: Grove Press.

Lappé, Frances Moore, Joseph Collins, and Peter Rosset, with Luis Esparza. 1998. *World Hunger: Twelve Myths*. 2nd ed. New York: Grove Press.

Lean, Geoffrey. 2008a. Multinationals Make Billions in Profit out of Growing Global Food Crisis. *The Independent*. <http://www.independent.co.uk/ environment/green-living/multinationals-make-billions-in-profit-out-of-growing -global-food-crisis-820855.html> (October 14, 2008).

Lean, Geoffrey. 2008b. Rising Prices Threaten Millions with Starvation, Despite Bumper Crops. The Independent. <http://www.independent.co.uk/environment/ green-living/multinationals-make-billions-in-profit-out-of-growing-global-food -crisis-820855.html> (accessed March 2, 2008).

McMichael, Philip. 2009. A Food Regime Genealogy. *Journal of Peasant Studies* 36 (1): 139–169.

OECD. 2001. Towards More Liberal Agricultural Trade. Office of Economic Cooperation and Development (OECD) Policy Brief. November. <http://www .oecd.org/dataoecd/39/16/2674624.pdf> (accessed May 25, 2010).

Polanyi, Karl. 1944. *The Great Transformation*. Boston: Beacon Press.

Reuters. 2009. One in Nine Americans on Food Stamps, USDA Says. Reuters, June 3, 2009. <http://www.reuters.com/article/domesticNews/ idUSTRE55270Y20090603> (accessed July 15, 2009).

Rossi, Andrea, and Yianna Lambrou. 2008. *Gender and Equity Issues in Liquid Biofuels Production: Minimizing the Risks to Maximize the Opportunities*. Rome: Food and Agricultural Organization of the United Nations.

Science Daily. 2009. Large Gulf of Mexico "Dead Zone" Predicted. *Science Daily*. <http://www.sciencedaily.com/releases/2009/06/090618124956.htm> (accessed July 12, 2009).

Stone, Dori. 2009. *Beyond the Fence: A Journey to the Roots of the Migration Crisis*. Oakland, CA: Food First Books.

USDA. 2008. *USDA Agricultural Projections to 2017*. Office of the Chief Economist, World Agricultural Outlook Board. Washington, DC: United States Department of Agriculture.

Vorley, Billy. 2003. Food Inc.: Corporate Concentration from Farm to Consumer. United Kingdom Food Group. <http://www.ukfg.org.uk/docs/UKFG-Foodinc -Nov03.pdf> (July 15, 2008).

Wiggins, Steve, and Stephanie Levy. 2008. *Rising Food Prices: A Global Crisis*. London: Overseas Development Institute.

15

Conclusion

Cultivating the Fertile Field of Food Justice

Alison Hope Alkon and Julian Agyeman

The central message of this book is that institutional racism intersects with an increasingly consolidated industrial agriculture to produce a variety of negative consequences for low-income people and people of color, and that an analysis of these processes can produce a broader critique of agribusiness than is currently offered by the food movement. Low-income people and people of color have been systematically denied access to the means of food production, and are often limited in their abilities to consume healthy foods. However, the food movement narrative ignores these injustices, an omission which reflects its adherents' race and class privilege. The cultivation of a food system that is both environmentally sustainable and socially just will require the creation of alliances between the food movement and the communities most harmed by current conditions. The food justice movement is laying the foundation for such coalition building. Our goals in this book have been to examine the conditions that produce food injustice, to celebrate the potential present in the food justice movement, and to highlight potential directions in which it might grow.

Food justice research is informed by a variety of interdisciplinary traditions including environmental justice, critical race theory, sustainable agriculture, and food studies. Bringing these literatures together produces new knowledge that can in turn inform each of them. To the environmental justice literature, food serves as an undertheorized environmental benefit to which access is deeply structured by race and class. Because of the intimate role food plays in the construction and performance of individual and collective identities, studies of food justice may also help the environmental justice literature to develop a deeper, more theoretically rich concept of race. Moreover, food justice offers critical race theorists an opportunity to better understand how environmental racism and environmental privilege can affect racial identity formation.

It also suggests that scholars of sustainable agriculture and food studies can better attend to the role of structural inequalities in producing patterns of food production and consumption. In other words, studies of justice need food, and studies of food need justice.

The resulting food justice scholarship is necessary not only to produce rich accounts of and explanations for various aspects of social life, but also to inform efforts to create a more just and sustainable society. Like its environmental justice and sustainable agriculture predecessors, the field of food justice scholarship is characterized by a deep engagement with activist movements and a calling to be useful and relevant to actors on the ground working to create a more just and sustainable food system. Our desire is not merely to better understand the effects of institutional racism and economic inequality on the food system, but also to help to create a broad, multiracial, and multiclass movement that can challenge the dominance of industrial agriculture and help to create something more sustainable and just. In order to do so, we aim to encourage the food movement to begin to see the low-income communities and communities of color most deeply harmed by industrial agriculture as potential allies. Additionally, we seek to highlight potentials and pitfalls within food justice activism. For this reason, our conclusion draws upon the previous chapters to offer a variety of pushes both from and for the food justice movement. That is, we seek to draw upon the concept of food justice to illuminate future pathways for the food movement, as well as to highlight potential directions for the food justice movement itself.

Lessons from the Food Justice Movement

The lessons described in this conclusion are not exhaustive of either the cases contained in this volume or of the food justice movement in general. The chapters in this volume do, however, contain some common insights that the food justice movement can offer to the food movement.

Moving beyond Colorblindness

Perhaps the most comprehensive lesson demonstrated by the chapters in this volume is that the food system is not racially neutral. The influence of race on the food system is reflected in the institutional policies that have decimated the food sovereignty of many communities of color, creating both poverty and hunger. By ignoring these histories, the food movement tends to take a colorblind and class-blind approach that fails to recognize the privileges enjoyed by many of its strongest supporters. This is true not only with regard to industrial agribusiness, but also the food movement's proposed solutions. For example, an understanding of what Norgaard, Reed, and Van Horn (chapter 2, this volume) call the "racialized environmental history" of U.S. agriculture might help the food movement to realize that its dominant social change strategy— encouraging economic support for small local and organic farm own- ers—requires the elevation of a group from which people of color have historically been excluded. Moreover, people of color whose families may have been or currently are exploited by white farm owners, which Guthman (2004) argues is common on organic as well as conventional farms, may be less inclined to join a movement that valorizes white farmers. Additionally, the food movement tends to argue for decentral- ized planning and local decision making. But discourses of local control, as DuPuis, Harrison, and Goodman argue (chapter 13, this volume), are not always read as community based and independent. They can also be aligned with racially exclusive and reactionary efforts to deny civil rights, particularly to African American and immigrant communities. In these examples, the food movement's failure to incorporate the histories and experiences of people of color into its narrative leads to the creation of ostensibly colorblind alternatives that subtly reflect white cultural histo- ries. The food movement then ignores, or even denies, the racialized implications of its narrative.

It is against this backdrop that the alternative stories told in these chapters become essential to the future of the food movement. Contem- porary food justice activism provides an alternative to the food move- ment's colorblind approach that focuses on both sustainability and justice. Food justice activists adopt many of the same tools beloved by the food movement—particularly small farms and community gardens— but also construct culturally specific narratives and foodways. Latino/a community gardeners in South Central LA and Seattle, for example, highlight the importance of corn, beans, and squash, as well as other staples they consider essential to their indigenous Latin American identi- ties. In addition, these projects highlight participants' immigrant identi- ties through transnational seed sharing. In this practice, gardeners code their plots of land as Latino/a immigrant spaces that reflect their own food histories and traditions. Similarly, the Pan African Orthodox Church and the Nation of Islam weave their farming traditions together with their goals of Black Nationalism. Here, explicitly racialized discourses of Black Nationalism code these farms as black spaces. Additionally, while the Growing Food and Justice Initiative (GFJI) creates a multiracial coalition, and thus does not invoke culturally specific foodways to frame its work, the central place accorded to "dismantling racism" marks it as a group seeking to facilitate food system transformation by and for people of color. These race-conscious projects create sustainable local food systems while providing an alternative to the colorblindness that characterizes an overwhelmingly white food movement.

A move beyond colorblindness is necessary if the food movement is to create alliances with low-income people and people of color in pursuit of just sustainability. Social movements such as the feminist and environmental/environmental justice movements have taught us that an understanding of various privileges and oppressions is essential to the creation of the kinds of multiracial and multiclass coalitions that will be necessary to take on the entrenched power of industrial agriculture. In order to create large coalitions, the food movement's predominantly white and middle-class adherents will have to examine the ways that white privilege shapes their life experiences, including their access to and the meanings they assign to local and organic food. Harper's and Guthman's chapters (10 and 12, respectively, this volume) join a growing body of scholarship on whiteness in the food movement (Slocum 2006, 2007; Alkon and McCullen 2011) to both argue for such a conversation and give examples of how it can begin.

Cultivating Ways of Being

A closely related point is that individuals and groups assign meanings to the foods they eat and the agricultural practices they perform, and that these meanings can be contested both within and among groups. The food movement's dominant narrative aligns the cultivation and consumption of local, organic food with a tightly bounded set of meanings. Food movement participation tends to be driven by desires for environmental sustainability, strong local communities, and personal health. Because its narrative is disproportionately told by white and middle-class individuals, including those with platforms such as the *New York Times*, these meanings can become widely repeated and accepted as objectively true. Through the food movement narrative, those who eat "properly" are lauded as well-disciplined, conscientious subjects while "industrial" eaters are depicted as apathetic to the needs of their own bodies, communities, and environments.

Much of the early work of environmental justice activists centered on issues of meaning. Throughout the twentieth century, environmentalists, who were largely affluent and white, tended to be most concerned about the protection of scenic landscapes and charismatic endangered species. In order to convince environmental organizations to support the work that low-income people and people of color were doing to protect the communities in which they lived, worked, and played, environmental justice activists needed to change the ways that environmentalists defined the environment. This need became obvious when, in the 1990s, large environmental organizations refused to support grassroots environmental justice groups because their concerns were not significantly "environmental" (Di Chiro 1996). Over the past twenty years, the environmental justice movement has largely succeeded in convincing environmentalists that wilderness is not the only environment worth saving. Together, environmental and environmental justice groups have worked to preserve and create urban green spaces, to restore urban waterways, and even to create urban organic farms, projects which are not only environmentally sustainable but also offer amenities to urban communities of color.

The chapters in this book suggest that the food movement might benefit from opening up to a similarly broad set of meanings, this time regarding food and cultural identity. Food justice activists often see food not only as a way to build community, but as a tool toward racial and economic liberation. This approach can be seen in McCutcheon's analysis of Black Nationalist foodways or in the Growing Food and Justice Initiative's emphasis on creating green jobs. Moreover, some working-class white populations align the production of local foods with reciprocity and affordability. Emphases on exchange and economy stand in direct contrast to the food movement's directive to pay, presumably to local farmers, the true cost of food. Each of these groups participate in the production of local food, but assign to their participation a variety of narratives not reflected by the food movement. Taken together, they suggest that the food movement might increase its potential to work with low-income people and people of color *who are already interested in the production and consumption of local food* if it were to create a more open dialogue about the various meanings people assign to the foods they consume. Such a dialogue might produce a broader vision of a just and sustainable food system.

Autonomy

Several of the chapters in this volume argue that low-income communities and communities of color need autonomy in order to create their own just and sustainable local food systems. With this autonomy, communities argue, not only can they feed themselves, but they can also create new narratives linking cultural traditions and racial identities to just and sustainable foodways. The most radical demand for autonomy is made by the Nation of Islam, which seeks the creation of a nation psychologically and geographically separate from whites. Less stringent desires for autonomy are held by the Pan African Orthodox Christian Church, which seeks only psychological separation, and the South Central and Seattle gardeners, who create autonomous cultural spaces that reflect their heritages and histories. Despite many differences, these cases are united by their desire for community self-sufficiency and to continue to provide food for their communities outside of—rather than by transforming—either the dominant food system or the food movement.

The Vegans of Color listserv and the Growing Food and Justice for All initiative can also be read as efforts toward autonomy. The former was created as a space in which vegans of color could speak about their experiences, and, in doing so, could collectively negotiate their relationships with the predominantly white vegan and animal rights movements. While comments from white vegans neither versed in nor dedicated to antiracism sometimes temporarily threaten this listserv's ability to operate outside the food movement, autonomy is often quickly reasserted by the responses of its regular contributors. Similarly, the Growing Food and Justice Initiative was born of a desire among food justice activists to build alliances among communities of color, without being subject to white activists' tendencies to minimize issues of race (Slocum 2006). These groups work not to create autonomous physical spaces in which food can be grown (though many of the groups aligned with the GFJI initiative do just that), but autonomous metaphorical spaces in which communities of color can build coherent and culturally resonant discourses connecting issues of food access and food sovereignty to various forms of racial and economic oppression. For these groups, autonomy is a necessary component of and step toward creating justice and sustainability for themselves.

Lessons for the Food Justice Movement

The chapters contained in this volume are, by and large, sympathetic to the goals and strategies of the food justice movement. Many of them use the movement as a standpoint from which to better understand issues located at the nexus of food, inequality, and environment. However, many chapters also contain both explicit and implicit pushes urging the food justice movement to be more reflective about its goals, strategies, and eventual chances for success. In the following section, we review and extend several of the chapter authors' comments concerning future directions for the food justice movement's theory and practice.

Race, Class, and Geography

Like many studying or working to address the manifestations of institutional racism, food justice movements and scholars have devoted little time to disentangling the effects of race, class, and geographic location. Several of the historical studies offered in this volume offer finely grained analyses of the interplays between these factors in the disenfranchisement of communities' food sovereignty and the establishment of food insecurity and hunger. However, the question remains as to how, in the present tense, to work from the nexus of multiple racial, spatial, and economic circumstances.

In an understandable and strategic response to a conservative political climate that seeks to reduce racial inequalities to economic ones, food justice activists have highlighted the role of institutional racism. This may, however, shift their visions away from other potential alliances based in class. With regard to this book, the need to untangle race and class is most clearly seen in McEntee's work (chapter 11, this volume) on white working-class notions of localism because the racial privilege experienced by those he studies is at odds with their marginal economic status. Clearly, McEntee's traditional localists enjoy participating in local food provisioning, and do so out of economic necessity, but couch their behaviors in a narrative quite different from that of the food movement. And yet, their story does not fit squarely within the analysis offered by the food justice movement either.

We are certainly not suggesting that the food justice movement, or scholars associated with it, must partition out the various oppressions of race and class. To the contrary, we recognize that all oppressions, but particularly those of race and class, are experienced and understood in tandem, and that structural racism is a process through which racial exclusions create and reify economic inequalities. Within the environmental justice literature, debating the relative effects of race and class often kept many scholars distracted from other, more nuanced analytical questions (Banerjee and Bell 2008). Our goal is not to call upon food justice activists or scholars to engage in a similar debate, but instead to begin a more proactive discussion of what members of white, working-class communities, many of whom are descended from farmers who could not withstand the increasing consolidation of agribusiness, might contribute to efforts to create a more just and sustainable food system. After all, the predominantly white town of Love Canal, New York, played an important role in the environmental justice movement, drawing national attention to toxic waste in residential communities and leading to the creation of the superfund law under which the Environmental Protection Agency can compel responsible parties to clean up hazardous waste (Levine 1982).

A third factor that affects issues of food access and food sovereignty is geography. To the degree that a national food justice movement has cohered, it has largely focused on urban areas with limited food access. However, many of the chapters in this volume reveal that hunger and food security can also be felt in rural areas, and indeed, national and international data on hunger depicts it largely as a rural phenomenon. The food justice movement's urban focus may constrain its ability to view rural communities facing similar hardships as allies in their struggle for just sustainability. Moreover, while the authors in this volume have presented issues of both urban and rural communities experiencing food insecurity and working toward food justice, this new body of work has not comparatively engaged these experiences. Rigorous comparative research could greatly illuminate the ways that urban and rural geographies of food sovereignty and food insecurity intersect with racial and economic inequalities.

A Place for Workers
In their explanation of the paradox of farmworker food insecurity, Brown and Getz (chapter 6, this volume) offer an excellent starting point to discuss issues of labor within the food justice movement. Their analysis mirrors the food justice movement's emphasis on low-income people of color's abilities to consume healthy foods. But such an analysis also suggests that labor is itself a fertile field that can nurture the movement for food justice.

Nationally, 75 percent of farmworkers are Mexican born (Center for Social Inclusion 2007) and in California, the nation's largest agricultural state, 97 percent are foreign born (Philpott 2008). On average, farmworkers garner half the wages paid to other manual labor occupations and experience high rates of work-related injuries (Center for Social Inclusion 2007). Indeed, farmworker deaths during heat waves in California's Central Valley, the site of Brown and Getz's study, occur with an alarming regularity. These brutal conditions, which certainly meet Ruth Wilson Gilmore's definition of racism as the "the state-sanctioned and/or extralegal production and exploitation of group-differentiated vulnerabilities to premature death" (2007, 247), are not restricted to California. In a case that has garnered significant national news, five residents of Immokelee, Florida, pled guilty to "enslaving Mexican and Guatemalan workers, brutalizing them and forcing them to work in farm fields" (Williams 2008). In addition, food processing plants in rural areas employ largely black or immigrant labor, and their inflation-adjusted wages have fallen by half in the past twenty-five years (Philpott 2008). Famous muckraking including Upton Sinclair's *The Jungle* ([1906] 1920) and Eric Schlosser's *Fast Food Nation* (2001) serve as chilling reminders of the kinds of past and present conditions that have been found in these workplaces.

And yet, neither the food movement nor the food justice movement has worked to build strong alliances with farmworkers and food processors, groups whose exploitation by agribusiness is certainly racialized, and, as Getz and Brown demonstrated, leads to hunger and food insecurity. The food justice movement has focused exclusively on low-income people and people of color as small producers, consumers, or those without access to local and organic food. But by reaching out to food-sector workers, they might find allies who understand the relationship between food and racial and economic exploitation from a different angle, and could offer additional visions of what a just and sustainable agriculture might look like.

With its vastly greater resources, the food movement has reached out to farmworkers (though not food processing workers) in limited ways, urging them to share their struggles at national conferences such as San Francisco's Slow Food Nation. However, because the movement's dominant narrative idealizes the small family farm and unprocessed foods, its writers and activists more often serve to metaphorically erase the presence of these workers. A less romantic and more realistic picture of U.S. farming might help the food movement to see this group of potential allies.

In the past few years, several important campaigns have emerged from farmworker communities. One example is the lawsuits filed against the California Department of Occupational Safety and Health Standards Board (OSHA) for failing to protect the safety of farm workers. OSHA, the lawsuit claims, does not have the capacity to inspect California's

35,000 farms for legal violations, and even when violations are found, fines are small and often are not collected. This lawsuit comes in the wake of farmworker deaths during the previously mentioned heat waves (United Farm Workers 2009). Another example is the Coalition of Immokalee Workers' Campaign for Fair Food, which raises wages and improves working conditions, and has reached agreements with fast food giants such as Taco Bell and McDonalds. These examples show that farmworkers, despite their extremely low wages and (predominantly) immigrant status, can be important participants in a movement to transform the food system. Parallels with the food justice movement are particularly strong in Immokalee, where the Campaign for Fair Food has recently argued that Chipotle,[1] a chain of Mexican-style restaurants with a mission to serve "food with integrity," must include not only environmental and animal rights concerns, but also human rights. Greater attention to labor within the food and food justice movements' emerging visions of just sustainability might lead to the creation of strong alliances with already politicized groups of workers.

Beyond Market-driven Strategies

Building these broad coalitions will require that the food justice movement "scale up" from its current form as a loose conglomeration of locally based programs to build the kind of power needed to fight against an agribusiness system in which increasingly consolidated corporate farms and food processors devastate consumers, workers, small-farm and business owners, and the environment. Doing so, however, requires that activists recognize the limitations of, though not necessarily abandon, the market-driven strategies that have dominated the movement thus far.

The food justice movement has largely adopted the food movement's position that the national government strongly supports agribusiness and that local and state governments are largely uninterested in creating a just and sustainable food system. Their initiatives, therefore, are often sponsored by foundation-funded nonprofits attempting to fulfill the basic human needs abandoned by an increasingly neoliberal state. Foundations, however, tend only to sponsor programs for a few years, requiring that they work toward financial self-sufficiency. To this end, food justice activists have had to emphasize the development of market gardens, community-supported agriculture programs, small-farm revenues, and other small businesses. In the urban setting, food from such programs is often sold to food desert residents at discounted prices, which is then subsidized by sales to wealthier patrons. This is problematic for several reasons. First, it echoes the food movement's message that individual consumption is a favorable route toward social change. Market-based solutions imply that food desert residents need only achieve the ability to purchase healthier food. This is in contrast to the food justice movement's broad, structural analysis of the causes of hunger and food insecurity in low-income communities of color. Second, market-based solutions are inherently undemocratic, as participation in them requires money. Market-based food justice programs are therefore uncomfortably split between the communities they seek to serve and the wealthier patrons who enable the financial success of their projects. Allen (2004) has convincingly argued that when food movement institutions prioritize economic success, they often undermine issues of social justice and inequality. Because products aligned with environmental imperatives tend to procure premium prices, the economic necessities of market-based projects can drive a wedge between the food justice movement's twin goals of justice and sustainability. Last, although food justice programs tend to offer fresh produce to their communities at subsidized costs, such food remains more expensive than the processed results of industrial monocultures. This puts food justice activists in the untenable position of trying to convince low-income people to spend more money on food than they otherwise would. Preliminary research in Oakland, California, has shown that many food-insecure people hold food justice programs in high regard, but still travel outside their food desert neighborhoods to purchase food from discount stores because their priority is to purchase the cheapest possible food (Alkon unpublished data; DeNuccio 2009).

Several of the chapters in this reader suggest that increased autonomy can be a pathway around the problem of market-based solutions. Borrowing from the food movement, food justice activists often seek to create spaces outside of, rather than in opposition to, an exploitative agribusiness system. Autonomy is often the goal of food justice projects such as those described earlier, which aim to employ and empower food desert residents, though the dictates of foundation funding tend to inspire their slippage into market-based strategies. Food justice projects that encourage low-income people of color to grow food for their families and communities, rather than for the market, offer an alternative to the problems described previously. The indigenous Latino immigrants working at the South Central Farm and Seattle Community gardens, for example, create their own just and sustainable food access and food sovereignty, alongside collective senses of community, history, and identity.

But the circumstances of the South Central Farm provide an important counterpoint to this argument. The South Central Farm was established in 1994 on land that the city of Los Angeles purchased through eminent domain to build a waste incinerator, which was then defeated by an array of environmental justice activists. Ten years later, under legal threat from the prior owner who argued that he had a right to repurchase, and despite the dedicated activism of the South Central farmers and their allies, the city sold the land. The cultural spaces established by the farmers were no longer autonomous when confronted by capital and state power (Lawson 2007).

Several of the chapters in this volume highlight more policy-oriented approaches to food justice, and implicitly push scholars and activists to think beyond current strategies. The Karuk Tribe, for example, has been party to nearly nine years of negotiations, lawsuits, and civil disobedience aimed at pressuring various government agencies and other stakeholders to remove four Klamath River Dams that block salmon runs (Karuk Tribe 2009a). An agreement to do so was reached in the fall of 2009 (Sullivan 2009). While it is merely a small step toward the reestablishment of Karuk food sovereignty, it demonstrates the potential for collective action in pursuit of food justice and just sustainability. The Karuk have also played important roles in other land management issues, such as lobbying for the increased regulation of a particularly environmentally damaging form of gold mining (Karuk Tribe 2009b). The Pigford and Keepseagle cases, in which black and Native American farmers collectively sued the USDA for racist lending practices, provide additional examples of broad, collective opposition to food injustice. Latino/a farmers in the U.S. Southwest, who faced remarkably similar circumstances, are currently attempting to bring a parallel suit (Goodwyn 2009). In each of these examples, communities that have been stripped of their food sovereignty have organized collectively to take it back.

Creating autonomous spaces and building power are not mutually exclusive goals. Indeed, the former can be an important pathway to create the latter. In contrast to strategies seeking to exist outside of state power are activists in the global South who develop autonomy in order to confront this power. Perhaps the strongest example is Brazil's Movimento dos Trabalhadores Rurais Sem Terra (Landless Workers Movement, or MST). The MST is best known for land occupations, in which collectives of landless farmers occupy privately owned but unoccupied lands throughout the country. Workers in these settlements do not simply hope to remain under the radar of capitalist landowners and state officials. They sue for property rights under a Brazilian law authorizing the government to expropriate land not serving a social function for the purpose of agrarian reform. Through this strategy, the MST has created more than a thousand settlements (Wolford 2007). Holt-Giménez (chapter 14, this volume) argues that the food justice movement has the potential to transform the U.S. food movement from its current reformist goals to a movement for food sovereignty that can stand with groups like the MST to confront not only agribusiness but also the globally unequal distribution of land and power. But even beginning to do so will require that food justice activists and scholars think much more broadly about the potential to draw-upon issues of food and hunger, labor, racial, and economic inequality, and environmental degradation to build a large and powerful collective movement. Creating local sustainable food systems in low-income communities of color can be an important step toward this goal, but should not become the goal itself.

Confronting the Food Movement

Just as the food movement can exist alongside industrial agribusiness, offering an alternative to it without confronting it, the food justice movement can be seen as merely offering a set of additional food provisioning choices. For this reason, Guthman (chapter 12, this volume) refers to food justice projects as "alternatives to the alternatives." This stands in stark contrast to the relationship between the environmental and environmental justice movements, in which direct confrontation by the latter led the former to revise its priorities in favor of both justice and sustainability. The food justice movement has made no similar claims, and has merely begun to work autonomously to create small farms and community gardens in low-income communities of color.

While working autonomously has other advantages for these food justice activists, it does not push the food movement to confront the race and class privileges that inform many of their discourses and practices. The food movement remains one in which both wealth, which is necessary to buy local, organic food, and whiteness, which creates greater resonance with the food movement narrative, continue to inform movement participation. Indeed, in some ways, the food justice movement gives the food movement the luxury of not having to worry about low-income communities and communities of color. Food movement activists can merely invoke the existence of food justice programs in order to argue that inequalities in the food system are being addressed. Additionally, by focusing its analysis only on the role of institutional racism, the food justice movement ignores the role that the food movement plays in increasing inequalities by improving conditions mainly for wealthy and white communities.

Moreover, it seems as if several aspects of the food movement, including those that make it so poorly equipped to address race and class, are being reproduced by the food justice movement. These include not only the market-based strategies already described, but also several aspects of the food narrative. For example, the blog of Growing Power, a prominent food justice organization, is replete with references to the "good food revolution." This phrase assumes there is some kind of universal consensus as to what kind of food is good without acknowledging the intimate relationship between food practices and racial identity. And while food justice activists often rightly insist on an empathetic approach to the structural constraints under which low-income people choose foods, their goal remains to introduce the "good food" of the food movement into their communities' diets. Growing Power's Will Allen, for example, understands that many descendents of sharecroppers "have some bad feelings about the farming life," but his answer is nonetheless to inspire them to grow food (Royte 2009). There seems to be a lack of alignment between the various aspects of the food movement narrative that drift into the food justice movement, and the conditions of racism and poverty that activists like Will Allen seek to address. Guthman argues that this lack of alignment between the food justice narrative and the experiences and histories of food-insecure people is one reason why many food justice programs struggle to engage the communities they intend to serve (chapter 12, this volume).

We recognize that many in the food justice movement are wary of alienating potential allies in the food movement. Indeed, this book seeks to contribute to the increased development of a dialogue, not a confrontation. Nonetheless, we believe that the food justice movement can better push the food movement to interrogate the ways its currently unacknowledged race and class privilege shape the movement's culture and political agenda. Doing so will greatly improve the abilities of these two movements to work together despite, and in full recognition of, the inequalities of wealth and power that divide their constituencies. Such collaboration is increasingly necessary if both movements are to move beyond supporting individual, local projects and begin to take on the collective challenge of organizing for structural change. For only through such collective action can a mass movement confront, and eventually transform, a destructive industrial agriculture into a just and sustainable food system.

Summary

The chapters contained in this collection are among the first seedlings sprouting in the field of food justice research. Chapter authors have offered a variety of stories that increase our understanding of how economic consolidation by industrial agriculture intersects with institutional racism to create spaces of hunger and food insecurity. This volume has offered glimpses into the historical and contemporary processes through which land-based communities are stripped of their food sovereignty, capital creates devaluated landscapes lacking access to healthy food, and agricultural workers are politically and economically marginalized. *Cultivating Food Justice* has also highlighted community-based responses, which include not only the local, nonprofit projects that comprise the core of the food justice movement, but also a variety of strategies ranging from supportive discursive communities to legal demands to influence current land-management practices. Last, this book offers a variety of theoretical and practical lessons, urging the food movement to acknowledge the role of institutional racism in the food system, and the food justice movement to engage with remaining issues of racial privilege, conflicting understandings of justice, and a growing global movement for food sovereignty. We hope that this kind of reflexive engagement might shed light on some of the seeds that lie beneath the surface and inspire readers to join the work of cultivating food justice.

Note

1. Chipotle was owned by McDonalds from 1998 to 2006.

References

Alkon, Alison, and Christie McCullen. 2011. Whiteness at Farmers Markets: Performances, Perpetuations . . . Contestations? *Antipode* 43.

Allen, Patricia. 2004. *Together at the Table: Sustainability and Sustenance in the American Agrifood System.* University Park: Pennsylvania State University Press.

Banerjee, Damayanti, and Michael M. Bell. 2008. Environmental Hazards. In *Encyclopedia of Race, Ethnicity, and Society*, ed. Richard T. Schafer., 394–397. Thousand Oaks, CA: Sage.

Center for Social Inclusion. 2007. Structural Racism and Our Food. <www .centerforsocialinclusion.org> (accessed October 30, 2009).

DeNuccio, Nicole. 2009. To In-food-ity and Beyond: Community Engagement in the Health for Oakland's People and the Environment (HOPE) Collaborative and Its Implications for Food Justice. Unpublished thesis, Department of Geography, University of California, Berkeley.

Di Chiro, Giovanna. 1996. Nature as Community: The Convergence of Environment and Social Justice. In *Uncommon Ground: Rethinking the Human Place in Nature*, ed. William Cronon, 298–320. New York: Norton.

Gilmore, Ruth Wilson. 2007. *Golden Gulag: Prisons, Surplus, Crisis, and Opposition in Globalizing California*. Berkeley: University of California Press.

Goodwyn, Wade. 2009. Hispanic Farmers Fight to Sue USDA. *All Things Considered, National Public Radio*. <http://www.npr.org/templates/story/story.php?storyId=113730694>. (accessed October 30, 2009).

Guthman, Julie. 2004. *Agrarian Dreams: The Paradox of Organic Farming in California*. Berkeley: University of California Press.

Karuk Tribe. 2009a. Klamath River Tribes and Fishermen Declare Mission Accomplished: Groups Succeed in Disrupting Warren Buffett's Woodstock of Capitalism. <http://karuk.us/press/press.php> (accessed October 30, 2009).

Karuk Tribe. 2009b. "Governor Signs Bill Banning In Stream Dredge Mining for Gold. <http://karuk.us/press/press.php> (accessed October 30, 2009).

Lawson, Laura. 2007. The South Central Farm: Dilemmas in Practicing the Public. *Cultural Geographies* 14 (4): 611–616.

Levine, Adeline Gordon. 1982. *Love Canal: Science, Politics and People*. Lanham, MD: Lexington.

Philpott, Tom. 2008. Schlosser: Food Industry Abuses Workers as Matter of Course. *Grist Magazine*. <http://www.grist.org/article/slow-food-nation-farmworkers-at-the-table/> (accessed October 30, 2009).

Royte, Elizabeth. 2009. Street Farmer. *The New York Times Magazine*. <http://www.nytimes.com/2009/07/05/magazine/05allen-t.html> (accessed November 1, 2009).

Schlosser, Eric. 2001. *Fast Food Nation: The Dark Side of the All-American Meal*. New York: Houghton Mifflin.

Sinclair, Upton. [1906] 1920. *The Jungle*. New York: Doubleday Page.

Slocum, Rachel. 2006. Anti-racist Practice and the Work of Community Food Organizations. *Antipode* 38 (2): 327–349.

Slocum, Rachel. 2007. Whiteness, Space and Alternative Food Practice. *Geoforum* 38 (3): 520–533.

Sullivan, Colin. 2009. Landmark Agreement to Remove 4 Klamath River Dams. *New York Times*. <http://www.nytimes.com/gwire/2009/09/30/30greenwire-landmark-agreement-to-remove-4-klamath-river-d-72992.html> (accessed October 31, 2009).

United Farm Workers. 2009. Landmark Lawsuit Accuses State of Failing to Protect Farm Workers from Heat-Related Death and Illness. <http://www.ufw.org/_board

.php?mode=view&b_code=cre_leg_back&b_no=5618> (accessed October 30, 2009).

Williams, Amy Bennett. 2008. Five Plead Guilty in Immokalee Slavery Case. <www.democraticunderground.com> (accessed October 30, 2009).

Wolford, Wendy. 2007. Neoliberalism and the Struggle for Land in Brazil. In *Neoliberal Environments: False Promises and Unnatural Consequences*, ed. Nik Heynen, James McCarthy, Scott Prudham, and Paul Robbins, 243–254. New York: Routledge.

Contributors

Julian Agyeman is professor and chair of urban and environmental policy and planning at Tufts University. He is author/editor of nine books including *Sustainable Communities and the Challenge of Environmental Justice* and *Environmental Inequities Beyond Borders: Local Perspectives on Global Injustices*.

Alison Hope Alkon is assistant professor of sociology at the University of the Pacific.

Sandy Brown is a PhD candidate in geography at the University of California, Berkeley.

E. Melanie DuPuis is professor of sociology at UC Santa Cruz. Her books include *Nature's Perfect Food: How Milk Became America's Drink* and *Smoke and Mirrors: The Politics and Culture of Air Pollution*.

Christy Getz is an associate Cooperative Extension Specialist in the Department of Environmental Science, Policy, and Management at UC Berkeley.

David Goodman is professor emeritus of environmental studies at UC Santa Cruz and visiting professor in the Department of Geography, Kings College London. He is coeditor of *Confronting the Coffee Crisis* and *Consuming Space: Placing Consumption in Perspective*.

Eleanor M. Green worked in the Archives and Museum at Delta State University where she also completed a master's degree in history education. She is now Delta Field Coordinator for The Partnership for a Healthy Mississippi.

John J. Green is associate professor of sociology at the University of Mississippi where he also directs the Center for Population Studies.

Julie Guthman is associate professor of community studies at UC Santa Cruz. She is author of *Agrarian Dreams: The Paradox of Organic Farming in California* and *Weighing In: Obesity, Food Justice, and the Limits of Capitalism*.

A. Breeze Harper is a PhD candidate in geography at the University of California, Davis. She is author of the novel *Scars* and editor of *Sistah Vegan: Black Female Vegans Speak on Food, Identity, Health and Society*.

Jill Lindsey Harrison is assistant professor of sociology at the University of Colorado Boulder. She is the author of *Pesticide Drift and the Pursuit of Environmental Justice*.

Eric Holt-Giménez is the executive director of Food First/Institute for Food and Development Policy. He is the author of *Food Rebellions! Crisis and the Hunger for Justice* and *Campesino a Campesino: Voices from Latin America's Farmer to Farmer Movement for Sustainable Agriculture.*

Anna M. Kleiner is associate professor of sociology in the Department of Sociology and Criminal Justice at Southeastern Louisiana University.

Teresa M. Mares is assistant professor of anthropology at the University of Vermont.

Nathan McClintock is assistant professor of urban studies and planning at Portland State University.

Priscilla McCutcheon is assistant professor of geography and African American studies at the University of Connecticut.

Jesse C. McEntee, PhD, is a postdoctoral fellow in sustainability studies in the Department of Urban and Environmental Policy and Planning at Tufts University.

Laura-Anne Minkoff-Zern is a PhD candidate in geography at the University of California, Berkeley.

Alfonso Morales is assistant professor of urban and regional planning at the University of Wisconsin. He is the editor of three books: *Renascent Pragmatism: Studies in Law and Social Science; Street Entrepreneurs: People, Place, & Politics in Local and Global Perspective;* and *Wealth Creation and Business Formation among Mexican-Americans: History, Circumstances and Prospects.*

Kari Marie Norgaard is assistant professor of sociology and environmental studies at the University of Oregon. She is the author of *Living in Denial: Climate Change, Emotions and Everyday Life.*

Nancy Lee Peluso is Henry J. Vaux Distinguished Professor of Forest Policy in the Department of Environmental Science, Policy and Management at UC Berkeley. Her books include *Taking Southeast Asia to Market: Commodities, People and Nature in a Neoliberal Age* and *Violent Environments.*

Devon G. Peña is professor of anthropology and American ethnic studies at the University of Washington. He is author/editor of four books including *Mexican Americans and the Environment: Tierra y Vida* and *The Terror of the Machine: Technology, Work, Gender and Ecology on the U.S.–Mexico Border.*

Ron Reed is a traditional Karuk dipnet fisherman and cultural biologist in the Fisheries Program of the Karuk Department of Natural Resources in Happy Camp, California.

Jennifer Sowerwine is a researcher in the Department of Environmental Science, Policy, and Management at the University of California, Berkeley.

Carolina Van Horn is an alumnus of Whitman College where she majored in environmental studies.

Index